Layered Double Hydroxides

Layered Double Hydroxides

Editors

Giuseppe Prestopino
Giuseppe Arrabito

MDPI • Basel • Beijing • Wuhan • Barcelona • Belgrade • Manchester • Tokyo • Cluj • Tianjin

Editors

Giuseppe Prestopino
Università di Roma
"Tor Vergata"
Italy

Giuseppe Arrabito
Università degli Studi
di Palermo
Italy

Editorial Office
MDPI
St. Alban-Anlage 66
4052 Basel, Switzerland

This is a reprint of articles from the Special Issue published online in the open access journal *Crystals* (ISSN 2073-4352) (available at: https://www.mdpi.com/journal/crystals/special_issues/layereddouble_hydroxide).

For citation purposes, cite each article independently as indicated on the article page online and as indicated below:

LastName, A.A.; LastName, B.B.; LastName, C.C. Article Title. *Journal Name* **Year**, *Volume Number*, Page Range.

ISBN 978-3-0365-0306-6 (Hbk)
ISBN 978-3-0365-0307-3 (PDF)

Cover image courtesy of Giuseppe Prestopino and Giuseppe Arrabito.

Contents

About the Editors

Giuseppe Prestopino was awarded his degree in Electronic Engineering from the University of Rome "Roma Tre" in 2004 and completed his Ph.D. in Microsystems Engineering at the University of Rome "Tor Vergata" in 2009, where he started his academic career. Currently, he has a permanent position at the "Tor Vergata" University, "Dipartimento di Ingegneria Industriale". The scientific interests of Giuseppe Prestopino have been mainly focused on the growth and characterization of synthetic single crystal diamond, and on the development of diamond-based devices. In particular, he contributed in developing diamond growth reactors based on microwave plasma-enhanced chemical vapor deposition for both intrinsic and boron-doped diamond films, and he specialized in numerous characterization techniques like time-resolved photoluminescence, X-ray diffraction, scanning electron microscopy, and many types of electrical measurements. He worked in the design and fabrication of many different diamond-based detectors, e.g., detectors for application in hadrontherapy dosimetry, radiation therapy in vivo dosimeters, diamond-based UV position sensitive detectors, and microDiamond dosimeters for external beam radiotherapy. Recently, Giuseppe Prestopino broadened his research interests to the field of nanostructured materials, and in particular to the synthesis and characterization of layered double hydroxide films and related composites, exploring new intriguing properties and applications of these exciting materials. Giuseppe Prestopino is the author of more than 80 publications in peer-reviewed journals. He has been the guest editor of two Special Issues of the Crystals journal by MDPI, namely "Layered Double Hydroxides" and "2D Materials: From Structures to Functions".

Giuseppe Arrabito was awarded his degree in Biomolecular Chemistry by the University of Catania in 2008 and his Ph.D. in Nanosciences from the Scuola Superiore di Catania in 2012. From April 2012 to April 2013, he was a post-doctoral researcher at the Max-Planck Institute for Molecular Physiology at the Technical University of Dortmund. From May 2013 to July 2014, he was a post-doctoral researcher at the Department of Electronic Engineering at the University of Rome Tor Vergata. Since August 2014, he is as a post-doctoral researcher at the Department of Physics and Chemistry—Emilio Segrè at the University of Palermo. He is the holder of two "Seal-of-Excellence" awards from EU projects. In 2020, he received the "Galileo Galilei Giovani" International Prize for scientific disciplines and the Bronze Award at the EIT Innovation Days 2020. The research interests of Giuseppe Arrabito are mainly directed towards the establishment of novel printing methodologies for the fabrication of life-like or life-inspired systems onto solid or liquid interfaces. He is a developer of innovative strategies for the assembly of ordered patterns of biomolecular systems (DNA, proteins) at different scales (from nano- up to the milli-scale), finding applications in biosensors, drug screening, single cell biology, cellular scaffolds and synthetic biology. He recently expanded his research interests towards piezoelectric and piezoresistive materials for the fabrication of wearable sensors, focusing his attention on ZnO nanowire synthesis by rational approaches and fullerene-based materials for bending sensors. Giuseppe Arrabito is the author of more than 27 publications in peer-reviewed journals. Two of his publications have been featured for journal cover pages. He is also the co-holder of a US patent and he is the editor of a book about DNA nanotechnology applications in bioanalysis and medicine.

Preface to "Layered Double Hydroxides"

Layered double hydroxides (LDHs) are an emerging class of clay-like inorganic layered compounds which have been extensively investigated in both academia and industry for their relevant applications in many different fields, including catalysis, drug delivery, flame retardants, nanomedicine, energy storage and conversion, anion exchangers, pollutant remediation as well as heavy metal ions absorption. These materials are characterized by a unique versatility of chemical composition and morphology and are the subject of significant interdisciplinary applications in physical and biology-related disciplines. The journal Crystals allows for the collections of sets of papers dealing with specific topics that are of interest for the readership of the journal. Some recent contributions of LDH-related research have been collected in the Special Issue entitled "Layered Double Hydroxides".

Giuseppe Prestopino, Giuseppe Arrabito
Editors

Editorial

Layered Double Hydroxides

Giuseppe Prestopino [1,*] and Giuseppe Arrabito [2,*]

[1] Department of Industrial Engineering, University of Rome "Tor Vergata", 00133 Rome, Italy
[2] Department of Physics and Chemistry Emilio Segrè, University of Palermo, 90128 Palermo, Italy
* Correspondence: giuseppe.prestopino@uniroma2.it (G.P.); giuseppedomenico.arrabito@unipa.it (G.A.)

Received: 13 November 2020; Accepted: 18 November 2020; Published: 19 November 2020

The impact of layered double hydroxides (LDHs) within the multidisciplinary fields of materials sciences, physics, chemistry, and biology is rapidly growing, given their easiness of synthesis, flexibility in composition, tunable biocompatibility and morphology. LDHs constitute a versatile platform for the realization of new classes of functional systems, showing unique enhanced surface effects and unprecedented properties for application in very different fields, namely, surface chemistry and catalysis, storage and triggered release of functional anions, flame retardants, drug delivery and nanomedicine, remediation, energy storage and conversion. These systems can be synthesized as self-assembled hierarchical nanosheet thin films by means of low temperature solution-based approaches, which are accessible by many laboratories and have the advantages of low cost, mild conditions, and environmental friendliness. In addition, the possibility of LDHs to be exfoliated into 2D nanosheets has been demonstrated to further improve their performance in many applications, as well as to be an attractive route to achieve building blocks for fabricating a wide plethora of hybrid functional architectures. LDHs are therefore a playground for exciting new research covering all of the most intriguing features of 2D materials and more. This Special Issue on "Layered Double Hydroxides" gathers a multidisciplinary collection of original contributions and review articles from authors with diverse scientific backgrounds and who employ LDHs for very different applications, permitting the demonstration of their versatility. Along with LDH-focused papers, this Special Issue also includes some research in which materials different to LDHs resulted in a convenient choice for selected purposes.

A study of particular interest is the report from Teixeira et al. [1], where the extreme flexibility of LDH matrices in changing both interlayer and metal components to tune their physicochemical properties is explored. The memory effect in thermally treated LDHs, i.e., the restoration of their lamellar structure by rehydration in aqueous solutions containing anions, is leveraged for the synthesis of rare earth Eu^{3+} doped luminescent LDHs intercalated with 1,3,5-benzenetricarboxylate anion. The latter acts as an anionic photosensitizer for Eu^{3+} ions, increasing the total observable luminescence by means of the so-called antenna effect. Such a combination of the two effects also provided a useful tool to monitor the rehydration process of the calcined LDHs.

This Special Issue has received many contributions from the field of pollutant remediation, highlighting the key role of LDH-based compounds for this particular application. Specifically, it is important to consider LDHs as outstanding candidates for selective adsorption of anionic contaminants, taking advantage of anion exchange with the interlayer anions, or anion trapping in the interlayer during rehydration of mixed metal oxides from calcined LDHs and subsequent reconstruction of the lamellar structure via the "memory effect". In this context, Dore et al. [2] showed a clear example of the usefulness of calcined LDHs, namely, mixed MgAlFe oxides and mixed ZnAl oxides from hydrotalcite-like and zaccagnaite-like compounds, respectively, to remove Sb(V), in the $Sb(OH)_6^-$ form, from aqueous solution. The authors also demonstrated the feasibility of LDH-based removal of $Sb(OH)_6^-$ from the slag drainage in an abandoned mine in Sardinia, Italy. Another excellent application in the field of water decontamination comes from Golban and coworkers [3], who proposed a new and convenient method to synthetize Mg_4Fe-LDHs from iron-containing acidic residual solution of the hot-dip galvanizing process, obtaining a material suitable for the effective decontamination of $MoO_4{}^{2-}$ from aqueous solutions. Similarly, the selective recovery of Cu(II) from metal mixtures was conveniently achieved by Yang et al. [4] with calcium alginate beads, which are well-known green sorbents for the biosorption of heavy metals. The fabricated alginate beads showed also excellent retainment of their properties after five cycles of sorption–desorption procedures.

Along with heavy metal pollution, antibiotics increasingly pose a serious concern for environmental water. Panplado et al. [5] demonstrated a simple strategy to very rapidly remove tetracycline (TC) antibiotic molecules from contaminated water. They propose an *in-situ* adsorption method which involves the utilization of Mg^{2+} and Al^{3+} containing LDH precursors to promote the precipitation of mixed metal hydroxides (MMHs), which act as fast sorbents for capturing TC from aqueous solution. The strong interactions between the charged MMH surface and the TC molecules, consisting of electrostatic attraction and hydrogen bonding, were leveraged to achieve instantaneous adsorption, which is superior to the use of LDH as sorbent in a conventional route.

This Special Issue also contains original contributions in the field of biology-related applications. Chang et al. [6] demonstrated the ability of Mg_2Fe-LDHs to significantly enhance the production of surfactin in bacterial cells of a *Bacillus subtilis* ATC 21,322 culture. Surfactin is a cyclic lipopeptide of seven amino acids and acts as an excellent biosurfactant. However, in order to obtain sufficient levels to allow for its commercial use, its production needs to be enhanced. Another interesting contribution is provided by Sun and coworkers [7], who leveraged solid lipid nanoparticles for efficient resveratrol loading, with the aim to obtain a substantial improvement of the mitochondrial function in mice, in comparison with control resveratrol supplementation in the absence of nanoparticle loading.

Finally, the role of LDHs as catalysts in relevant organic chemistry reactions and their emerging application in cellular biology were extensively reviewed by Arrabito et al. [8], highlighting the conspicuous studies focusing on the synthesis, characterization, and applications of LDH-based systems. In a second review article [9], the same authors reviewed the role of LDHs in the scenario of bioinspired nanomaterials research and applications thereof. This work provides a possible link between the role of LDHs in the origin of life and the formation of pre-biotic molecules, to inspire the fabrication of artificial LDH-based compartments that mimic prebiotic assemblies. The design of LDH-based systems for life-like and life-inspired devices was also reviewed.

In summary, the present Special Issue on "Layered Double Hydroxides" can be considered as a status report that gathers and reviews different contributions summarizing the progress of many different LDH-related research and applications in the past several years.

Funding: This research received no external funding.

Conflicts of Interest: The authors declare no conflict of interest.

References

1. Teixeira, A.; Morais, A.; Silva, I.; Breynaert, E.; Mustafa, D. Luminescent Layered Double Hydroxides Intercalated with an Anionic Photosensitizer via the Memory Effect. *Crystals* **2019**, *9*, 153. [CrossRef]
2. Dore, E.; Frau, F.; Cidu, R. Antimonate Removal from Polluted Mining Water by Calcined Layered Double Hydroxides. *Crystals* **2019**, *9*, 410. [CrossRef]
3. Golban, A.; Lupa, L.; Cocheci, L.; Pode, R. Synthesis of MgFe Layered Double Hydroxide from Iron-Containing Acidic Residual Solution and Its Adsorption Performance. *Crystals* **2019**, *9*, 514. [CrossRef]
4. Yang, N.; Wang, R.; Rao, P.; Yan, L.; Zhang, W.; Wang, J.; Chai, F. The Fabrication of Calcium Alginate Beads as a Green Sorbent for Selective Recovery of Cu(II) from Metal Mixtures. *Crystals* **2019**, *9*, 255. [CrossRef]
5. Panplado, K.; Subsadsana, M.; Srijaranai, S.; Sansuk, S. Rapid Removal and Efficient Recovery of Tetracycline Antibiotics in Aqueous Solution Using Layered Double Hydroxide Components in an In Situ-Adsorption Process. *Crystals* **2019**, *9*, 342. [CrossRef]
6. Chang, P.-H.; Li, S.-Y.; Juang, T.-Y.; Liu, Y.-C. Mg-Fe Layered Double Hydroxides Enhance Surfactin Production in Bacterial Cells. *Crystals* **2019**, *9*, 355. [CrossRef]

7. Sun, J.; Zhou, Y.; Su, Y.; Li, S.; Dong, J.; He, Q.; Cao, Y.; Lu, T.; Qin, L. Resveratrol-Loaded Solid Lipid Nanoparticle Supplementation Ameliorates Physical Fatigue by Improving Mitochondrial Quality Control. *Crystals* **2019**, *9*, 559. [CrossRef]
8. Arrabito, G.; Bonasera, A.; Prestopino, G.; Orsini, A.; Mattoccia, A.; Martinelli, E.; Pignataro, B.; Medaglia, P.G. Layered Double Hydroxides: A Toolbox for Chemistry and Biology. *Crystals* **2019**, *9*, 361. [CrossRef]
9. Arrabito, G.; Pezzilli, R.; Prestopino, G.; Medaglia, P.G. Layered Double Hydroxides in Bioinspired Nanotechnology. *Crystals* **2020**, *10*, 602. [CrossRef]

Article

Resveratrol-Loaded Solid Lipid Nanoparticle Supplementation Ameliorates Physical Fatigue by Improving Mitochondrial Quality Control

Jingyu Sun [1,†], Yunhe Zhou [1,†], Yajuan Su [2], Sheng Li [2], Jingmei Dong [1], Qing He [2], Yang Cao [1], Tianfeng Lu [1,*] and Lili Qin [1,*]

1 Sports and Health Research Center, Department of Physical Education, Tongji University, Shanghai 200092, China; jysun@tongji.edu.cn (J.S.); maggie211@tongji.edu.cn (Y.Z.); djm1969@tongji.edu.cn (J.D.); caoyang@tongji.edu.cn (Y.C.)

2 School of Life Sciences and Technology, Tongji University, Shanghai 200092, China; 18338691923@163.com (Y.S.); lisheng1209@163.com (S.L.); 1831521@tongji.edu.cn (Q.H.)

* Correspondence: sytyltf@126.com (T.L.); qinlili@tongji.edu.cn (L.Q.); Tel.: +86-21-6598-5242 (T.L.); +86-21-6598-1711 (L.Q.)

† These authors contributed equally to this article.

Received: 2 September 2019; Accepted: 21 October 2019; Published: 25 October 2019

Abstract: Resveratrol (RSV) has various pharmacological effects; however, few studies have directly addressed the possible antifatigue effects of long-term endurance exercise. The clinical use of RSV is limited by its poor water solubility and extremely short plasma half-life. Solid lipid nanoparticles (SLNs) are considered as reasonable drug delivery systems to overcome some of these drawbacks and expand its applications. In this study, RSV-SLNs were successfully prepared through emulsification and low-temperature solidification. Results showed that RSV-SLN supplementation effectively enhanced endurance performance. RSV-SLN supplementation might enhance mitochondrial function by ameliorating mitochondrial quality control (QC), which was superior to RSV application. These results revealed an unexpected role of RSV-SLN compared with RSV in terms of linking nutrient deprivation to mitochondrial oxidant production through mitochondrial QC. A mitochondrion-mediated pathway was likely involved in RSV-SLN, thereby improving endurance performance. Overall, this study highlighted new possibilities for anti-physical-fatigue strategies.

Keywords: resveratrol; solid lipid nanoparticles; endurance exercise; mitochondrial nutrients; mitochondrial quality control

---

1. Introduction

Physical fatigue and mental fatigue are two main aspects of fatigue. Physical fatigue is often accompanied by the deterioration of physical function [1]. Exhaustive exercise-induced mitochondrial dysfunction may result in physical fatigue [2]. Mitochondrial nutrients protect organelles from chronic and repeated exercise or excessive fatigue-induced damage and maintain metabolic homeostasis. Therefore, scientists are actively exploring natural products to reduce oxidative damage caused by exercise and fight against physical fatigue [3]. Resveratrol (3-5-4′-trihydroxy-trans-stilbene, RSV), a polyphenol compound, has various pharmacological effects, including improvement of mitochondrial function, prevention of obesity and obesity-related diseases [4], suppression of inflammation [5], and protection against oxidative stress [6]. However, to our knowledge, few studies have directly addressed the possible anti-physical-fatigue effects of RSV. The pharmacokinetic properties of RSV are less favorable because of the poor water solubility of RSV. Beyond that, RSV metabolism is rapid and extensive [7], and its plasma half-life is short [8]. A reasonable strategy is needed to propose and expand the applications of RSV to resolve some of these problems. Solid lipid nanoparticles (SLNs) are considered as an efficient drug delivery system because of their good physicochemical properties [9,10]. SLNs, which can be metabolized by many organisms, can modulate drug release [11–13]. SLNs can efficiently protect encapsulated resveratrol drugs in the biological environment and improve their physiochemical properties [14–16]. Hence, this study evaluated the effect of RSV-loaded SLN (RSV-SLN) supplementation on exercise performance to explore its possible mechanisms.

Maintaining mitochondrial function involves mitochondrial biogenesis, mitophagy, fusion, and fission. The integration of this series of processes reflects mitochondrial quality control (QC). We hypothesized that chronic fatigue states induced by excessive endurance exercise might be linked to injured mitochondrial QC. This study was the first to explore the effects of RSV-loaded SLNs on gene regulation that involves mitochondrial QC following excessive endurance exercise in mice. These data improved the understanding of the role of RSV in a nanometer form as mitochondrial nutrition during successive sessions of prolonged endurance exercise.

2. Materials and Methods

2.1. Preparation of RSV-SLNs

RSV-SLNs were produced by emulsification and low-temperature solidification. In brief, RSV (150 mg) (Aladdin Industrial Corporation, Shanghai, China), lecithin (100 mg), and stearic acid (200 mg) (Shanghai Chemical Reagent Company, Shanghai, China) were dissolved in 10 mL of chloroform in glass bottles as an organic phase through ultrasound. Myrj 52 (Sigma-Aldrich Co., St Louis, MO, USA) was dissolved in 30 mL of distilled water and heated to 75 ± 2 °C in a water bath as an aqueous phase. Under 1000 rpm mechanical stirring, the organic phase was injected into the hot water phase, and the solution was kept at 75 °C at the same stirring speed to remove organic solvents. Approximately 5 mL of condensed solvent remained after the organic solvent was removed. The condensed solvent was then mixed with the same amount of cold water (0 °C to 2 °C) and stirred for 2 h. The resultant suspension was centrifuged at 20,000 rpm (Avanti J25centrifuge, JA 25.50 rotor; Beckman Coulter, Palo Alto, CA, USA) to remove the supernatant. Afterward, the pellets were suspended in ultrapure water, refrigerated, and freeze-dried.

2.2. Characterization byTransmission Electron Microscopy (TEM), Scanning Electron Microscopy (SEM), and Zetasizer of RSV-SLN

For TEM, 5 mL of each sample was placed on carbon formvar-coated 400-mesh spacing grids, left to become adsorbed for 5 min, negatively stained with 2% sodium phosphotungstate for 45 s, and allowed to dry. The grid was visualized by using a JEM 1400 electron microscope (JEOL-1230, Tokyo, Japan) at 80 kV. For the SEM, the samples were coated with aurum for 6 min using an Ion Sputter (JFC-1100, JEOL Ltd, Tokyo, Japan), and the thickness of cladding material was thinner than 20 nm. Finally, after vacuuming, the shape and surface morphology of the samples were observed under S-4800 (Hitachi, Tokyo, Japan) scanning electron microscopes at an accelerating voltage of 20 kV. The magnification of the SEM images was 10,000×. The particle size and the zeta potential were determined at 25 °C by photon correlation spectroscopy (Zetasizer Nano ZS, Malvern Instruments, Malvern, UK). For each sample, the measurements were repeated thrice.

2.3. X-Ray Powder Diffraction (XRD) Analysis

X-ray diffraction (XRD) patterns of pure RSV, RSV-SLN, and SLN were performed in order to characterize their crystallographic structure. The patterns were carried out with 'X' pert PRO, PANalytical instrument (Westborough, MA, USA), using Cu–Ka rays with a voltage of 40 kV and a current of 30 mA, over the 2θranges 5–60°, with a step width 0.05°and a scan time of 2.0 s per step.

2.4. Fourier-Transform Infrared (FTIR) Spectra of RSV-SLNs

FTIR spectra were obtained on a CARY 50 spectrophotometer. Potassium bromide disc technique was employed to obtain the FTIR spectra of RSV, SLN-RSV, and SLN by the standard KBr disk method (sample/KBr = 1/100). The samples were ground gently with anhydrous KBr and compressed to form pellets. The spectrum was recorded in the range of 500–4000 cm^{-1} using TEN-SOR27(Bruker Co., Ettlingen, Germany).

2.5. Animals

Eight-week-old male C57BL/6J mice were purchased from Shanghai Laboratory Animal Research Center (SLAC, Shanghai, China) and kept in a controlled environment (12 h/12 h light/dark cycle, 08:00–20:00, temperature: 23 ± 2 °C, humidity: 60% ± 5%). They were randomly divided into four groups 1 week after acclimatization: (1) sedentary control group (SC; $n = 6$); (2) endurance exercise (EE; $n = 6$); (3) endurance exercise combined resveratrol supplementation (EE + RSV; $n = 6$); and (4) endurance exercise combined with RSV-SLN supplementation (EE + RSV-SLN; $n = 6$). During the whole experiment, all the mice had free access to purified water and food and were weighed weekly. RSV and RSV-SLN were administered orally in EE + RSV group and EE + RSV-SLN group, respectively. RSV was administered at a dose of 25 mg/kg [4]. The amount of RSV-SLN was adjusted to be equal to that of the RSV treatment group. The SC and EE groups were performed with physiological saline solution as vehicle. Each treatment was administered once a day for 6 days/week for 8 weeks and performed 1 h before exercise. All the animal experimental protocols in this study were approved by the Animal Care and Use Committee of Shanghai Model Biology Research Center (Approval number SRCMO-IACUC No. 20140002).

2.6. Exercise Protocol

The whole training process was conducted on a motor treadmill (Jiangsu Saiangsi Biologic Technology Co., Jiangsu, China). After acclimating for 1 week, the mice exercised at low-moderate intensity for 8 weeks (speed initially at 10 m/min, 120 min per day), and the speed was gradually increased to 20 m/min until exhaustion. The exhaustive distance of the mice in each group was recorded. The mice in the control group were exposed to noise and handling, which were similar to those in the EE group, the EE + RSV group, and the EE + RSV-SLN group, to regulate exercise-associated stress.

2.7. Indirect Calorimetry

Indirect calorimetry was administered by computer-controlled automatic systems (Oxymax/ CLAMS-SC, Comprehensive Lab Animal Monitoring System, Columbus Instruments). The mice were tested in separate chambers to provide free water and ad libitum access to food. The velocity of indoor air passing through the chamber was 0.5 L/min. The exhaust gas of the combustion chamber was sampled for 1 min at an interval of 12 min. Oxygen consumption and carbon dioxide production were estimated using O_2 and CO_2 sensors. Respiratory exchange rate (RER) was measured in terms of the volume of oxygen consumption and carbon dioxide production ($VO_2 = ViO_2i - VoO_2o$; $VCO_2 = VoCO_2o - ViCO_2i$; $RER = VCO_2/VO_2$) [17]. Measurements were collected immediately after the endurance exercise challenge for 2 days.

2.8. Tissue Sampling

At the end of each manipulation, all the mice were subjected to fasting overnight and anesthetized by intraperitoneally injecting 2% sodium pentobarbital (6.5 mg/100 g body weight). The gastrocnemius (GAS) muscle tissues were completely excised and weighed individually. A portion of the GAS muscle tissue was kept for TEM or homogenized immediately to determine mitochondrial respiratory chain enzymes. The remaining portions were stored at −80 °C until further analysis.

2.9. Ultrastructural Changes of Skeletal Muscle Tissues

The cross-sectional GAS muscle tissues were immobilized in a fixed buffer (2% glutaraldehyde, 0.1 M sodium cacodylate, 0.5% polyformaldehyde, 3 mM $CaCl_2$, and 0.1 M sucrose) for 4 h at 4 °C, rinsed in PBS three times for 15 min at each time, dehydrated in ethanol, then acetone, and inserted in LX-112 (Ladd, Burlington, VT, USA). The tissues were then cut into 60–80 nm sections. The slides were stained by double staining with uranium lead (2% uranyl acetate and lead citrate saturated aqueous solution) and examined at 80 kV by using a Tecnai 10 TEM (TECNAI G2 F20 S-TWIN, FEI, Oregon, USA). Digital images were captured with cameras (Olympus Soft Imaging Solutions, GmbH, Munster, Germany).

2.10. Mitochondrial Respiratory Chain Enzyme Assays

Citrate synthase (CS) activity, cytochrome (Cyt) c content, and adenosine triphosphatase (ATPase) activity in skeletal muscle were measured using the corresponding microplate assay kits (Nanjing Jiancheng Bioengineering Institute, Jiangsu, China) in accordance with the manufacturers' instructions. Absorbance was recorded with a TECAN microplate reader (TECAN, Sunrise, Mannedorf, Switzerland).

2.11. Mitochondrial DNA (mtDNA) Content

Quantitative real-time RT-PCR was performed to determine the ratio of a mitochondrial gene to a nuclear gene and to examine the mtDNA content in each sample in accordance with previously described methods with some modifications [18,19]. Mitochondrial NADH dehydrogenase subunit 1 (*ND1*) served as the mitochondrial mark, and platelet endothelial cell adhesion molecule-1 (*PECAM-1*) functioned as the nuclear reference mark (primer sequences in Table 1). Total DNA was extracted using a QIAamp DNA Mini kit in accordance with the manufacturer's instructions. A melting curve was obtained to ensure specific amplification, and the standard curve method was used for relative quantification. The ratio of mitochondrial *ND1* to *PECAM-1* was then calculated.

Table 1. List of primers used in the study.

Gene	Forward Primer (5′–3′)	Reverse Primer (5′–3′)
ND1	CCTATCACCCTTGCCATCAT	GAGGCTGTTGCTTGTGTGAC
PECAM-1	ATGGAAAGCCTGCCATCATG	TCCTTGTTGTTCAGCATCAC
COX II	TTCAACACACTCTATCACTGGC	AGAAGCGTTTGCGGTACTCAT
COX IV	TCACTGCGCTCGTTCTGATT	TGGCCTTCATGTCCAGCATT
CPT-1M	GCACACCAGGCAGTAGCTTT	CAGGAGTTGATTCCAGACAGGTA
CD36	ATGGGCTGTGATCGGAACTG	TTTGCCACGTCATCTGGGTTT
PGC-1α	TATGGAGTGACATAGAGTGTGCT	CCACTTCAATCCACCCAGAAAG
NRF1	AGCACGGAGTGACCCAAAC	TGTACGTGGCTACATGGACCT
Tfam	ATTCCGAAGTGTTTTTCCAGCA	TCTGAAAGTTTTGCATCTGGGT
Bnip3	TCCTGGGTAGAACTGCACTTC	GCTGGGCATCCAACAGTATTT
Beclin-1	ATGGAGGGGTCTAAGGCGTC	TCCTCTCCTGAGTTAGCCTCT
NIX	ATGTCTCACTTAGTCGAGCCG	CTCATGCTGTGCATCCAGGA
β-actin	ATTGCTGACAGGATGCAGAA	GCTGATCCACATCTGCTGGAA

2.12. RNA Extraction and Semiquantitative RT-PCR

Total RNA was isolated from the skeletal muscle tissues of each mouse by using Trizol (Invitrogen, Carlsbad, CA, USA) and reverse transcribed with a Superscript II kit (Invitrogen) in accordance with the manufacturer's recommendation. Table 1 shows the forward (F) and reverse (R) primers of mouse genes. PCR was conducted under the following conditions: 10 min at 94 °C, 30–35 cycles at 94 °C (30 s), 55 °C (30 s), 72 °C (1 min), and 10 min of incubation at 72 °C. The mRNA levels were normalized to that of β-actin mRNA and quantified using the $2^{-\Delta\Delta Ct}$ method.

2.13. Data and Statistical Analyses

Data were expressed as mean ± SEM. Differences in means were analyzed through one-way ANOVA. Significant level was set at $p < 0.05$ (two sided). Data were examined using SPSS 19.0 (Chicago, IL, USA).

3. Results

3.1. Characterization of SLN and RSV-SLN

TEM images showed that SLN and RSV-SLN were both spherical in shape with smooth surfaces (Figure 1A,B). The discrete spheres were solid particles and had no aggregations. SEM analysis showed the smooth surface and the spherical morphology of the prepared SLN (Figure 1C) and RSV-SLN (Figure 1D), and these findings were consistent with previous reports [20]. Figure 1E showed the equivalent mean hydrodynamic diameter of RSV-SLN was about 112.5 ± 10.3 nm with a narrow particle size distribution. Meanwhile, the zeta potential of RSV-SLN was −24.7 mV with a narrow polydispersity index (PDI = 0.36 ± 0.02) (Figure 1F).

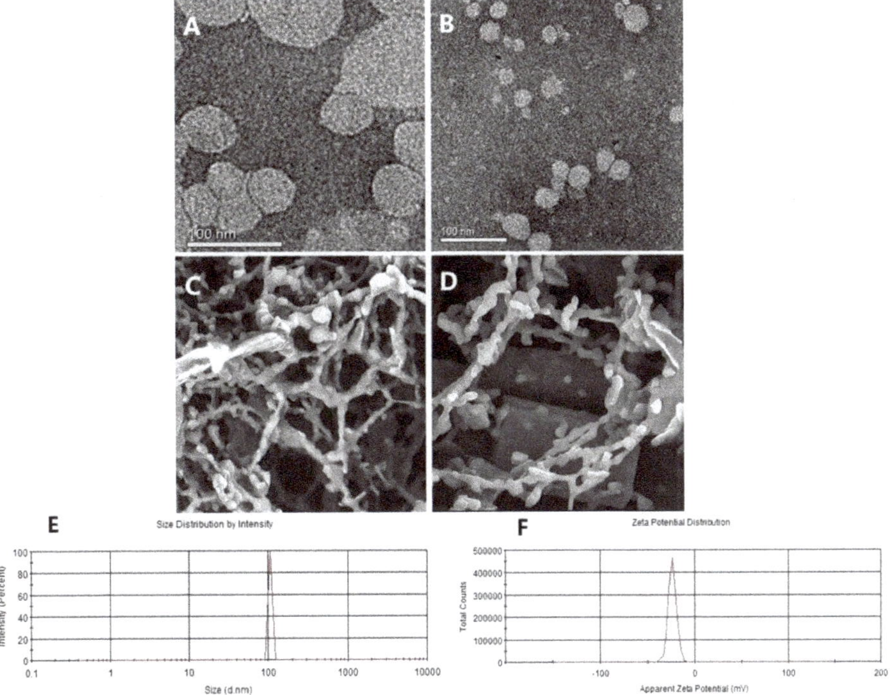

Figure 1. Characterization of RSV-SLN and SLN. TEM images of (**A**) SLN and (**B**) RSV-SLN after they were stained with one drop of 2% phosphotungstic acid. SEM images of (**C**) SLN and (**D**) RSV-SLN. (**E**,**F**) The size distribution and zeta potential for RSV-SLN. RSV: Resveratrol; SLN: solid lipid nanoparticles.

3.2. X-Ray Diffraction Analysis of RSV-SLN

XRD studies were performed in order to characterize drug status inside the SLN. As shown in Figure 2, the XRD pattern of the SLN showed the peaks at 2θ value of 21.55° and 24.05°. The diffraction pattern of pure RSV showed different peaks at 2θ value of 16.36°, 19.18°, 22.67°, 23.02°, and 27.67°, indicating highly crystalline nature structures. In addition, as for the XRD pattern of RSV-SLN, some similar characteristic peaks were also observed, which suggests that the RSV was present in crystalline state in SLN [14,20].

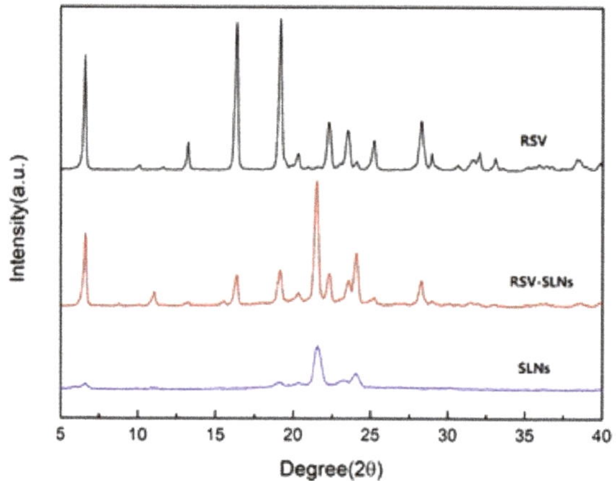

Figure 2. Powder X-ray diffraction patterns for RSV, RSV-SLN, and SLN.

3.3. Fourier-Transform Infrared Raman (FTIR) Spectroscopy of RSV-SLN

The FTIR spectra of RSV, RSV-SLN, and SLN (Figure 3) displayed no peak shifting and no loss of RSV peaks. In the FT-IR spectra, pure RSV has characteristic peaks around 1587 cm^{-1} representing benzene skeleton vibration. Peaks seen around 830 cm^{-1} represent the bending vibration of C=C–H. The spectra of RSV-SLN showed a part of the functional characteristic peaks of RSV, probably because of RSV molecular dispersion or entrapment within SLN [16]. Moreover, no additional peaks were observed in the RSV-SLN spectra, indicating that the loading of RSV into SLN did not change the nature of SLN. Therefore, this study confirmed that no interaction occurred between RSV and the solid lipid component of SLN, indicating that RSV was compatible with SLN formulations; this finding was consistent with previous reports [21].

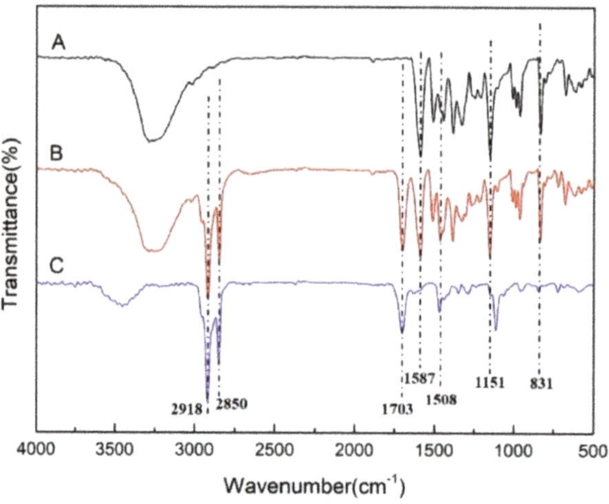

Figure 3. Fourier-transform infrared Raman (FTIR) spectra of RSV, RSV-SLN, and SLN; (**A**) FTIR spectra for RSV; (**B**) FTIR spectra for RSV-SLN; (**C**) FTIR spectra for SLN.

3.4. *Effect of RSV-SLN Supplementation on Endurance Performance and Muscle Energy Utilization*

In this study, we observed the beneficial effects of RSV and RSV-SLN supplementation on excessive endurance exercise challenges and measured other biochemical indicators after 8 weeks of experimental intervention. The running distance to exhaustion significantly increased after RSV-SLN was administered (Figure 4A). The average exhaustive running distance of the EE + RSV-SLN group was 8617.78 m, which was longer by 28.7% than that of the EE group. These results suggested that RSV-SLN supplementation could effectively enhance endurance performance.

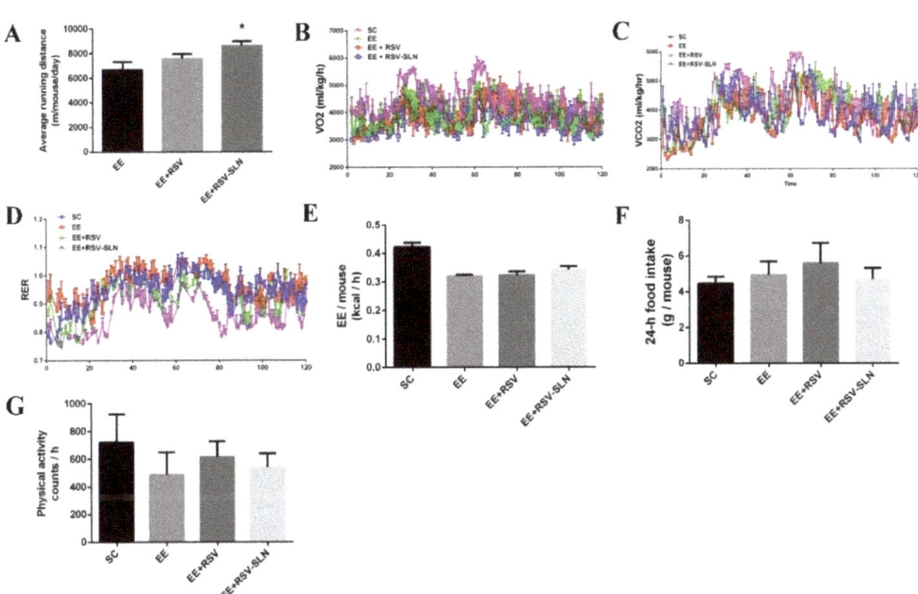

Figure 4. RSV-SLN supplementation enhances endurance performance. (**A**) Average running distances to the point of exhaustion. (**B**) Oxygen consumption (VO_2). (**C**) Carbon dioxide production (VCO_2). (**D**) Respiratory exchange ratio (RER). (**E**) Energy expenditure (EE). (**F**) Food intake for 24 h. (**G**) Physical activity of mice from four groups. * $p < 0.05$ vs. SC (sedentary control) group.

Muscle energy utilization is essential for endurance exercise performance [22]. As such, we evaluated the muscle energy utilization of RSV-SLN-supplemented mice by measuring their oxygen utilization (Figure 4B), CO_2 generation (Figure 4C), and respiratory exchange ratio (Figure 4D). During low-moderate intensity exercise, the main energy source is fat. Consistent with this observation, our results showed that the RER decreased in the EE + RSV-SLN group. This result was accompanied by the enhancement of endurance exercise tolerance (Figure 4A). These data suggested that the amount of fat consumed by the mice with RSV-SLN supplementation was higher, and their dependence on lipid metabolism was greater than that of the EE controls because a lower RER typically reflects a substrate shift favoring fat metabolism. No significant difference was found in other indicators (Figure 4E–G).

3.5. *Effect of RSV-SLN Supplementation on the Muscular Ultrastructural Changes of Mice that Underwent Excessive Endurance Exercise*

Ultrastructural changes are shown in Figure 5A,B. The muscle fibers in the SC group showed normal characteristics under TEM images. The M line and the Z line were arranged orderly, and the mitochondrion ridge was clearly observed (Figure 5A). After excessive endurance exercise was completed, the M line and the Z line were arranged disorderly, and the mitochondria were damaged, swollen, and vacuolized (Figure 5B). However, the damaged ultrastructure was significantly improved after RSV and RSV-SLN treatments, especially RSV-SLN, were administered. The disordered M and Z lines were relieved, and the number of morphologically normal mitochondria increased (Figure 5C,D).

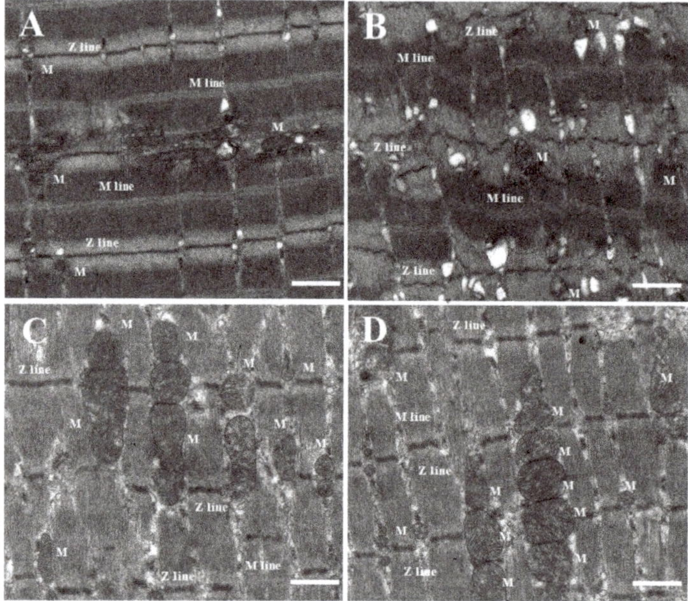

Figure 5. RSV-SLN alleviates the ultrastructural damage of skeletal muscles by excessive endurance exercise. Ultrastructure changes were shown via transmission electron microscopy from four groups of mice. Mt indicates mitochondria. The scale bars represent 500 nm in skeletal muscle tissues.

3.6. Effect of RSV-SLN Supplementation on Mitochondrial Function and Mitochondrial Long-Chain FA Translocase in Response to Excessive Endurance Exercise

We assume that, after RSV-SLN supplementation was administered, endurance performance was likely improved through the enhancement of mitochondrial function and mitochondrial long-chain FA translocase expression. As a result, energy insufficiency induced by long-term endurance exercise was alleviated. Enzymatic assays for mitochondrial respiratory chain complexes have been widely used to estimate mitochondrial function [23]. In our study, mitochondrial function was assessed by measuring the ATPase activity, Cyt c content, CS activity, and mRNA contents of mitochondrial complex IV (*COX IV*) and mitochondrial complex II (*COX II*) in the skeletal muscle tissue. Our results showed that long-term endurance exercise resulted in a decrease in the ATPase activity ($p < 0.05$), the CS activity ($p < 0.05$), and the transcription level of *COX IV* ($p < 0.05$) (Figure 6A,B,D), which induced an impairment in mitochondrial function. As expected, the CS activity and *COX IV* mRNA significantly increased after RSV-SLN was administered (Figure 6B,D), whereas the increase in the levels of ATPase activity, Cyt c content, and *COX II* mRNA was unnoticeable (Figure 6A,C,E). However, no changes in the mRNA levels of *CPT-1M* (a prominent isoform specific to skeletal muscle) and *CD36* were observed in response to excessive exercise with or without RSV-SLN application (Figure 6F,G).

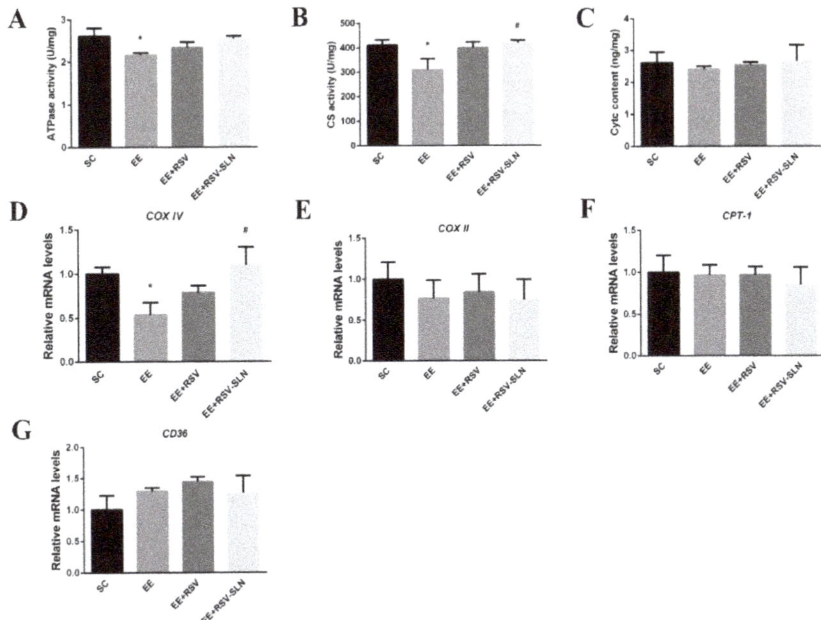

Figure 6. RSV-SLN alleviates mitochondrial dysfunction but does not alter mitochondrial long-chain FA translocase during excessive endurance exercise. (**A**) ATPase activity. (**B**) CS activity. (**C**) Cyt c content. (**D**,**E**) mRNA expression levels of *COX IV* and *COX II*. (**F**,**G**) mRNA expression levels of *CPT-1* and *CD36* in the skeletal muscle from four groups of mice. * $p < 0.05$ vs. SC group; # $p < 0.05$ vs. EE group. FA: fatty acid; CS: citrate synthase; Cyt c: cytochrome c; COX IV: mitochondrial complex IV; COX II: mitochondrial complex II.

3.7. Effect of RSV-SLN Supplementation on Mitochondrial QC in Response to Excessive Endurance Exercise

The above results prompted us to explore the molecular mechanism involved in the amelioration of mitochondrial function through RSV-SLN supplementation in response to excessive endurance exercise. The healthy mitochondrial population is dynamic, and it shows a variable turnover rate in an active transcriptional process that links mitochondrial biogenesis to the degradation of damaged and senescent mitochondria via mitophagy [24]. Mitochondrial biogenesis and mitophagy were investigated to address mitochondrial function via the stimulation of the mitochondrial QC. Our results indicated that the mitochondrial DNA copy number decreased in mice that were subjected to excessive endurance exercise (Figure 7A), and this observation was consistent with the results on CS activity.

The activity of CS is a marker of mitochondrial content [25]. Furthermore, we measured the peroxisome proliferator receptor gamma coactivator-1 alpha (*PGC-1α*), which is the main regulator of mitochondrial biogenesis, and its downstream transcription factors, namely, mitochondrial transcription factor A (*Tfam*) and nuclear respiratory factor 1 (*NRF-1*). Interestingly, the mRNA levels of *PGC-1α*, *Tfam*, and *NRF-1* were significantly upregulated (Figure 7B–D) and incompatible with mtDNA content. As such, we aimed to explore the status of mitophagy. Consequently, we found that excessive endurance exercise significantly increased the expression levels of genes, including Bcl-2/adenovirus E1B 19 kD-interacting protein 3 (*Bnip3*), and *Beclin-1*, involved in mitophagy (Figure 7E,F). These results indicated that the increase in mitophagy, but not the suppression of mitochondrial biogenesis, resulted in the reduced mitochondrial content.

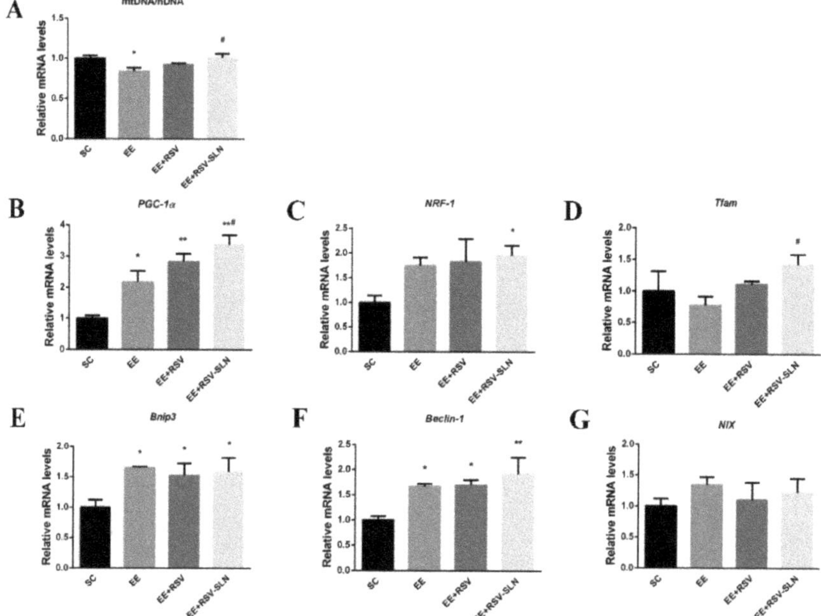

Figure 7. RSV-SLN elevates mitochondrial content by upregulating mitochondrial biogenesis and enhancing mitophagy capacity during excessive endurance exercise. (**A**) Mitochondrial DNA copy number (mtDNA). (**B–D**) The mRNA expression profile of mitochondrial biogenesis, *PGC-1α*, *NRF1*, and *Tfam*. (**E–G**) The mRNA expression profile of mitophagy, *Bnip3*, *Beclin-1*, and *NIX* in the skeletal muscle from four groups of mice. * $p < 0.05$, and ** $p < 0.01$ vs. SC group; # $p < 0.05$ vs. EE group.

In accordance with the above observations, the mtDNA content of the mice exhausted through exercise following RSV-SLN administration increased (Figure 7A). Consistent with the increased abundance of mitochondria, the transcript levels of *PGC-1α* and *Tfam* were significantly higher in the EE + RSV-SLN group than in the EE group. However, no changes were found in the transcription levels of *Bnip3* and *Beclin-1* or the mitophagic receptor NIP3-like protein X (*NIX*) in the skeletal muscle compared with those in the EE group, indicating that the increase in mitochondrial biogenesis contributed to the reversal of the mtDNA content but not to mitophagy. Altogether, our results implied a precise control between the production of functional mitochondria and the selective elimination of dysfunctional ones via RSV-SLN supplementation in mice that were subjected to excessive endurance exercise.

4. Discussion

In recent years, nanomaterials have been more used as drug delivery to improve their availability and efficiency. For instance, solid lipid nanoparticles have been introduced as an attractive alternative to traditional drug delivery systems. They are composed of physiological lipids and present many potential advantages, such as low toxicity, targeting ability, and bioavailability.

In this study, RSV-SLN was successfully prepared by emulsification and low-temperature solidification method. As observed via TEM and SEM, the RSV-SLN was spherical in shape with a narrow distribution. The mean particle size of RSV-SLNs with a nominal hydrodynamic diameter were about 112.5 ± 10.3 nm. The particle size with 100 nm range could improve cellular drug uptake and retention. The particle size measured by PCS was larger than those estimated by TEM. This is because PCS is used for hydrodynamic diameter examination, while TEM is used for characterization of particles in dried states. The zeta potential of RSV-SLN was −24.7 mV, which could avoid the aggregation and keep the system a good physical dispersion. The XRD and FTIR spectra of the RSV-SLN revealed the drug status in the SLN, which confirmed that the RSV drug retained its chemical properties and successfully encapsulated in the SLN.

Exhaustive exercise produces physical stress, which transiently disrupts homeostasis [26], and the working skeletal muscles are the most directly affected organ during physical activity [27]. Excessive exercise can cause excessive accumulation of reactive oxygen species (ROS), which may lead to the imbalance between oxidative intermediates and antioxidant systems, and develop oxidative stress in muscles [28,29]. Some nutrients can limit mitochondrial ROS generation and relieve severe oxidative stress [3]. Effective ROS scavengers not only reduce mitochondrial damage but also interfere with the redox signaling involved in mitochondrial quality control. Therefore, scientists have been trying to investigate natural antioxidant extracts to relieve excessive oxidative damage induced by exercise and improve mitochondrial function.

RSV has been widely explored because of the discovery of antioxidant effects [30]. RSV protects mitochondrial function and mass against stress-induced oxidative damage [31]. However, its pharmacokinetic properties are unfavorable. Thus, incorporating RSV into nano formulation may be a useful and viable approach. As it is reported, RSV-SLN can display a sustained release effect, which will help to improve the practical delivery of RSV [16]. RSV molecules are highly stable when they are entrapped in SLN, thereby protecting RSV from degradation in biochemical reactions [32]. Additionally, the potential penetrating ability and cell uptake capacity of RSV-SLN affect the tissue accumulation of RSV that may be responsible for the effects of RSV-SLN in muscle tissue [15]. We mainly explored the improved protective effects and underlying mechanism of RSV-SLN on endurance performance and focused on the mitochondrial QC process. Our results showed that RSV-SLN applications significantly extended the running distance as a duration performance indicator. The three possibilities were as follows: first, RSV-SLN supplementation contributed to a substrate shift in favor of fat metabolism, which may be conducive to long-term energy that supplies working muscles. Second, RSV-SLN, as mitochondrial nutrition supplementation, can improve mitochondrial function through the indicators of ATPase and CS activity, thereby protecting against energy insufficiency during prolonged endurance exercise; this finding also agreed with our previous findings [33]. Finally, the application of RSV-SLN enhanced mitochondrial quality and quantity to match the increased energy requirements during long-term endurance exercise.

High mitochondrial content is closely related to endurance performance and muscular oxidative capacity. This study showed that RSV-SLN administration could reverse the decrease in the mtDNA content in response to excessive endurance exercise, and this decrease was parallel to CS activity and *COX IV* mRNA levels. Studies have indicated that the number of mitochondria is maintained through mitochondrial biogenesis and autophagy [34]. Once this cycle is disrupted, cells not only become susceptible to loss of energy regulation but also become subjected to oxidative damage secondary to dysfunctional mitochondria. However, the underlying role of RSV-SLN in improving the number of mitochondria during prolonged endurance exercise remains unclear.

A relatively small number of DNA-binding transcriptional regulators, such as *PGC-1α*, *NRF-1*, and *Tfam*, are essential for mitochondrial biogenesis [35]. Other studies have indicated that RSV administration can upregulate mitochondrial biogenesis by increasing *PGC-1α* expression levels [33]. Similarly, despite the exhaustive training, RSV-SLN supplementation could still be positive and cause an increase in the mRNA expression levels of the upstream markers of mitochondrial biogenesis (*PGC-1α* and *Tfam*) and downstream targets (CS and *COX IV*). Interestingly, in the EE group, the increase in the mRNA expression of mitochondrial proliferation regulators did not translate into an increase in mtDNA. The cause of these changes might be related to the aggravation of mitochondrial damage and mitophagy. Thus, mitophagy was often regarded as another center mechanism of mitochondrial quality and quantity control.

Different mitophagy effectors, including the mitophagy receptors NIX, Bnip3, and PINK1/Parkin pathway, are involved in the selective clearance of mitochondria [36,37]. Bnip3 is an important mitochondrial redox sensor, which leads to cell death when cells are subjected to severe oxidative stress [38]. Under conditions such as hypoxia, Bnip3 and NIX may interact with Bcl-2 and release Beclin-1, resulting in the autophagic removal of mitochondria [36]. Our study showed that the transcriptional levels of *Bnip3* and *Beclin-1* were obviously upregulated in EE mice, suggesting that mitophagy that is induced by long-term endurance exercise might be dependent on the Bnip3/Beclin-1 pathway. We also observed a robust increase in the mRNA expression levels of *Bnip3* and *Beclin-1* after RSV or RSV-SLN supplementation when the mice were subjected to excessive exercise, indicating that RSV and RSV-SLN applications contributed to the clearing of damaged mitochondria. Taken together, our data demonstrated that the integration of mitochondrial biogenesis and mitophagy might be necessary to maintain energy supply during excessive endurance exercise; moreover, RSV-SLN could also partially relieve the exhaustive-exercise-induced mitochondrial response in vivo. Thus, we propose the probable mechanism as follows: first, drugs loaded in nanoparticle carriers can improve cellular drug uptake and retention via endocytosis in the cell membrane [39]. Second, the nanoparticle system may target mitochondrial sites via passive or active mechanisms that likely improve mitochondrial function [40]. Overall, we provided a novel perspective of RSV nanomaterial supplementation in improving endurance performance through a mitochondrion-mediated pathway (Figure 8).

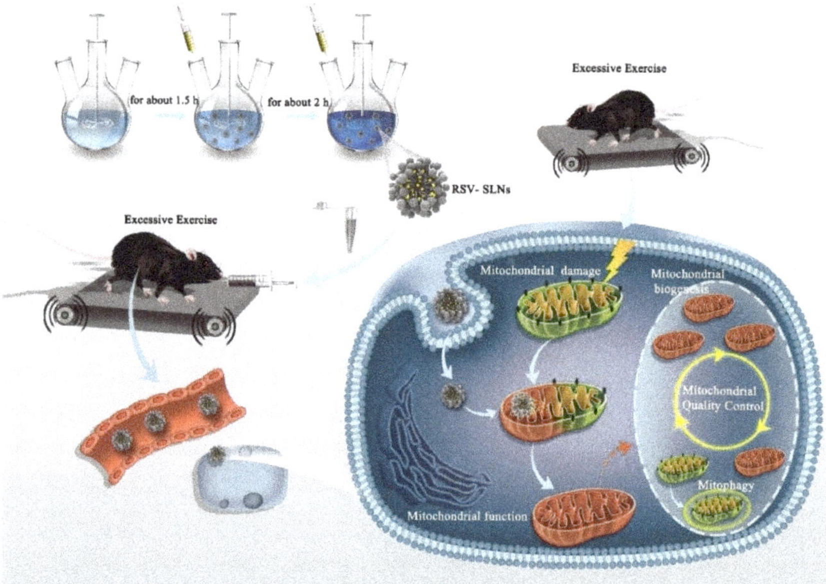

Figure 8. Schematic shows the mechanisms underlying RSV-SLNs to improve endurance performance by modulating mitochondrial quality control. RSV-SLNs were prepared through emulsification and low-temperature solidification. Results indicated that RSV-SLNs were superior to RSV because permeability and retention of the former were better than those of the latter. The possible biological mechanism is as follows: RSV-SLNs, as mitochondrial nutrients, can alleviate excessive endurance exercise-produced mitochondrial stress and improve mitochondrial function by ameliorating mitochondrial QC, thereby enhancing endurance performance and fatigue recovery.

5. Conclusions

In summary, we revealed that an exhausting endurance exercise induces a decrease in mitochondrial quantity and quality through an imbalance between mitochondrial biogenesis and mitophagy, thereby impairing mitochondrial function. In turn, this impaired mitochondrial function leads to an insufficient energy supply. However, RSV-SLN administration might enhance mitochondrial function by ameliorating mitochondrial QC, thereby improving endurance performance, which was superior to RSV application. These results revealed an unexpected role for RSV-SLN compared with RSV in terms of linking nutrient deprivation to mitochondrial oxidant production through mitochondrial QC; moreover, a possible mechanism involved in improving endurance performance was identified. This study highlighted new possibilities for anti-physical-fatigue strategies.

Author Contributions: Conceptualization, J.S., T.L. and L.Q.; Data curation, Y.Z., Y.S. and Y.C.; Formal analysis, J.D.; Funding acquisition, J.S., Y.Z. and L.Q.; Methodology, S.L. and Q.H.; Project administration, L.Q.; Supervision, T.L.; Writing—original draft, J.S. and L.Q.; Writing—review and editing, T.L.

Funding: This research was funded by the National Natural Science Foundation of China (grant number 31600966, 31771313, and 31401019), the scientific fitness guidance project of National Sports General Administration of China (grant number 2017B026), Tengfei project of the Shanghai Municipal Sports Bureau (grant number 16T006) and the Fundamental Research Fund for the Central Universities of Tongji University.

Conflicts of Interest: The authors declare no conflict of interest.

References

1. Tanaka, M.; Baba, Y.; Kataoka, Y.; Kinbara, N.; Sagesaka, Y.M.; Kakuda, T.; Watanabe, Y. Effects of (-)—Epigallocatechin gallate in liver of an animal model of combined (physical and mental) fatigue. *Nutrition* **2008**, *24*, 599–603. [CrossRef] [PubMed]
2. Filler, K.; Lyon, D.; Bennett, J.; McCain, N.; Elswick, R.; Lukkahatai, N.; Saligan, L.N. Association of Mitochondrial Dysfunction and Fatigue: A Review of the Literature. *BBA Clin.* **2014**, *1*, 12–23. [CrossRef] [PubMed]
3. Huang, C.C.; Hsu, M.C.; Huang, W.C.; Yang, H.R.; Hou, C.C. Triterpenoid-Rich Extract from Antrodia camphorata Improves Physical Fatigue and Exercise Performance in Mice. *Evid.-Based Complement. Alternat. Med.* **2012**, *2012*, 1–8.
4. Sun, J.; Zhang, C.; Kim, M.; Su, Y.; Qin, L.; Dong, J.; Zhou, Y.; Ding, S. Early potential effects of resveratrol supplementation on skeletal muscle adaptation involved in exercise-induced weight loss in obese mice. *BMB Rep.* **2018**, *51*, 200–205. [CrossRef] [PubMed]
5. Busch, F.; Mobasheri, A.; Shayan, P.; Lueders, C.; Stahlmann, R.; Shakibaei, M. Resveratrol modulates interleukin-1beta-induced phosphatidylinositol 3-kinase and nuclear factor kappaB signaling pathways in human tenocytes. *J. Biol. Chem.* **2012**, *287*, 38050–38063. [CrossRef] [PubMed]
6. Zhao, H.; Niu, Q.; Li, X.; Liu, T.; Xu, Y.; Han, H.; Wang, W.; Fan, N.; Tian, Q.; Zhang, H.; et al. Long-term resveratrol consumption protects ovariectomized rats chronically treated with D-galactose from developing memory decline without effects on the uterus. *Brain Res.* **2012**, *1467*, 67–80. [CrossRef] [PubMed]
7. Walle, T.; Hsieh, F.; DeLegge, M.H.; Oatis, J.E.; Walle, U.K. High absorption but very low bioavailability of oral resveratrol in humans. *Drug Metab. Dispos.* **2004**, *32*, 1377–1382. [CrossRef]
8. Neves, A.R.; Lucio, M.; Lima, J.L.; Reis, S. Resveratrol in medicinal chemistry: A critical review of its pharmacokinetics, drug-delivery, and membrane interactions. *Curr. Med. Chem.* **2012**, *19*, 1663–1681. [CrossRef]
9. Doktorovova, S.; Silva, A.M.; Gaivão, I.; Souto, E.B.; Teixeira, J.P.; Martins-Lopes, P. Comet assay reveals no genotoxicity risk of cationic solid lipid nanoparticles. *J. Appl. Toxicol.* **2014**, *34*, 395–403. [CrossRef]
10. Doktorovova, S.; Souto, E.B.; Silva, A.M. Nanotoxicology applied to solid lipid nanoparticles and nanostructured lipid carriers—A systematic review of in vitro data. *Eur. J. Pharm. Biopharm.* **2014**, *87*, 1–18. [CrossRef]
11. Souto, E.B.; Doktorovová, S. Chapter 6—Solid lipid nanoparticle formulations pharmacokinetic and biopharmaceutical aspects in drug delivery. *Methods Enzymol.* **2009**, *464*, 105–129. [PubMed]

12. Souto, E.B.; Müller, R.H. Lipid nanoparticles: Effect on bioavailability and pharmacokinetic changes. *Handb. Exp. Pharmacol.* **2010**, *197*, 115–141.
13. Teixeira, M.C.; Carbone, C.; Souto, E.B. Beyond liposomes: Recent advances on lipid based nanostructures for poorly soluble/poorly permeable drug delivery. *Prog. Lipid Res.* **2017**, *68*, 1–11. [CrossRef] [PubMed]
14. Jose, S.; Anju, S.S.; Cinu, T.A.; Aleykutty, N.A.; Thomas, S.; Souto, E.B. In vivo pharmacokinetics and biodistribution of resveratrol-loaded solid lipid nanoparticles for brain delivery. *Int. J. Pharm.* **2014**, *474*, 6–13. [CrossRef]
15. Mohseni, R.; ArabSadeghabadi, Z.; Ziamajidi, N.; Abbasalipourkabir, R.; RezaeiFarimani, A. Oral Administration of Resveratrol-Loaded Solid Lipid Nanoparticle Improves Insulin Resistance through Targeting Expression of SNARE Proteins in Adipose and Muscle Tissue in Rats with Type 2 Diabetes. *Nanoscale Res. Lett.* **2019**, *14*, 227. [CrossRef]
16. Gumireddy, A.; Christman, R.; Kumari, D.; Tiwari, A.; North, E.J.; Chauhan, H. Preparation, Characterization, and In vitro Evaluation of Curcumin- and Resveratrol-Loaded Solid Lipid Nanoparticles. *AAPS PharmSciTech* **2019**, *20*, 145. [CrossRef]
17. Kleiber, M. Metabolic turnover rate: A physiological meaning of the metabolic rate per unit body weight. *J. Theor. Biol.* **1975**, *53*, 199–204. [CrossRef]
18. Zhang, Q.; Zheng, J.; Qiu, J.; Wu, X.; Xu, Y.; Shen, W.; Sun, M. ALDH2 restores exhaustive exercise-induced mitochondrial dysfunction in skeletal muscle. *Biochem. Biophys. Res. Commun.* **2017**, *485*, 753–760. [CrossRef]
19. Pieters, N.; Koppen, G.; Smeets, K.; Napierska, D.; Plusquin, M.; De Prins, S.; Van De Weghe, H.; Nelen, V.; Cox, B.; Cuypers, A.; et al. Decreased mitochondrial DNA content in association with exposure to polycyclic aromatic hydrocarbons in house dust during wintertime: From a population enquiry to cell culture. *PLoS ONE* **2013**, *8*, e63208. [CrossRef]
20. Wang, W.; Zhang, L.; Chen, T.; Guo, W.; Bao, X.; Wang, D.; Ren, B.; Wang, H.; Li, Y.; Wang, Y.; et al. Anticancer Effects of Resveratrol-Loaded Solid Lipid Nanoparticles on Human Breast Cancer Cells. *Molecules* **2017**, *22*, 1814. [CrossRef]
21. Ramalingam, P.; Ko, Y.T. Improved oral delivery of resveratrol from N-trimethyl chitosan-g-palmitic acid surface-modified solid lipid nanoparticles. *Colloids Surf. B Biointerfaces* **2016**, *139*, 52–61. [CrossRef] [PubMed]
22. Fu, T.; Xu, Z.; Liu, L.; Guo, Q.; Wu, H.; Liang, X.; Zhou, D.; Xiao, L.; Liu, L.; Liu, Y.; et al. Mitophagy Directs Muscle-Adipose Crosstalk to Alleviate Dietary Obesity. *Cell Rep.* **2018**, *23*, 1357–1372. [CrossRef] [PubMed]
23. Lobo-Jarne, T.; Ugalde, C. Respiratory chain supercomplexes: Structures, function and biogenesis. *Semin. Cell Dev. Biol.* **2018**, *76*, 179–190. [CrossRef] [PubMed]
24. Palikaras, K.; Tavernarakis, N. Mitochondrial homeostasis: The interplay between mitophagy and mitochondrial biogenesis. *Exp. Gerontol.* **2014**, *56*, 182–188. [CrossRef] [PubMed]
25. Larsen, S.; Nielsen, J.; Hansen, C.N.; Nielsen, L.B.; Wibrand, F.; Stride, N.; Schroder, H.D.; Boushel, R.; Helge, J.W.; Dela, F.; et al. Biomarkers of mitochondrial content in skeletal muscle of healthy young human subjects. *J. Physiol.* **2012**, *590*, 3349–3360. [CrossRef] [PubMed]
26. Mastorakos, G.; Pavlatou, M. Exercise as a stress model and the interplay between the hypothalamus-pituitary-adrenal and the hypothalamus-pituitary-thyroid axes. *Horm. Metab. Res.* **2005**, *37*, 577–584. [CrossRef] [PubMed]
27. Hoene, M.; Franken, H.; Fritsche, L.; Lehmann, R.; Pohl, A.K.; Haring, H.U.; Zell, A.; Schleicher, E.D.; Weigert, C. Activation of the mitogen-activated protein kinase (MAPK) signalling pathway in the liver of mice is related to plasma glucose levels after acute exercise. *Diabetologia* **2010**, *53*, 1131–1141. [CrossRef]
28. Fogarty, M.C.; Hughes, C.M.; Burke, G.; Brown, J.C.; Trinick, T.R.; Duly, E.; Bailey, D.M.; Davison, G.W. Exercise-induced lipid peroxidation: Implications for deoxyribonucleic acid damage and systemic free radical generation. *Environ. Mol. Mutagen.* **2011**, *52*, 35–42. [CrossRef]
29. Alessio, H.M.; Hagerman, A.E.; Fulkerson, B.K.; Ambrose, J.; Rice, R.E.; Wiley, R.L. Generation of reactive oxygen species after exhaustive aerobic and isometric exercise. *Med. Sci. Sports Exerc.* **2000**, *32*, 1576–1581. [CrossRef]
30. Haohao, Z.; Guijun, Q.; Juan, Z.; Wen, K.; Lulu, C. Resveratrol improves high-fat diet induced insulin resistance by rebalancing subsarcolemmal mitochondrial oxidation and antioxidantion. *J. Physiol. Biochem.* **2015**, *71*, 121–131. [CrossRef]

31. Sheu, S.J.; Liu, N.C.; Ou, C.C.; Bee, Y.S.; Chen, S.C.; Lin, H.C.; Chan, J.Y. Resveratrol stimulates mitochondrial bioenergetics to protect retinal pigment epithelial cells from oxidative damage. *Investig. Ophthalmol. Vis. Sci.* **2013**, *54*, 6426–6438. [CrossRef] [PubMed]
32. Qin, L.; Wang, W.; You, S.; Dong, J.; Zhou, Y.; Wang, J. In vitro antioxidant activity and in vivo antifatigue effect of layered double hydroxide nanoparticles as delivery vehicles for folic acid. *Int. J. Nanomed.* **2014**, *9*, 5701–5710. [CrossRef] [PubMed]
33. Lagouge, M.; Argmann, C.; Gerhart-Hines, Z.; Meziane, H.; Lerin, C.; Daussin, F.; Messadeq, N.; Milne, J.; Lambert, P.; Elliott, P.; et al. Resveratrol improves mitochondrial function and protects against metabolic disease by activating SIRT1 and PGC-1alpha. *Cell* **2006**, *127*, 1109–1122. [CrossRef] [PubMed]
34. Gottlieb, R.A.; Stotland, A. MitoTimer: A novel protein for monitoring mitochondrial turnover in the heart. *J. Mol. Med.* **2015**, *93*, 271–278. [CrossRef] [PubMed]
35. Scarpulla, R.C. Transcriptional paradigms in mammalian mitochondrial biogenesis and function. *Physiol. Rev.* **2008**, *88*, 611–638. [CrossRef] [PubMed]
36. Parikh, S.M.; Yang, Y.; He, L.; Tang, C.; Zhan, M.; Dong, Z. Mitochondrial function and disturbances in the septic kidney. *Semin. Nephrol.* **2015**, *35*, 108–119. [CrossRef] [PubMed]
37. Wei, H.; Liu, L.; Chen, Q. Selective removal of mitochondria via mitophagy: Distinct pathways for different mitochondrial stresses. *Biochim. Biophys. Acta* **2015**, *1853*, 2784–2790. [CrossRef]
38. Kubli, D.A.; Quinsay, M.N.; Huang, C.; Lee, Y.; Gustafsson, A.B. Bnip3 functions as a mitochondrial sensor of oxidative stress during myocardial ischemia and reperfusion. *Am. J. Physiol. Heart Circ. Physiol.* **2008**, *295*, H2025–H2031. [CrossRef]
39. Wang, J.; Wang, H.; Zhu, R.; Liu, Q.; Fei, J.; Wang, S. Anti-inflammatory activity of curcumin-loaded solid lipid nanoparticles in IL-1beta transgenic mice subjected to the lipopolysaccharide-induced sepsis. *Biomaterials* **2015**, *53*, 475–483. [CrossRef]
40. Choi, J.Y.; Gupta, B.; Ramasamy, T.; Jeong, J.H.; Jin, S.G.; Choi, H.G.; Yong, C.S.; Kim, J.O. PEGylated polyaminoacid-capped mesoporous silica nanoparticles for mitochondria-targeted delivery of celastrol in solid tumors. *Colloids Surf. B Biointerfaces* **2018**, *165*, 56–66. [CrossRef]

Article

Synthesis of MgFe Layered Double Hydroxide from Iron-Containing Acidic Residual Solution and Its Adsorption Performance

Alin Golban, Lavinia Lupa *, Laura Cocheci * and Rodica Pode

Politehnica University of Timisoara, Faculty of Industrial Chemistry and Environmental Engineering, 6 Vasile Parvan Blvd., Timisoara 300223, Romania; rodica.pode@upt.ro

* Correspondence: lavinia.lupa@upt.ro; laura.cocheci@upt.ro; Tel.: +402-5640-4159

Received: 19 August 2019; Accepted: 1 October 2019; Published: 3 October 2019

Abstract: The paper presents a new method of layered double hydroxide (LDH) synthesis starting from secondary sources, namely acidic residual solutions. The iron content of the acidic solution resulting from the pickling step of the hot-dip galvanizing process make it suitable to be used as an iron precursor in LDH synthesis. Here, Mg_4Fe–LDH synthesized through the newly proposed method presented structural and morphological characteristics similar to the properties of layered double hydroxides synthesized from analytical-grade reagents. Moreover, the as-synthesized LDH and its calcined product presented efficient adsorption properties in the removal process of Mo(VI) from aqueous solutions. The adsorption studies are discussed from the equilibrium, kinetic, and thermodynamic points of view. The proposed novel technologies present both economic and environmental protection benefits.

Keywords: iron precursor; acidic residual solution; LDH synthesis; Mo(VI) adsorption

1. Introduction

Concerns over water resource pollution are continuously increasing, which has intensified research efforts regarding water decontamination. Because issues arise when the contaminants are in trace amounts but still at a concentration which exceeds the maximum admitted values, one of the most studied water treatment methods is adsorption. The adsorption process has gained researchers' interest due to its simple operating conditions and the versatile types of adsorbent materials which exist on the market [1–3]. Among the multitude of adsorbent materials, a considerable amount of attention has been paid to layered double hydroxide (LDH) compounds. The general formula of an LDH is $[M^{II}_{1-x}M^{III}_{x}(OH)_2]^{x+}.[A^{n-}_{x/n} \cdot mH_2O]^{x-}$, where M^{II} is a divalent cation, M^{III} is a trivalent cation, and A^{n-} is an anion. Their lamellar structure is based on brucite-like sheets, where some divalent cations are replaced with trivalent cations, resulting in some positively charged layers that contain between them various anions such as CO_3^{2-}, Cl^-, NO_3^-, or even organic anions [4,5]. This structure increases their adsorptive properties, especially if the contaminant is in the form of oxyanions [6–8]. Due to the positive charge of the brucite-like lamellar layers, one of the main properties of synthetized LDHs is anionic exchange. If the LDH is thermal treated at temperatures up to 600 °C, when the obtained mixed metal oxides are immersed in aqueous solution, they are able to rehydrate and restore the lamellar structure of the LDH while retaining the anions present in solution in order to provide a neutral LDH molecule. This property is referred to as the "memory effect" and it is most often utilized to treat water containing undesirable anions or to introduce various anions into the LDH structure. From the multitude of Me^{2+}/Me^{3+} LDH types, much focus has been directed toward the Mg^{2+}/Fe^{3+} pair. A literature search revealed that Mg/Fe–LDH has been used for phosphate removal from aqueous solutions [9–11]; as adsorbent materials of various heavy metals, such as Cr, As, Pb, Zn, Cu, Se, Sb, and so forth [12–16]; for treatment of aqueous solutions contaminated with different reactive dyes [17–19]; or as a catalyst [20–22].

Presently, there is a focus on obtaining cheaper adsorbent materials that still retain properties similar to those obtained from pure chemicals. Therefore, the tendency is to replace some raw materials with secondary sources of various metals. This helps reduce production costs as well as protect the environment by recycling and recovering various waste products. To this end, some researchers have studied the possibility of obtaining LDHs by using, as a precursor of various metal ions, different industrial wastes, such as fly ash, zinc ash, furnace slag, aluminum slag, and so forth. [23–27]. In most cases, the wastes were used as precursors for magnesium, zinc, or aluminum ions. However, even though the obtained LDHs presented very good properties similar to LDHs obtained from reagents and the process helped the environment by recycling wastes, there were no significant reductions in the production cost. In all of these cases, the precursor was a solid waste, which first needed to be brought into solution. This required the use of some acid solutions and added an extra step in the LDH production process.

Therefore, in the present study, we present a new Mg/Fe–LDH obtained from secondary sources using, for the first time, an acidic residual solution resulting from the pickling step in the hot-dip galvanizing process as an iron precursor. In a previous study, we demonstrated that an Mg/Fe–LDH obtained using a secondary source as an iron precursor (iron sludge resulting from the hot-dip galvanizing process) presented properties similar to those obtained from reagents [28]. The novelty of this method is that, by directly using an acidic waste solution as an iron precursor, two steps are avoided in the process of obtaining LDHs. First, the acidic solution from the hot-dip galvanizing process is not neutralized in order to obtain the sludge, and second, the hydrometallurgical leaching of sludge is avoided when obtaining the iron precursor. In order to determine the efficiency of using the acidic residual solution as an iron precursor in the process of obtaining Mg/Fe–LDH, the LDH was analyzed and used as an adsorbent material in the removal process of Mo(VI) as molybdate ($MO_4{}^{2-}$) from aqueous solutions.

Due to the extensive use of Mo(VI) in many practices, molybdate anions ($MO_4{}^{2-}$), the most common oxyanions of Mo(VI), can be found in various waste waters, such as mining waters, scrubber effluent of municipal solid waste incinerators, waste waters from the production of stainless-steel or cast iron alloys, and waste water from the production of various pigments and catalysts for high-temperature chemical processes [29–31]. If molybdenum is present in drinking water at a level higher than 0.07 mg/L, which is the maximum value admitted by the World Health Organization (WHO) [32], it could be toxic to animals and humans [33]. Therefore, finding an efficient method to remove oxyanions from aqueous solutions is still a challenge, taking into account the fact that the traditional method of waste water treatment, precipitation, can remove only cations and not oxyanions [30]. There are few papers which report the removal of Mo(VI) from aqueous solutions involving an ionic exchange mechanism using various LDHs synthesized from pure reagents [7,8,34]. The purpose of this study was to compare the adsorption performance of the new, synthesized LDH from a secondary iron source with those reported in the literature.

2. Experimental

2.1. Mg/Fe–LDH Synthesis and Characterization

The acidic residual solution was received from a local hot-dip galvanizing plant and was subjected to chemical analysis in order to determine the metal ion concentrations. The concentration of Fe^{2+} was determined through titration with $KMnO_4$. The concentrations of total iron and other ions present in solution were determined by atomic absorption spectrometry using a Varian SpectrAA 280 FS spectrophotometer (Agilent, Santa Clara, CA, USA).

Mg/Fe–LDH was synthesized using the coprecipitation method at low oversaturation [4]: 200 mL of 1 M solution containing iron residual solution and magnesium nitrate, at a Mg:Fe molar ratio of 4:1, was added dropwise under continuous stirring to 100 mL of 1 M Na_2CO_3 solution. The pH of the suspension was maintained in the range of 10–11 using a 2 M NaOH solution. The suspension was aged at 70 °C for 20 h, then filtered and washed with distilled water several times until reaching a pH of 7. The obtained slurry was dried overnight, then milled and sieved in order to obtain particles with dimensions less than 90 µm, which were used in this study. A portion of the obtained Mg_4Fe–LDH was calcined at 450 °C at a heating rate of 10 °C/min. It was maintained at this temperature for 3 h using a Nabertherm oven. The obtained material was named Mg_4Fe-450.

The as-synthesized sample and the calcined one were characterized by powder X-ray diffraction (XRD), scanning electron microscopy (SEM), X-ray dispersion analysis (EDX), and specific surface area and pore volume. The RX diffractograms were recorded using a Rigaku Ultima IV X-ray diffractometer (Rigaku Analytical Devices Inc., Wilmington, MA, USA). SEM images were recorded using a Quanta FEG 250 microscope (FEI Company, Hillsboro, OR, USA) equipped with a ZAF-type EDX quantifier (FEI Company). A Micromeritics ASAP 2020 instrument (Micromeritics, Norcross, GA, USA) was used to determine the specific surface area and pore volume of the studied samples.

In order to determine the adsorption performance of the obtained materials, both samples—the as-synthesized LDH (Mg_4Fe) and the calcined one (Mg_4Fe-450)—were used in experiments to remove Mo(VI) from aqueous solutions.

The resulting adsorbent after Mo(VI) adsorption was analyzed using an FEI Tecnai F20 G2 TWIN TEM (FEI Company) at an accelerating voltage of 200 kV in bright field mode in order to determine the morphology.

2.2. Adsorption Studies

Both materials, the as-synthesized and the calcined samples, were used in the removal process of Mo(VI) anions from aqueous solutions. The adsorption process was conducted in batch mode using a Julabo SW23 shaker for sample shaking at a constant rotation speed (200 rpm). The adsorption performance of the obtained materials developed in the removal process of Mo(VI) anions from aqueous solutions was studied taking into account the influence of the Mo(VI) initial concentrations (15–200 mg/L), stirring time (5–240 min), and temperature (20, 35, and 50 °C). For all the adsorption studies, the solid:liquid ratio between the adsorbent and Mo(VI)-containing aqueous solutions was 0.025 g in 25 mL (i.e., solid:liquid ratio (S:L) = 1 g/L). The initial pH of the solution was adjusted to pH = 6.0 ± 0.5. According to the literature review and our previous studies [2,35], Mo(VI) adsorption onto LDHs is maximal at this pH due to the fact that, at this value, Mo(VI) is found in the solution under MoO_4^{2-}.

After each adsorption experiment, the samples were filtered in order to separate the phases and to analyze the residual concentration of Mo(VI). The Mo(VI) concentration before and after adsorption was determined through atomic absorption spectrometry using a Varian SpectrAA 280 FS spectrometer.

The mass balance presented in Equation (1) was used for the calculation of Mo(VI) uptake on 1 g of adsorbent:

$$q_e = \frac{(C_0 - C_e) \cdot V}{m} \tag{1}$$

where q_e is the adsorbed quantity of Mo(VI), expressed in milligrams per 1 g of studied adsorbent; C_o and C_e represent the Mo(VI) concentration of the aqueous solutions before adsorption and after the established equilibrium (mg/L); V is the volume of the Mo(VI)-containing aqueous solution (L); and m is the mass of the Mg_4Fe–LDH (g) used in the experiments.

3. Results and Discussions

3.1. Material Characterizations

It is estimated that, in the European Union, almost 300,000 m^3 of acidic residual solution is produced annually in the hot-dip galvanizing industry. [36] Due to its corrosive nature and high concentration of metal ions, this solution is considered a toxic waste and requires treatment before dumping. The treatment processes for spent pickling solutions can be classified as neutralization treatments, treatments for acid recovery, and treatments for metal recovery. Neutralization treatments consist of adding NaOH or $Ca(OH)_2$ when the metal ions are precipitated, and the resulting sludge is dried and considered a solid waste. Treatment processes for acid recovery include evaporation, membrane processes (membrane dialysis, membrane electrolysis, and membrane distillation), and processes with fluidized beds (Ruthner or spray-roasting processes). Metal ion recovery can employ ion exchange, crystallization, or solvent extraction. The physicochemical characterization of the acidic residual solution resulting from the pickling step of the hot-dip galvanizing process is presented in Table 1. It can be observed that the residual solution contains a high concentration of iron (65 g/L), which makes it suitable to be used as an iron precursor in LDH synthesis. The concentration of iron ions in the residual solutions can be considered constant because these acidic solutions resulting from the pickling step in the hot-dip galvanizing industry are continuously analyzed, and when they achieve the maximum value of iron ions and the minimum value of acidity, they are removed from the process and can be used as a secondary source of iron ions for LDH manufacturing applications. Besides iron ions, the residual solution also contains other metal ions, but in a smaller concentration; their presence does not inhibit the synthesis of LDHs.

Table 1. Physicochemical characterization of the acidic residual solution.

No.	Parameter	Value
1.	pH	2.15
2.	Fe	65.0 g/L
3.	Cu	57.4 mg/L
4.	Cd	0.453 mg/L
5.	Pb	15.2 mg/L
6.	Zn	565 mg/L
7.	Cl^-	81.7 g/L
8.	TOC	68 mg/L

The XRD patterns of the obtained adsorbent are presented in Figure 1. It can be observed that the Mg_4Fe obtained from a secondary source of iron showed a unique main phase which corresponded to pyroaurite. The cell parameters a and c were calculated using Bragg's law, assuming a rhombohedral symmetry of crystallization. The cell parameter a represents the cation–cation distance within the brucite-like layer ($a = 2 \cdot d_{(110)} = 3.11$ Å) and c represents the adjacent distance of the hydroxide layer ($c = 3 \cdot d_{(003)} = 23.9$ Å), where $d = \lambda/2 \sin \theta$ and $\lambda = 1.54056$ Å. The obtained results are in good agreement with other results reported by several authors [19,28,37]. The impurities present in the acidic residual solution were under 5% and they did not lead to the formation of secondary phases in the Mg_4Fe–LDH structure. In the case of the sample calcined at 450 °C, the LDH structure was modified, and the RX diffractogram was specific to a poor crystalline phase corresponding to MgO (periclase), with Fe^{3+} probably dispersed in the structure. Due to smaller Fe^{3+} than Mg^{2+} concentration utilized in the starting solution and the low temperature of the thermal treatment (450 °C) of Mg_4Fe, it was difficult to achieve the formation of iron oxide species such as magnesioferrite ($MgFe_2O_4$) or maghemite (Fe_2O_3). At this calcination temperature, the iron oxides were amorphous and the crystals were in the course of forming; therefore, they could not be identified by the RX analysis. These findings are also in agreement with other results reported in the literature [37,38]. For this reason, it is expected that this sample has the highest adsorption capacity in the removal process of Mo(VI) anions from aqueous solutions [19,28].

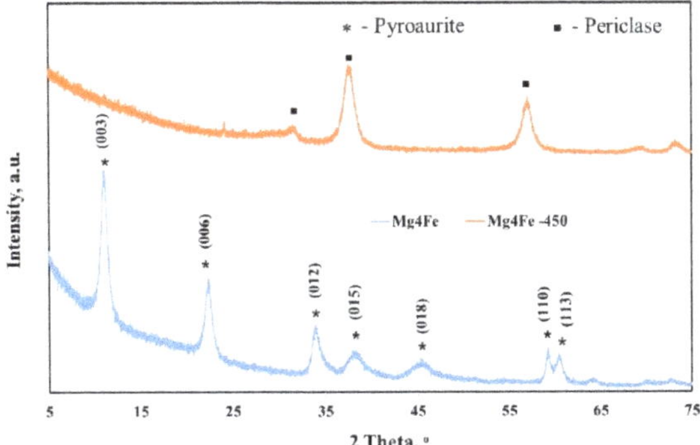

Figure 1. XRD pattern of the synthesized materials.

The morphology of the synthesized samples can be observed from the SEM images presented in Figure 2. The Mg_4Fe–LDH presented an aerated structure of fluffy particles. Through calcination, the sample became more amorphous, and the surface had aspects of cotton flowers.

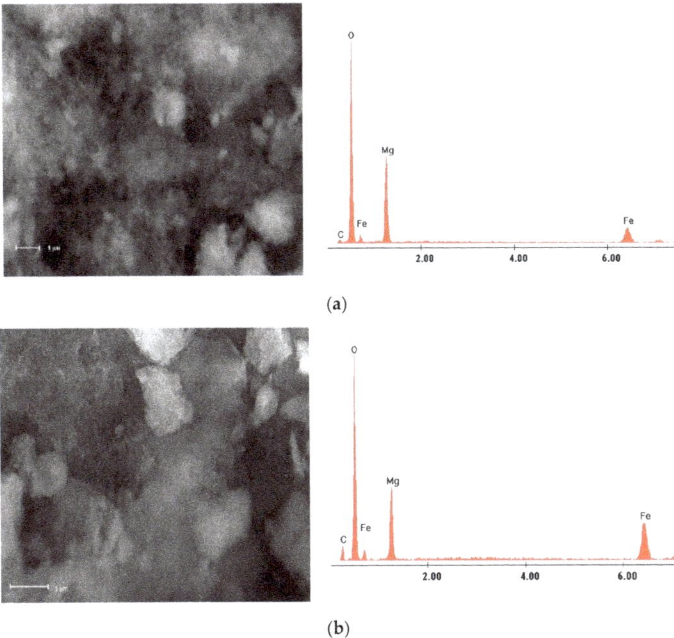

Figure 2. SEM images and X-ray dispersion analysis (EDX) spectra of the synthesized samples: (**a**) Mg_4Fe and (**b**) Mg_4Fe–450.

The EDX spectra presented in Figure 2 show the peaks of the characteristic elements of the studied materials. No characteristic peaks appeared for the impurities present in the precursor solution. This indicates that there was a negligible quantity of impurities.

The molar ratio between Mg and Fe ions from the studied samples together with their BET specific surface area and pore volume are presented in Table 2.

Table 2. Physical and chemical properties of the studied samples.

Parameters		Mg_4Fe	Mg_4Fe–450
Molar ratio	Theoretical	4	4
	Experimental	3.98	4.08
S_{BET} (m^2/g)		101	0.536
V_p (cm^3/g)		192	0.802
Cd, %		0.16	0.08
Pb, %		0.12	0.22
Zn, %		1.19	1.59

The impurities present in the spent pickling solution were analyzed in the washing solutions of the LDH in the as-synthesized solid samples and the calcined one and in the solutions after Mo(VI) adsorptions. It can be observed from the chemical analysis that impurities could be found in the solid synthesized samples but in a concentration under 2%, this being the reason why these do not appear in the EDX and RX analyses, as they are under the detection limit of 5%. Impurities could not be detected in the washing waters of the synthesized Mg_4Fe–LDH and in the residual solutions after Mo(VI) adsorption. This demonstrates that the impurities were not released during the adsorption process from the solid support.

3.2. Equilibrium Studies

The equilibrium isotherms representing the dependence of the adsorption capacities developed by the Mg_4Fe and Mg_4Fe-450 as a function of the Mo(VI) concentrations at equilibrium are presented in Figure 3. Increasing the initial concentration of Mo(VI) increased the active sites available for adsorption and, therefore, increased the adsorption capacities of both the studied materials. Mg_4Fe developed an experimental maximum adsorption capacity in the removal process of Mo(VI) from aqueous solutions of 39.9 mg/g. Further, due to the memory effect, the calcined samples exhibited a higher experimental maximum adsorption capacity of almost 50% (q_e = 52.8 mg/g). For aqueous solutions with Mo(VI) initial concentrations of ≤30 mg/L, the removal degree of Mo using a S:L ratio of 1 g/L was higher than 90%. For aqueous solutions with higher Mo(VI) initial concentrations, in order to obtain higher removal degrees, it was necessary to increase the S:L ratio. Other studies reported in the literature also present higher adsorption capacities for calcined samples compared with synthesized samples [18,28].

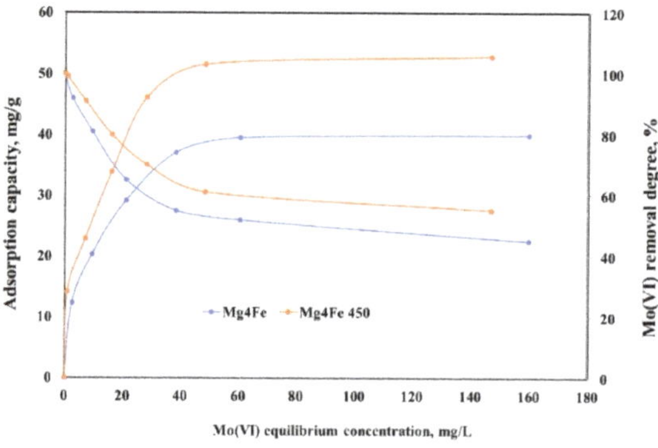

Figure 3. Equilibrium isotherms of Mo(VI) adsorption onto Mg_4Fe and Mg_4Fe-450. S:L = 1 g/L, t = 60 min, pH = 6, T = 20 °C.

Every adsorption study aims to determine the maximum adsorption capacity of the studied adsorbents and the equilibrium coefficient in order to achieve the design. Therefore, several isotherms, such as Langmuir, Freundlich, Temkin, and Dubinin–Radushkevich (DR) isotherms in their linear form, have been employed for this purpose [2,18,39]. The Langmuir isotherm supposes that the adsorption process takes place in a single layer on a uniform surface containing equivalent sites of the studied adsorbents. The Freundlich isotherm is used to express the affinity of the studied adsorbent to the retained pollutant. If the adsorption data fit the Temkin isotherm well, this means that the adsorbent surface is heterogeneous. In order to design the equilibrium adsorption regarding Mo(VI) removal using Mg_4Fe and Mg_4Fe-450, the linear graphs of the mentioned isotherms were plotted (Figure 4) and the obtained equilibrium isotherm parameters together with the regression coefficients are presented in Table 3.

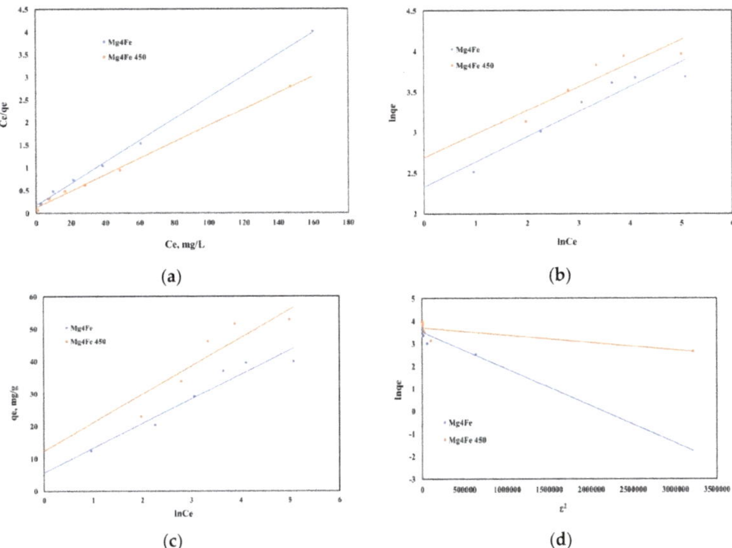

Figure 4. Equilibrium isotherms of Mo(VI) adsorption onto the studied adsorbent: (**a**) Langmuir, (**b**) Freundlich, (**c**) Temkin, and (**d**) Dubinin–Radushkevich.

Table 3. Equilibrium sorption isotherm parameters for Mo(VI) adsorption onto Mg_4Fe and Mg_4Fe-450.

Type of Isotherm	Parameter	Adsorbent	
		Mg_4Fe	**Mg_4Fe-450**
Langmuir	K_L, L/mg	0.137	0.151
	q_m calc, mg/g	42.1	55.2
	R^2	0.9981	0.9964
Freundlich	K_F, mg/g	9.33	14.7
	1/n	0.3086	0.2908
	R^2	0.9112	0.9330
Temkin	K_T, L/g	1.006	1.016
	b_T, J/mol	875	758
	R^2	0.9286	0.9074
Dubinin–Radushkevich	K_{ad}, mol^2/kJ^2	2×10^{-6}	3×10^{-7}
	q_s, mg/g	33.2	40.04
	R^2	0.7725	0.6644

Both studied materials presented affinity for Mo(VI) ion removal (1/n parameters were below unity).

By comparing the equilibrium isotherm parameters presented in Table 3, it can be concluded that Mo(VI) removal from the aqueous solutions occurred as a monolayer at the uniform surfaces of the Mg_4Fe and Mg_4Fe-450 materials because the Langmuir isotherm obtained the highest regression coefficients (closed to unity). Further, there was no significant difference between the maximum adsorption capacity calculated from its plot and those determined experimentally.

The lowest regression coefficients were obtained for the DR plots. At the same time, different values were obtained for the monomolecular adsorption capacities q_s for the studied materials and the values experimentally obtained in the adsorption process of Mo(VI) removal from aqueous solutions. The results suggest that the DR plot cannot be used to model the adsorption data of MoO_4^{2-} onto Mg_4Fe and Mg_4Fe-450 [18,39].

3.3. Kinetic Studies

Kinetic studies were used to determine the optimum time necessary to establish the equilibrium between the Mg_4Fe–LDH and the Mo(VI) anions. The experiments regarding the various stirring times were conducted at three different temperatures (Figure 5). Mo(VI) removal occurred quite quickly in the first minutes of contact between the adsorbent and adsorbate, especially when the calcined sample was used. After 60 min of stirring, the adsorption capacity increased slowly, so it can be considered that equilibrium was achieved in 60 min at all the studied temperatures for both adsorbent materials. The temperature increase led to a slight increase of the adsorption capacity of Mg_4Fe and Mg_4Fe-450 in the removal process of Mo(VI).

The well-known models of pseudo-first-order, pseudo-second-order, and intraparticle diffusion were used to simulate the kinetics of Mo(VI) adsorption onto the studied materials [18,19,40]. The representations of their linear plots are presented in Figures 6–8, and the kinetic parameters together with the obtained correlation coefficients are summarized in Table 4. From the linear representation of the experimental data according to the pseudo-first-order and pseudo-second-order kinetic models (Figures 6 and 7) and the interpretation of their parameters (Table 4), it can be concluded that Mo(VI) adsorption onto Mg_4Fe and Mg_4Fe-450, separately, is best described by the pseudo-second-order kinetic model. In this case, for all three temperatures, correlation coefficients were obtained that were close to 1, and the calculated adsorption capacities were similar to those determined experimentally.

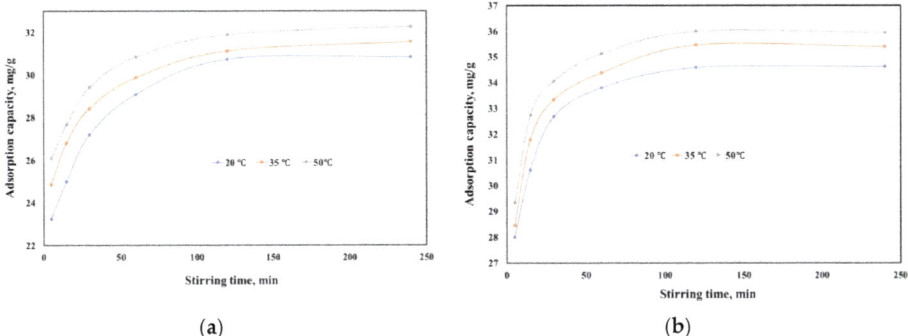

(a) (b)

Figure 5. Effect of contact time on the adsorption capacity of (**a**) Mg_4Fe and (**b**) Mg_4Fe-450 at three different temperatures in the removal process of MoO_4^{2-} from aqueous solutions. S:L = 1 g/L, C_0 = 50 mg/L Mo(VI), pH = 6.

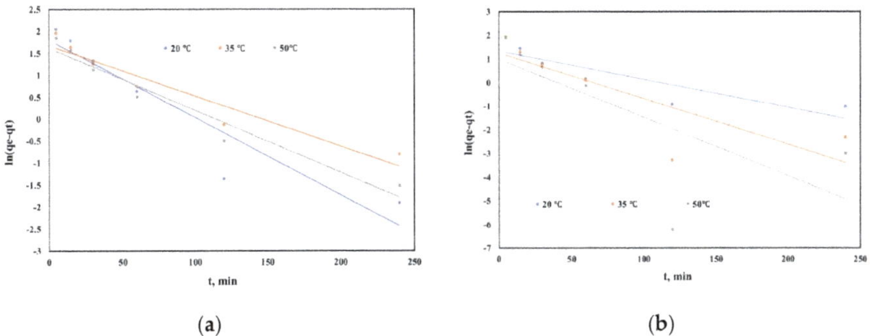

(a) (b)

Figure 6. Pseudo-first-order kinetic model of MoO_4^{2-} adsorption onto (**a**) Mg_4Fe and (**b**) Mg_4Fe-450.

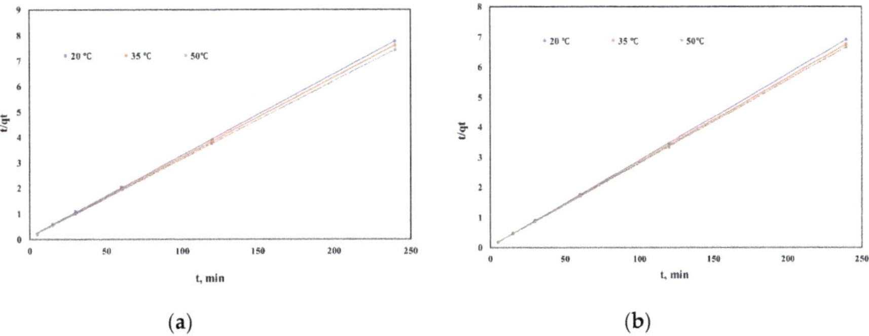

(a) (b)

Figure 7. Pseudo-second-order kinetic model of MoO_4^{2-} adsorption onto (**a**) Mg_4Fe and (**b**) Mg_4Fe-450.

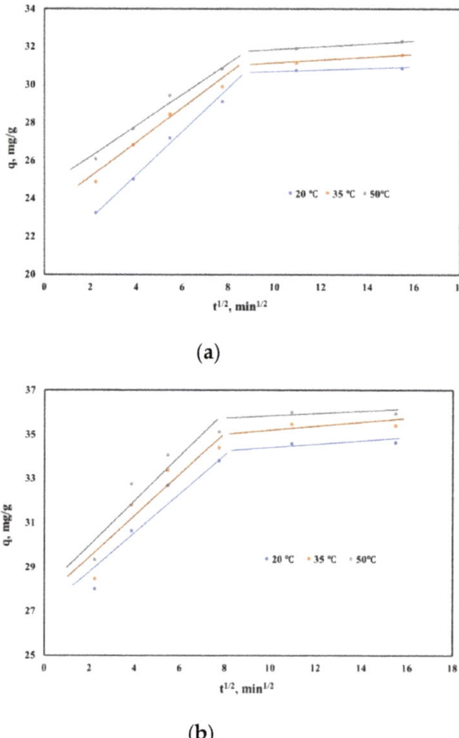

(a)

(b)

Figure 8. Intraparticle diffusion model of MoO_4^{2-} adsorption onto (**a**) Mg_4Fe and (**b**) Mg_4Fe-450.

Table 4. Kinetic model parameters for Mo(VI) adsorption onto Mg_4Fe and Mg_4Fe-450.

Model	Parameter	Adsorbent					
		Mg_4Fe			Mg_4Fe-450		
		Temperature			Temperature		
		20 °C	35 °C	50 °C	20 °C	35 °C	50 °C
	$q_{e\ exp}$, mg/g	31	32	32.5	35	35.5	36
Pseudo-first-order	K_1, min^{-1}	0.0176	0.0115	0.0142	0.0120	0.0194	0.0247
	q_e calc, mg/g	6.095	5.37	5.04	3.91	3.59	2.72
	R^2	0.8887	0.9189	0.9478	0.7772	0.6894	0.5104
Pseudo-second-order	K_2, min/(mg/g)	0.00934	0.0109	0.0119	0.0137	0.0163	0.0185
	q_e calc, mg/g	31.4	31.8	32.6	35	35.7	36.2
	R^2	0.9998	0.9999	0.9999	0.9999	0.9999	0.9999
Intraparticle diffusion	K_{int}	1.082	0.909	0.874	1.051	1.038	1.004
	R^2	0.9905	0.9777	0.9850	0.9356	0.8901	0.8780

The pseudo-second-order kinetic model can be used to simulate the experimental data regarding MoO_4^{2-} adsorption onto the studied materials; this means that the process is controlled by a chemical sorption [19,40,41].

Through the simulation of the experimental data according to the intraparticle diffusion model (Figure 8), it can be observed that the straight line does not pass through origin, indicating that the rate-limiting step for Mo(VI) adsorption is not the intraparticle diffusion. Also, the straight line presents a fragmentation after a while, suggesting that the adsorption process is more complex. The fast removal of Mo(VI) in the first minutes of contact between the adsorbent and adsorbate was controlled by the film diffusion when the studied adsorbent surfaces were covered with MoO_4^{2-} anions. After the surface coverage, the transportation of Mo(VI) inside the adsorbent particles occurred, as suggested by the second straight line obtained through the representation of q versus $t^{1/2}$ (Figure 8) [19].

3.4. Thermodynamic Studies

From the linear plot representation of the Arrhenius equation ($\ln(K_2)$ versus 1/T, Figure 9), the activation energy of Mo(VI) adsorption onto the studied materials was determined. The activation energy value confers information regarding the type of adsorption process (physical, when E < 4.2 kJ/mol, or chemical, when E > 4.2 kJ/mol) [33]. The activation energy determined for MoO_4^{2-} sorption onto Mg_4Fe and M_4Fe-450 was determined to be 6.37 and 7.89 kJ/mol, respectively, indicating chemisorption (Table 5).

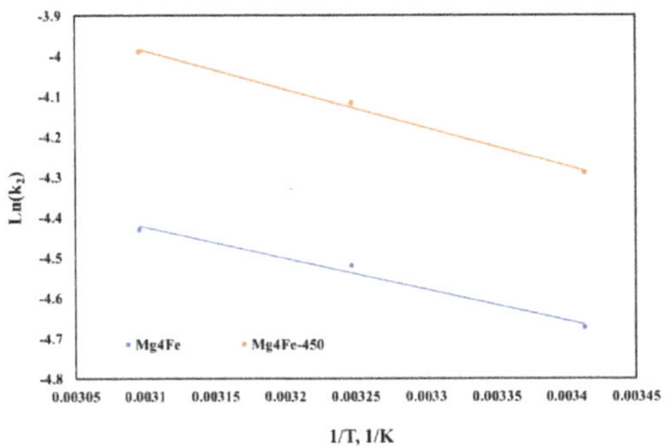

Figure 9. Arrhenius plot of MoO_4^{2-} adsorption onto the studied materials.

Table 5. Thermodynamic parameters for Mo(VI) adsorption onto Mg_4Fe and Mg_4Fe-450.

Parameter		Adsorbent	
		Mg_4Fe	Mg_4Fe-450
E, kJ/mol		6.37	7.89
ΔH^o, kJ/mol^{-1}		3.35	2.14
ΔS^o, J/(mol · K)		15.3	14.4
R^2		0.9869	0.9997
ΔG^o, kJ/mol	293 K	−1.13	−2.15
	308 K	−1.36	−2.30
	323 K	−1.59	−2.51

The thermodynamic equilibrium constant K_d, defined as the ratio between the Mo(VI) concentration in the solid phase and the solution at equilibrium, was used to calculate the Gibbs free energy (ΔG^o) at the studied temperatures and for the van 't Hoff representation in order to determine the thermodynamic parameters enthalpy (ΔH^o) and entropy (ΔS^o) for MoO_4^{2-} adsorption onto the studied materials [19,40,41]. The plot of the van 't Hoff representation is shown in Figure 10 and the thermodynamic parameters are listed in Table 5.

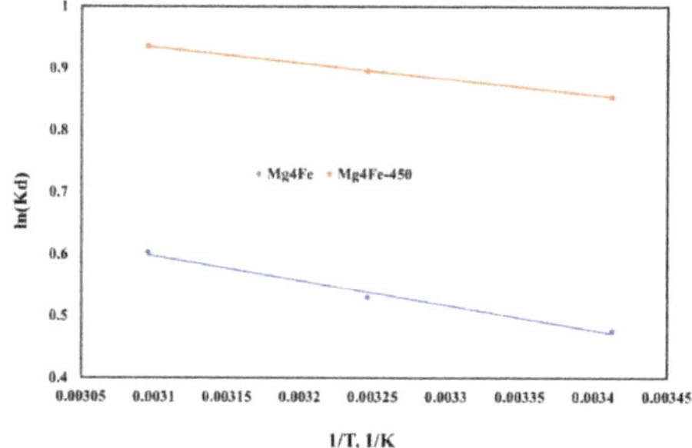

Figure 10. Van 't Hoff plot of MoO_4^{2-} adsorption onto the studied materials.

The MoO_4^{2-} adsorption onto the Mg_4Fe and Mg_4Fe-450 materials was an endothermic process, due to the obtained positive value for ΔH^o, and spontaneous because the Gibbs free energy presented negative values, which decreased as the temperature increased. Also, the positive values for ΔH^o and ΔS^o suggest that, during the adsorption process, the solid–liquid interface increases the randomness and the adsorption takes place due to chemical interactions [41].

3.5. Adsorption Performance

The adsorption performances of the synthesized materials developed in the removal process of MoO_4^{2-} adsorption from aqueous solutions were compared with the adsorption capacities developed by similar materials that were reported in the literature. The results are presented in Table 6, and it can be observed that Mg/Fe–LDH obtained from secondary sources can be efficiently used as an adsorbent material for treatment of water with dissolved Mo(VI).

Table 6. The comparison between the adsorption capacities of similar adsorbents developed for the treatment processes of aqueous solutions containing MoO_4^{2-} anions.

Adsorbent	q_m mg/g	References
Zn_3Al–CO_3-LDH	13.7	[2]
Zn_3Al-CLDH	39.8	
$ZnAl_3$–Cl-LDH	114.9	[8]
$ZnAl_3$–CO_3-LDH	<10	
$MgFeSO_4$-type HT-LDH	15.5	[42]
Mg_4Fe	42.1	Present paper
Mg_4Fe-450	55.2	

After adsorption, the solid materials were recovered by filtration and were subjected to XRD analysis. The XRD patterns of the Mg_4Fe and Mg_4Fe-450 are presented in Figure 11.

The as-synthesized material, Mg_4Fe, did not suffer any change in its crystalline structure after adsorption, with the unit cell parameters being a = 3.11 Å and c = 24.0 Å (compared to the unit cell parameters before adsorption: a = 3.11 Å and c = 23.9 Å). The 003 reflection was almost the same for Mg_4Fe before adsorption ($d_{(003)}$ = 7.967 Å) and after adsorption ($d_{(003)}$ = 8.000 Å). This demonstrates that the adsorption of Mo(VI), as molybdate anions, was performed on the LDH surface and not through anion exchange, considering that carbonate is the most difficult to replace among the anions present in the LDH interlayer gallery. On the other hand, as shown in Figure 11, the calcined material, Mg_4Fe-450, regained its layered double hydroxide structure, and all the characteristic peaks of pyroaurite were present in the diffractogram. The unit cell parameters of Mg_4Fe-450 after adsorption (a = 3.12 Å and c = 24.7 Å) suggest that the rehydration process of the calcined material (the memory effect) was developed through adsorption of Mo(VI) as molybdate anions from the solution and incorporated in the interlayer space of LDH. Furthermore, $d_{(003)}$ for Mg_4Fe-450 after adsorption increased to 8.233 Å. This increase in basal spacing indicates that molybdate anions were intercalated into the interlayer spaces of the reformed LDH. During Mo(VI) adsorption onto the calcined sample, instead of carbonate ions from the atmosphere, Mo(VI) anions were intercalated from the solutions due to the fact that the anion volume of Mo(VI) is higher than that of carbonate ($V_{MoO4}{}^{2-}$ = 0.088 nm^3; $V_{CO3}{}^{2-}$ = 0.061 nm^3) [43]. The retention of Mo(VI) through the memory effect between the interlayer gallery and also through adsorption onto the layer of the Mg_4Fe-450 surface explain the higher adsorption capacity developed by this material compared with its precursor. These results agree with other results reported in the literature [34,44].

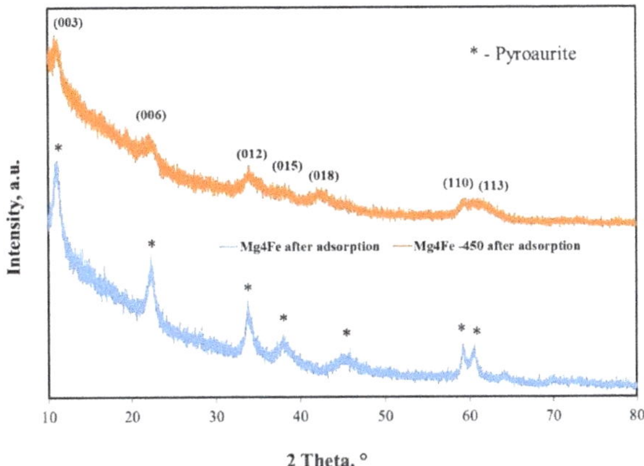

Figure 11. XRD patterns of the materials after Mo(VI) adsorption.

The morphology of the samples after Mo(VI) adsorption can be observed from the TEM images presented in Figure 12. The specific hexagonal morphology with ultrathin layers for the LDH samples can be observed in the presented TEM images. Due to their small crystallite size and the fact that it is difficult to isolate a unique crystal, the formation of some aggregates can be observed. The calcined sample, through the memory effect after Mo(VI) adsorption, returned to the hexagonal shape specific to LDH due to reformation of the layer structures.

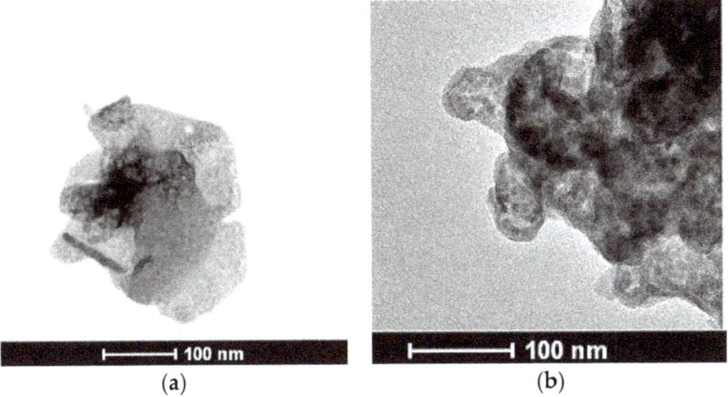

Figure 12. TEM images of the materials after Mo(VI) adsorption: (**a**) Mg_4Fe and (**b**) Mg_4Fe-450.

4. Conclusions

This paper reports the successful synthesis of a new Mg_4Fe–LDH using, as iron precursor, a secondary source, namely, the acidic residual solution resulting from the pickling step of the hot-dip galvanizing process. The obtained LDH presented similar properties to those obtained from pure reagents. The impurities present in the residual solutions, besides the iron ions, did not interfere with the structure of the obtained LDH. The obtained Mg_4Fe and the calcined product presented efficient adsorption properties in the removal process of MoO_4^{2-} from aqueous solutions. The proposed method of obtaining LDH presents multiple benefits: (1) it decreases the cost of obtaining LDH; (2) it decreases the cost of acidic residual solution neutralization; (3) it reduces waste discharge into the environment, and (4) it minimizes the use of raw materials.

Author Contributions: Conceptualization, L.L. and R.P.; methodology, L.C.; formal analysis, A.G. and L.L.; investigation, A.G., L.L. and L.C.; writing—original draft preparation, L.L. and R.P.; writing—review and editing, L.L. and L.C.; visualization, L.C.; supervision, R.P.

Funding: This research received no external funding.

Acknowledgments: The studies were done during the PhD program from the Doctoral School of the University Politehnica Timisoara.

Conflicts of Interest: The authors declare no conflict of interest.

References

1. Dore, E.; Frau, F.; Cidu, R. Antimonate Removal from Polluted Mining Water by Calcined Layered Double Hydroxides. *Crystals* **2019**, *9*, 410. [CrossRef]
2. Cocheci, L.; Lupa, L.; Lazau, R.; Voda, R.; Pode, R. Zinc recovery from waste zinc ash—A new "green" route for the preparation of Zn-Al layered double hydroxide used for molybdate retention. *J. Alloys Compd.* **2019**, *787*, 332–343. [CrossRef]
3. Lupa, L.; Cocheci, L.; Pode, R.; Hulka, I. Phenol adsorption using Aliquat 336 functionalized Zn-Al layered double hydroxide. *Sep. Pur. Technol.* **2018**, *196*, 82–95. [CrossRef]
4. Cavani, F.; Trifiro, F.; Vaccari, A. Hydrotalcite-type anionic clays: Preparation, properties and applications. *Catal. Today* **1991**, *11*, 173–301. [CrossRef]
5. Asiabi, H.; Yamini, Y.; Shamsayei, M.; Tahmasebi, E. Highly selective and efficient removal and extraction of heavy metals by layered double hydroxides intercalated with the diphenylamine-4-sulfonate: A comparative study. *Chem. Eng. J.* **2017**, *323*, 212–223. [CrossRef]
6. Carriazo, D.; del Arco, M.; Martin, C.; Rives, V. A comparative study between chloride and calcined carbonate hydrotalcites as adsorbents for Cr (VI). *Appl. Clay Sci.* **2007**, *37*, 231–239. [CrossRef]
7. Ardau, C.; Frau, F.; Dore, E.; Lattanzi, P. Molybdate sorption by Zn-Al sulphate layered double hydroxides. *Appl. Clay Sci.* **2012**, *65–66*, 128–133. [CrossRef]
8. Koilraj, P.; Srinivasan, K. ZnAl layered double hydroxides as potential molybdate sorbents and valorise the exchanged sorbent for catalytic wet peroxide oxidation of phenol. *Ind. Eng. Chem. Res.* **2013**, *52*, 7373–7381. [CrossRef]
9. Ashekuzzaman, S.M.; Jiang, J.Q. Strategic phosphate removal/recovery by a re-usable Mg-Fe-Cl layered double hydroxide. *Process. Saf. Environ. Prot.* **2017**, *107*, 454–462. [CrossRef]
10. Wan, S.; Wang, S.; Li, Y.; Gao, B. Functionalizing biochar with Mg-Al and Mg-Fe layered double hydroxides for removal of phosphate from aqueous solutions. *J. Ind. Eng. Chem.* **2017**, *47*, 246–253. [CrossRef]
11. Ogata, F.; Nagai, N.; Kishida, M.; Nakamura, T.; Kawasaki, N. Interaction between phosphate ions and Fe-Mg type hydrotalcite for purification of wastewater. *J. Environ. Chem. Eng.* **2019**, *7*, 102897. [CrossRef]
12. Matsuik, J.; Rybka, K. Removal of chromates and sulphates by Mg/Fe LDH and heterostructured LDH/Halloysite materials: Efficiency, selectivity, and stability of adsorbents in single and multi-element systems. *Materials* **2019**, *12*, 1373. [CrossRef] [PubMed]
13. Caporale, A.G.; Pigna, M.; Azam, S.M.G.G.; Sommella, A.; Rao, M.A.; Violante, A. Effect of competing ligands on the sorption/desorption of arsenite on/from Mg-Fe layered double hydroxides (Mg-Fe-LDH). *Chem. Eng. J.* **2013**, *225*, 704–709. [CrossRef]

14. Xie, Y.; Yuan, X.; Wu, Z.; Zeng, G.; Jiang, L.; Peng, X.; Li, H. Adsorption behaviour and mechanism of Mg/Fe layered double hydroxide with Fe_3O_4-carbon sphere on the removal of Pb(II) and Cu(II). *J. Coll. Interf. Sci.* **2019**, *536*, 440–455. [CrossRef]
15. Das, J.; Sairam Patra, B.; Baliarsingh, N.; Parida, K.M. Calcined Mg-Fe-CO_3 LDH as and adsorbent for the removal of selenite. *J. Coll. Interf. Sci.* **2007**, *316*, 216–223. [CrossRef]
16. Hudcova, B.; Erben, M.; Vitkova, M.; Komarek, M. Antimonate adsorption onto Mg-Fe layered double hydroxides in aqueous solutions at different pH values: Coupling surface complexation modelling with solid-state analyses. *Chemosphere* **2019**, *229*, 236–246. [CrossRef]
17. Ahmed, I.M.; Gasser, M.S. Adsorption study of anionic reactive dye from aqueous solution to Mg-Fe-CO_3 layered double hydroxides (LDH). *Appl. Surf. Sci.* **2012**, *259*, 650–656. [CrossRef]
18. Ayawei, N.; Angaye, S.S.; Wankasi, D. Mg/Fe layered double hydroxide as a novel adsorbent for the removal of Congo red. *Int. J. Appl. Sci. Technol.* **2017**, *7*, 83–92.
19. Abdelkader, N.B.-H.; Bentouami, A.; Derriche, Z.; Bettahar, N.; de Ménorval, L.-C. Synthesis and characterization of Mg–Fe layer double hydroxides and its application on adsorption of Orange G from aqueous solution. *Chem. Eng. J.* **2011**, *169*, 231–238. [CrossRef]
20. Zhang, S.; Jiao, Q.; Wang, C.; Yu, H.; Zhao, Y.; Li, H.; Wu, Q. In situ synthesis of Mg/Fe LDO/carbon nanohelix composites as adsorbing materials. *J. All. Comp.* **2016**, *658*, 505–512. [CrossRef]
21. Tahir, N.; Abdelssadek, Z.; Halliche, D.; Saadi, A.; Chebout, R.; Cherifi, O.; Bachari, K. Mg-Fe-hydrotalcite as catalyst for the benzylation of benzene and other aromatics by benzyl chloride reactions. *Surf. Inter. Anal.* **2008**, *40*, 254–258. [CrossRef]
22. Chang, P.S.; Li, S.Y.; Juang, T.Y.; Liu, Y.C. Mg-Fe layered double hydroxides enhance surfactin production in bacterial cells. *Crystals* **2019**, *9*, 355. [CrossRef]
23. Muriithi, G.N.; Petrik, L.F.; Gitari, W.M.; Doucet, F.J. Synthesis and characterization of hydrotalcite from South Africa Coal fly ash. *Powder Technol.* **2017**, *312*, 299–309. [CrossRef]
24. Cocheci, L.; Lupa, L.; Gheju, M.; Golban, A.; Lazau, R.; Pode, R. Zn–Al–CO_3 layered double hydroxides prepared from a waste of hot-dip galvanizing process. *Clean Technol. Environ. Policy* **2018**, *20*, 1105–1112. [CrossRef]
25. Kuwahara, Y.; Yamashita, H. Synthesis of Ca-based layered double hydroxide from blast furnace slag and its catalytic applications. *ISIJ Intern.* **2015**, *7*, 1531–1537. [CrossRef]
26. Galindo, R.; Lopez-Delgato, A.; Padilla, I.; Yates, M. Hydrotalcite-like compounds: A way to recover a hazardous waste in the aluminium tertiary industry. *Appl. Clay Sci.* **2014**, *95*, 41–49. [CrossRef]
27. Murayama, N.; Maekawa, I.; Ushiro, H.; Miyoshi, T.; Shibata, J.; Valix, M. Synthesis of various layered double hydroxides using aluminium dross generated in aluminium recycling process. *Int. J. Miner. Process.* **2012**, *110–111*, 46–52. [CrossRef]
28. Golban, A.; Cocheci, L.; Lazau, R.; Lupa, L.; Pode, R. Iron ions reclaiming from sludge resulted from hot-dip galvanizing process, as Mg_3Fe-layered double hydroxide used in the degradation process of organic dyes. *Des. Water Treat.* **2018**, *131*, 317–327. [CrossRef]
29. Panayotova, M.; Panayotov, V. An Electrochemical Method for Decreasing the Concentration of Sulfate and Molybdenum Ions in Industrial Wastewater. *J. Environ. Sci. Health Part A-Toxic/Hazard. Subst. Environ. Eng.* **2004**, *39*, 173–183. [CrossRef]
30. Lievens, P.; Block, C.; Cornelis, G.; Vandecasteele, C.; De Voogd, J.C.; Van Brecht, A. Mo, Sb and Se Removal from Scrubber Effluent of a Waste Incinerator. In *Water Treatment Technologies for the Removal of High-Toxicity Pollutants*; Vaclavikova, M., Vitale, K., Gallios, G.P., Ivanicova, L., Eds.; Springer: Dordrecht, The Netherlands, 2010.
31. Verbinnen, B.; Block, C.; Hannes, D.; Lievens, P.; Vaclavikova, M.; Stefusova, K.; Gallios, G.; Vandecasteele, C. Removal of molybdate anions from water by adsorption on zeolite-supported magnetite. *Water Environ. Res.* **2012**, *84*, 753–760. [CrossRef]
32. World Health Organization (WHO). Guidelines for Drinking-Water Quality, Fourth Edition. 2011. Available online: http://whqlibdoc.who.int/publications/2011/9789241548151_eng.pdf (accessed on 2 September 2019).
33. Underwood, E.J. *Trace Elements in Human and Animal Nutrition*, 3rd ed.; Academic Press: London, UK, 1971.
34. Palmer, S.J.; Soisonard, A.; Frost, R.I. Determination of the mechanism(s) for the inclusion of arsenate, vanadate or molybdate anions into hydrotalcites with variable cationic ratio. *J. Colloid Interface Sci.* **2009**, *329*, 404–409. [CrossRef] [PubMed]

35. Tkac, P.; Paulenova, A. Speciation of molybdenum (VI) in aqueous and organic phases of selected extraction systems. *Sep. Sci. Technol.* **2008**, *43*, 2641–2657. [CrossRef]
36. Pietrelli, L.; Ferro, S.; Vocciante, M. Raw materials recovery from spent hydrochloric acid-based galvanizing wastewater. *Chem. Eng. J.* **2018**, *341*, 539–546. [CrossRef]
37. Ferreira, O.P.; Alves, O.L.; Gouveia, D.X.; Souza Filho, A.G.; de Paiva, J.A.C.; Mendes Filho, J. Thermal decomposition and structural Reconstruction effect on Mg–Fe based hydrotalcite compounds. *J. Solid State Chem.* **2004**, *177*, 3058–3069. [CrossRef]
38. Elmoubarki, R.; Mahjoubi, F.Z.; Elhalil, A.; Tounsadi, H.; Abdennouri, M.; Sadiq, M.; Qourzal, S.; Zouhri, A.; Barka, N. Ni/Fe and Mg/Fe layered double hydroxides and their calcined derivatives: Preparation, characterization and application on textile dyes removal. *J. Mat. Res. Technol.* **2017**, *6*, 271–283. [CrossRef]
39. Dada, A.O.; Olalekan, A.P.; Olatunya, A.M.; Dada, O. Langmuir, Freundlich, Temkin and Dubinin–Radushkevich isotherms studies of equilibrium sorption of Zn^{2+} unto phosphoric acid modified rice husk. *IOSR-JAC* **2012**, *3*, 38–45.
40. Ertugay, N.; Malkoc, E. Adsorption isotherm, kinetic, and thermodynamic studies for methylene blue from aqueous solution by needles of Pinus sylvestris L. *Pol. J. Environ. Stud.* **2014**, *23*, 1995–2006.
41. Zhang, L.; Huang, T.; Zhang, M.; Guo, X.; Yuan, Z. Studies on the capability and behavior of adsorption of thallium on nano-Al_2O_3. *J. Hazard. Mater.* **2008**, *157*, 352–357. [CrossRef]
42. Paikaray, S.; Hendry, M.J.; Essilfie-Dughan, J. Controls on arsenate, molybdate, and selenate uptake by hydrotalcite-like layered double hydroxides. *Chem. Geol.* **2013**, *345*, 130–138. [CrossRef]
43. Jenkins, H.D.B.; Roobottom, H.K.; Passmore, J.; Glasser, L. Relationships among ionic lattice energies, molecular (formula unit) volumes, and thermochemical radii. *Inorg. Chem.* **1999**, *38*, 3609–3620. [CrossRef]
44. Smith, H.D.; Parkinson, G.M.; Hart, R.D. In situ absorption of molybdate and vanadate during precipitation of hydrotalcite from sodium aluminate solutions. *J. Cryst. Grow.* **2005**, *275*, e1665–e1671. [CrossRef]

Article

Antimonate Removal from Polluted Mining Water by Calcined Layered Double Hydroxides

Elisabetta Dore *, Franco Frau and Rosa Cidu

Department of Chemical and Geological Sciences, University of Cagliari, 09042 Monserrato, Cagliari, Italy
* Correspondence: elisabettadore@yahoo.it

Received: 20 June 2019; Accepted: 1 August 2019; Published: 6 August 2019

Abstract: Calcined layered double hydroxides (LDHs) can be used to remove Sb(V), in the $Sb(OH)_6^-$ form, from aqueous solutions. Sorption batch experiments showed that the mixed MgAlFe oxides, obtained from calcined hydrotalcite-like compound (3HT-cal), removed $Sb(OH)_6^-$ through the formation of a non-LDH brandholzite-like compound, whereas the mixed ZnAl oxides, resulting from calcined zaccagnaite-like compound (2ZC-cal), trapped $Sb(OH)_6^-$ in the interlayer during the formation of a Sb(V)-bearing LDH (the zincalstibite-like compound). The competition effect of coexistent anions on $Sb(OH)_6^-$ removal was $HAsO_4^{2-} >> HCO_3^- \geq SO_4^{2-}$ for 2ZC-cal and $HAsO_4^{2-} >> HCO_3^- >> SO_4^{2-}$ for 3HT-cal. Considering the importance of assessing the practical use of calcined LDHs, batch experiments were also carried out with a slag drainage affected by serious Sb(V) pollution (Sb = 9900 μg/L) sampled at the abandoned Su Suergiu mine (Sardinia, Italy). Results showed that, due to the complex chemical composition of the slag drainage, dissolved $Sb(OH)_6^-$ was removed by intercalation in the interlayer of carbonate LDHs rather than through the formation of brandholzite-like or zincalstibite-like compounds. Both 2ZC-cal and 3HT-cal efficiently removed very high percentages (up to 90–99%) of Sb(V) from the Su Suergiu mine drainage, and thus can have a potential application for real polluted waters.

Keywords: layered double hydroxides; antimonate uptake; mine water; brandholzite; zincalstibite

---

1. Introduction

Antimony (Sb) is an element widely present in the environment as a result of both natural processes and anthropogenic sources [1]. Due to its potential risk for human health, the World Health Organization has set the guideline value for drinking water at 20 μg/L of Sb [2], while the European Community has established 5 μg/L [3]. The Sb concentration in uncontaminated freshwater is usually lower than drinking water limits [4,5], however considerable higher concentrations (up to mg/L) can be related to both natural sources and anthropogenic activities [1,4,6–8]. In natural environments, Sb is generally present in the trivalent Sb(III) and pentavalent Sb(V) oxidation states, with the Sb(III) species being ten times more toxic than the Sb(V) ones [9]. In aqueous solution, Sb(III) and Sb(V) prevail, respectively, under reducing and oxidizing conditions as antimonous acid H_3SbO_3 and antimonic acid H_3SbO_4 and their dissociation products, with the $Sb(OH)_6^-$ anion being the most common and stable aqueous species in a wide range of natural pH values [1,9].

Among the techniques suitable for the abatement of Sb concentration in the solution, such as coagulation-flocculation [10], electrochemical methods [11,12] and membrane separation [13,14], the adsorption is considered a low cost and effective method [9]. Several studies reported that metal hydroxides and oxohydroxides (e.g., MnOOH, $Al(OH)_3$, FeOOH) are good Sb removers, however they result more efficient for Sb(III) than Sb(V) under slightly acid to acid conditions [15–18]. Also, nano-TiO_2 electroactive carbon nanotube (CNT) filter and ZrO_2-carbon nanofibers (ZNC) were tested, respectively, for Sb(III) and simultaneous Sb(III) and Sb(V) removal; in particular, it was reported that the adsorption capacity of Sb(III) is much higher than Sb(V) in ZNC [19,20]. Recent works showed that the acidic conditions are also favorable for the Sb(V) removal from solution by La and Ce-doped magnetic biochars [21,22]. With respect to the other sorbents, layered double hydroxides (LDHs) show the advantage of being able to remove hazardous anions from solution at circumneutral pH, and can therefore be potential removers for $Sb(OH)_6^-$ at the circumneutral pH and oxidizing conditions usually found in the environment [23].

The LDHs are minerals with general formula $[M^{2+}{}_{1-x}M^{3+}{}_x(OH)_2](A^{n-})_{x/n}{\cdot}mH_2O$, where M^{2+} and M^{3+} are respectively bivalent and trivalent metals (Mg^{2+}, Zn^{2+}, Ca^{2+}, Al^{3+}, Fe^{3+}, etc.), A^{n-} are anions (Cl^-, $SO_4{}^{2-}$, $CO_3{}^{2-}$, etc.) and x is the $M^{3+}/(M^{2+} + M^{3+})$ molar ratio ($0.20 \leq x \leq 0.33$). The LDHs structure consists of octahedral brucite-like layers positively charged due to the partial substitution of M^{2+} by M^{3+}, stacked along the *c* axis and intercalated with interlayer anions, which neutralize the positive charge, and variable quantity of water molecules [24,25]. The LDHs can successfully remove anionic contaminants from solution through the anion exchange with the interlayer anions, or by trapping anions in the interlayer region during the reconstruction of the layered structure by the rehydration of mixed metal oxides obtained from LDHs calcination (the so called "memory effect") [23,26–29].

Previous authors reported that LDHs with different compositions are potential $Sb(OH)_6^-$ removers: nitrate and chloride bearing Mg-Al LDHs, both untreated and calcined, efficiently remove $Sb(OH)_6^-$ from solution through the formation of a brandholzite-like compound [30,31], and the removal capacity can be improved by doping LDHs with Fe^{2+} [32]; sulfate bearing Zn-Al and Zn-Fe(III) LDHs uptake $Sb(OH)_6^-$ from solution by anion exchange [33,34] and also Fe-Mn LDHs obtained by electro-coagulation process result good removers [35]. Although interest in the use of LDHs for Sb removal from aqueous solutions has increased in recent years, to the best of our knowledge their practical use with real polluted water has not been investigated yet. Therefore, the aim of this study was to assess the $Sb(OH)_6^-$ removal capacity of calcined LDHs from real water affected by serious Sb pollution.

In our previous work we showed that calcined synthetic LDHs with composition like hydrotalcite (with formula $Mg_6(Al_{0.5}Fe_{0.5})_2(CO_3)(OH)_{16}{\cdot}4H_2O$) and zaccagnaite (with formula $Zn_4Al_2(CO_3)(OH)_{12}{\cdot}3H_2O$) remove $Sb(OH)_6^-$ from solution, respectively, through the formation of a brandholzite-like phase (a non-LDH mineral with general formula $Mg[Sb(OH)_6]_2{\cdot}6H_2O$) and a zincalstibite-like compound (an LDH mineral with general formula $Zn_2Al(OH)_6[Sb(OH)_6]$) [36]. In this work we used calcined hydrotalcite-like and zaccagnaite-like compounds to carry out batch experiments with coexistent anions in solution to evaluate their competition effect on $Sb(OH)_6^-$ removal. We successively assessed the practical use of these sorbents with real water by sorption batch experiments performed with the drainage water flowing out from the foundry slag impoundments at the abandoned mine of Su Suergiu (Sardinia, Italy), which is affected by serious Sb pollution [6,37].

2. Materials and Methods

2.1. LDHs Synthesis and Calcination

Synthetic hydrotalcite $Mg_6(Al_{0.5}Fe_{0.5})_2(CO_3)(OH)_{16}{\cdot}4H_2O$ and zaccagnaite $Zn_4Al_2(CO_3)(OH)_{12}{\cdot}3H_2O$ were prepared with a coprecipitation method at constant pH [36]. Depending on composition, a solution (0.2 M) with the desired metals was prepared by dissolving in ultrapure water (Millipore, Milli-Q©, 18.2 MΩ cm) appropriate amounts of $Mg(NO_3)_2{\cdot}6H_2O$, $Al(NO_3)_3{\cdot}9H_2O$, $Fe(NO_3)_3{\cdot}9H_2O$ and $Zn(NO_3)_2{\cdot}6H_2O$. All reagents were of analytical grade (ACS-for analysis, CARLO ERBA Reagents S.r.l., Cornaredo (MI), Italy) and were used without further purification. The so-obtained metal solution was dropped into a reactor containing a Na_2CO_3 solution (0.05 M), under stirring (500 rpm), and the precipitation was induced at constant pH (ranging between 9.5 and 10.5) by adding dropwise a NaOH (0.5 M) solution. After 24 h of aging at 65 °C, the solids were recovered through filtration (30 μm pore size cellulose filter, Whatman Plc, Little Chalfont, Buckinghamshire, UK), washed with deionized water and dried at room temperature. Calcination was performed at 450 °C for 4 h. The M^{2+}/M^{3+} molar ratios of synthetic hydrotalcite and zaccagnaite, and their calcined products, were close to those of the starting solutions (Table 1).

From now on, synthetic hydrotalcite is termed 3HT and zaccagnaite is 2ZC, the numbers before the labels indicate the M^{2+}/M^{3+} molar ratio; moreover, the suffix -CO_3 will be used for the untreated carbonate LDHs and the suffix -cal for the calcined LDHs.

Table 1. Chemical composition of synthetic untreated carbonate (-CO_3) and calcined (-cal) hydrotalcite-like (3HT) and zaccagnaite-like (2ZC) compounds.

Sample	Zn	Al	M^{2+}/M^{3+}	Sample	Mg	Al	Fe^{3+}	M^{2+}/M^{3+}
	mmol/g	mmol/g	Molar Ratio		mmol/g	mmol/g	mmol/g	Molar Ratio
2ZC-CO_3	6.8	3.2	2.1	3HT-CO_3	7.4	1.3	1.4	2.8
2ZC-cal	9.0	4.1	2.2	3HT-cal	12.5	2.2	2.3	2.8

2.2. Sorption Experiments

2.2.1. Effect of Coexistent Anions

In batch experiments, the coexistent anions were selected taking into account the chemical composition of Su Suergiu mine drainage [6,37]. The experimental solutions were prepared dissolving appropriate amounts of $KSb(OH)_6$ and Na_2SO_4, $NaHCO_3$ or $Na_2HAsO_4{\cdot}7H_2O$ (ACS-for analysis, CARLO ERBA Reagents S.r.l., Cornaredo (MI), Italy) in ultrapure water. To perform the experiments, 0.1 g of 2ZC-cal or 3HT-cal was suspended, for 48 h under stirring, in 400 mL of solution containing equal concentrations (about 1 mmol/L) of dissolved $Sb(OH)_6^-$ and one competitor at a time. Experiments with only dissolved $Sb(OH)_6^-$ without competitors were also carried out to compare the results. Before the addition of the sorbents and during the experiments the pH of solutions was monitored and a portion of solution was withdrawn and acidified with HNO_3 1% *v*/*v* for chemical analysis of Sb, As, S, Mg, Zn, Al and Fe by inductively coupled plasma optical emission spectroscopy (ICP-OES, ARL Fisons 3520, Waltham, MA, USA). At the end of the reaction time the solids were recovered through filtration (0.45 μm pore size polycarbonate filters, Whatman Plc, Little Chalfont, Buckinghamshire, UK), washed with distilled water and dried at room temperature for mineralogical characterization.

2.2.2. Sorption Experiments with Su Suergiu Mine Drainage

The real water for sorption experiments is a slag drainage sampled at the abandoned Su Suergiu mine (Sardinia, Italy), at the sampling point named SU1 (Supplementary Materials Figure S1) as reported by Cidu et al. [37]. The physical and chemical parameters were determined at the sampling site using the sampling protocol described in Cidu et al. [37]. After sampling, the slag drainage (from now on SU1) was stored in HDPE bottles at 4 °C, and batch experiments were carried out, at room temperature (25 °C), less than 24 h after the sampling.

Different amounts of 3HT-cal or 2ZC-cal, equal to 0.1, 0.25, 0.5 and 1 g, were suspended in 400 ml of SU1 under stirring for 24 h. During the experiments the pH of solutions was monitored. At the end of reaction time, the solids were separated from solution through filtration, washed with distilled water and dried at room conditions for mineralogical characterization. The solutions recovered after the experiments were stored in two different aliquots: one aliquot was unacidified for analysis of major ions by ion chromatography (IC, Dionex ICS3000, ThermoFisher SCIENTIFIC, Waltham, MA, USA); a second aliquot was acidified for trace elements analysis (Sb, As, Fe, Zn and Al) by inductively coupled plasma mass spectrometry (ICP-MS, quadrupole, PerkinElmer SCIEX ELAN DRC-e, Waltham, MA, USA) with Rh as internal standard; concentrations of Sb > 1000 μg/L were also determined by ICP-OES.

2.3. Mineralogical Characterization

Mineralogical characterization of synthetic LDHs and their calcined products before and after the experiments was performed by collecting XRD patterns in the 5–80° 2θ angular range on an automated Panalytical X'pert Pro diffractometer (PANalytical, Almelo, Netherlands), with Ni-filter Cu-Kα_1 radiation (λ = 1.54060 Å), operating at 40 kV and 40 mA, using the X'Celerator detector.

3. Results

3.1. *Effect of Coexistent Anions on Sb(V) Removal*

3.1.1. Sorptive Competition

The results of competition experiments showed that 2ZC-cal was slightly more effective than 3HT-cal (Figure 1). At the end of the experiments without competitors, 2ZC-cal removed 85% of $Sb(OH)_6^-$ whereas 3HT-cal reached 72%. The coexistence of SO_4^{2-} and HCO_3^- slightly affected the $Sb(OH)_6^-$ uptake by 2ZC-cal, whereas in the experiments with 3HT-cal, the percentage of $Sb(OH)_6^-$ removed did not vary significantly in the presence of SO_4^{2-} but decreased up to 50% with coexistent HCO_3^- (Figure 1, Tables 2 and 3).

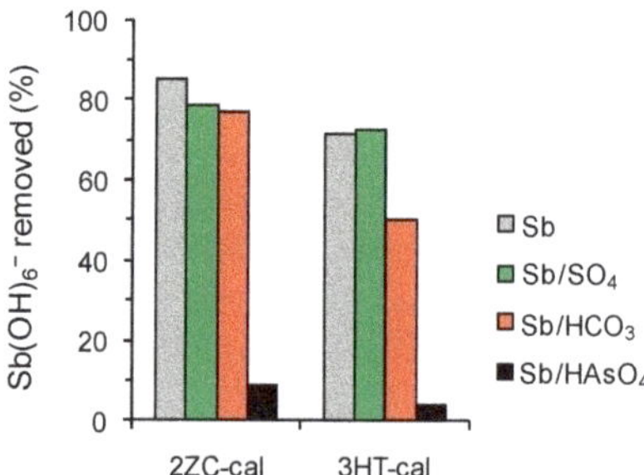

Figure 1. Percentage of $Sb(OH)_6^-$ removed from solution by 2ZC-cal and 3HT-cal at the end of the experiments without competitors (Sb) and with HCO_3^- (Sb/HCO$_3$), SO_4^{2-} (Sb/SO$_4$) or $HAsO_4^{2-}$ (Sb/HAsO$_4$) as coexistent anions.

The $HAsO_4^{2-}$ anion was the strongest competitor in the experiments with both 2ZC-cal and 3HT-cal, with percentages of $Sb(OH)_6^-$ removed lower than 10%. Moreover, the $HAsO_4^{2-}$ concentration at the end of the experiment with 3HT-cal markedly decreased by about 60% (Table 3), showing a strong affinity of $HAsO_4^{2-}$ for the interlayer region of 3HT.

Table 2. Solution pH values and dissolved ions determined before and at the end of the sorption experiments performed with 2ZC-cal without competitors (Sb), and with coexistent HCO_3^- (Sb/HCO$_3$), SO_4^{2-} (Sb/SO$_4$) or $HAsO_4^{2-}$ (Sb/HAsO$_4$). (A^{n-} = HCO_3^-, SO_4^{2-} or $HAsO_4^{2-}$; na = not analyzed; dl = detection limit; dl_{Zn} = 0.2 μmol/L; dl_{Al} = 7 μmol/L).

Sample	Experiment	Time	pH	$Sb(OH)_6^-$	$Sb(OH)_6^-$ removed	A^{n-}	A^{n-} removed	Zn	Al
		h		mmol/L	%	mmol/L	%	mmol/L	
2ZC-cal	Sb	0	5.1	1.02				0	0
		48	8.5	0.15	85			<dl	<dl
	Sb/HCO$_3$	0	8.2	1.03		1.09		0	0
		48	8.7	0.24	77	na	na	<dl	<dl
	Sb/SO$_4$	0	5.2	1.01		1.08		0	0
		48	9.5	0.22	79	1.04	4	<dl	<dl
	Sb/HAsO$_4$	0	8.6	1.02		1.03		0	0
		48	8.1	0.93	8.6	0.87	16	0.02	<dl

The effect of coexistent anions on $Sb(OH)_6^-$ uptake resulted to be $HAsO_4^{2-} >> HCO_3^- \geq SO_4^{2-}$ for 2ZC-cal and, in partial agreement with previous work [30], $HAsO_4^{2-} >> HCO_3^- >> SO_4^{2-}$ for 3HT-cal.

At the end of the experiments with 3HT-cal, slight concentrations of Mg were determined, whereas the concentrations of Al and Fe were always below the corresponding detection limits (Table 3).

Table 3. Solution pH values and dissolved ions determined before and at the end of the sorption experiments performed with 3HT-cal without competitors (Sb), and with coexistent HCO_3^- (Sb/HCO_3), SO_4^{2-} (Sb/SO_4) or $HAsO_4^{2-}$ (Sb/$HAsO_4$). (A^{n-} = HCO_3^-, SO_4^{2-} or $HAsO_4^{2-}$; na = not analyzed; dl = detection limit; dl_{Al} = 7 μmol/L; dl_{Fe} = 1 μmol/L).

Sample	Experiment	Time	pH	$Sb(OH)_6^-$	$Sb(OH)_6^-$ removed	A^{n-}	A^{n-} removed	Mg	Al	Fe
				mmol/L	%	mmol/L	%	mmol/L		
3HT-cal	Sb	0	5.2	1.03				0	0	0
		48	9.3	0.29	72			0.20	<dl	<dl
	Sb/HCO_3	0	8.3	1.02		1.05		0	0	0
		48	9.5	0.51	50	na	na	0.22	<dl	<dl
	Sb/SO_4	0	5.1	1.03		1.08		0	0	0
		48	9.1	0.28	73	1.06	2	0.33	<dl	<dl
	Sb/$HAsO_4$	0	8.6	1.03		1.02		0	0	0
		48	9.8	0.99	3.7	0.43	57	0.19	<dl	<dl

3.1.2. Kinetics

In the experiments with both 2ZC-cal and 3HT-cal the solution pH values increased sharply (up to about 11) after the addition of sorbents (Figure 2a,b), and decreased after 48 h in the range of 8.1–9.8 (Tables 2 and 3; Figure 2a,b). Most of the $Sb(OH)_6^-$ was removed within the first six hours (Figure 2c,d) indicating that $Sb(OH)_6^-$ uptake occurred mainly during the reconstruction of the lamellar structure of LDHs [36] as schematized in reaction (1):

$$M_{1-x}^{2+}M_x^{3+}(OH)_2\,(CO_3)_{\frac{x}{2}} \overset{calcination}{\rightarrow} M_{1-x}^{2+}M_x^{3+}O_{1+(\frac{x}{2})} + xSb(OH)_6^- \overset{reconstruction}{\rightarrow} M_{1-x}^{2+}M_x^{3+}(OH)_2(Sb(OH)_6)_x + xOH^- \quad (1)$$

The $Sb(OH)_6^-$ removal as a function of time was studied through the pseudo-first order [38] and the pseudo-second order kinetic models [39].

The $Sb(OH)_6^-$ sorption capacity has been calculated through the Formula (2):

$$q_t = (C_0 - C_t) \cdot V/W \quad (2)$$

where the sorption capacity (q_t) is the amount of $Sb(OH)_6^-$ sorbed per unit of sorbent (mmol/g) at the reaction time t (h), C_0 and C_t are the $Sb(OH)_6^-$ concentrations in solution (mmol/L) before the addition of the sorbent and at the reaction time t, V is the volume of solution (L) and W the weight of sorbent (g).

The pseudo-first order and the pseudo-second order equations are expressed as follows:

$$pseudo-first\ order\ kinetic\ model \qquad k_1 = \frac{2.303}{t} \cdot log\frac{q_e}{q_e - q_t} \quad (3)$$

$$pseudo-second\ order\ kinetic\ model \qquad k_2 = \frac{1}{t} \cdot \frac{q_t}{q_e(q_e - q_t)} \quad (4)$$

where q_e is the $Sb(OH)_6^-$ sorption capacity at equilibrium (mmol/g), k_1 (1/h) and k_2 (g/mmol h) are the rate constant of sorption. These equations expressed in the linear form result as follows:

$$pseudo-first\ order\ kinetic\ model \qquad \frac{dq_t}{dt} = k_1\,(q_e - q_t) \quad (5)$$

$$\text{pseudo} - \text{second order kinetic model} \qquad \frac{dq_t}{dt} = k_2\,(q_e - q_t)^2 \tag{6}$$

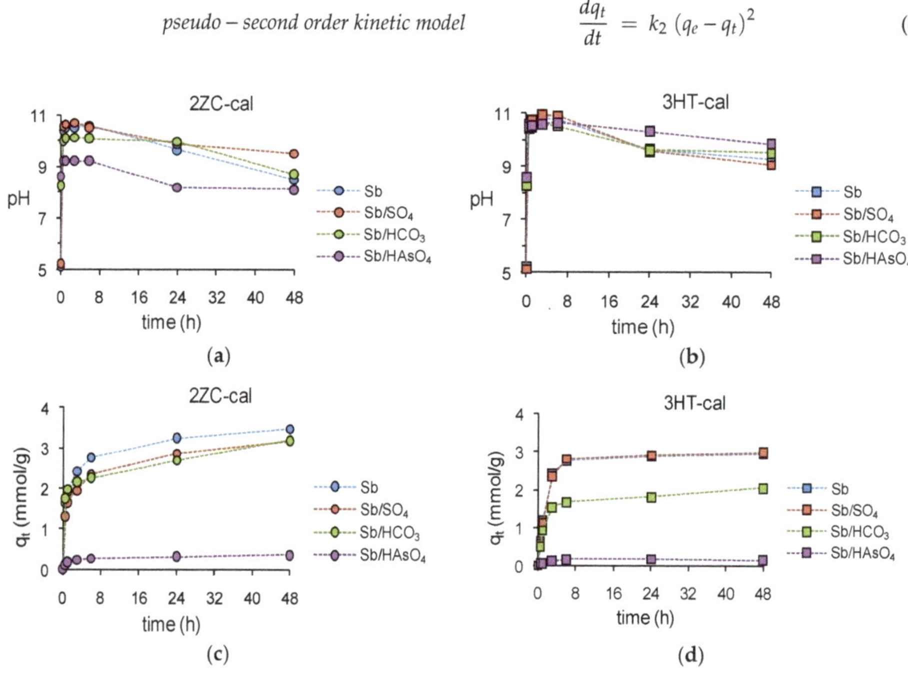

Figure 2. The solution pH values determined as a function of time during the sorption experiments without competitors and with coexistent anions performed with (**a**) 2ZC-cal and (**b**) 3HT-cal. The $Sb(OH)_6^-$ sorption capacity (q_t) as a function of time in the experiments performed with (**c**) 2ZC-cal and (**d**) 3HT-cal without competitors and with coexistent anions.

To verify the applicability of kinetic models at the sorption system, the experimental data were plotted as $\log(q_e - q_t)$ vs. time for the pseudo-first order (Supplementary Material Figure S2) and t/q_t vs. time (Supplementary Material Figure S3) for the pseudo-second order kinetic model. If the plots give a linear correlation, then the theoretical sorption capacity at equilibrium and the rate constants can be calculated from the slope and the intercept of the straight lines. The good fit of the data, the r^2 values close to the unit and the good agreement between the experimental sorption capacity at equilibrium (q_e) and the theoretical sorption capacity (q_{calc}) indicated that the sorption system is better described by the pseudo-second order kinetic model (Table 4), suggesting that the $Sb(OH)_6^-$ uptake by both 2ZC-cal and 3HT-cal might principally occur by chemisorption.

Table 4. The $Sb(OH)_6^-$ sorption capacity of 2ZC-cal and 3HT-cal determined at equilibrium from the experimental data (q_e) and calculated from the kinetic models (q_{calc}).

Sample	Experiment		Pseudo-First Order Kinetic Model			Pseudo-Second Order Kinetic Model		
		q_e mmol/g	q_{calc} mmol/g	k_1 1/h	r^2	q_{calc} mmol/g	k_2 g/mmol h	r^2
2ZC-cal	Sb	3.47	1.86	0.093	0.848	3.52	0.235	0.999
	Sb/HCO_3	3.18	1.59	0.052	0.631	3.19	0.196	0.992
	Sb/SO_4	3.17	1.86	0.080	0.856	3.23	0.186	0.997
	$Sb/HAsO_4$	0.35	4.78	0.068	0.754	0.36	1.568	0.992
3HT-cal	Sb	2.94	1.51	0.143	0.761	3.02	0.272	0.999
	Sb/HCO_3	2.04	1.13	0.074	0.641	2.07	0.303	0.997
	Sb/SO_4	2.99	1.58	0.137	0.738	3.08	0.231	0.999
	$Sb/HAsO_4$	0.15				0.16	12.9	0.995

3.1.3. Characterization of Sorbents

The XRD pattern of 2ZC-CO_3 (Figure 3a) showed the characteristic basal reflections (003) and (006) attributable to a zaccagnaite-like compound [40].

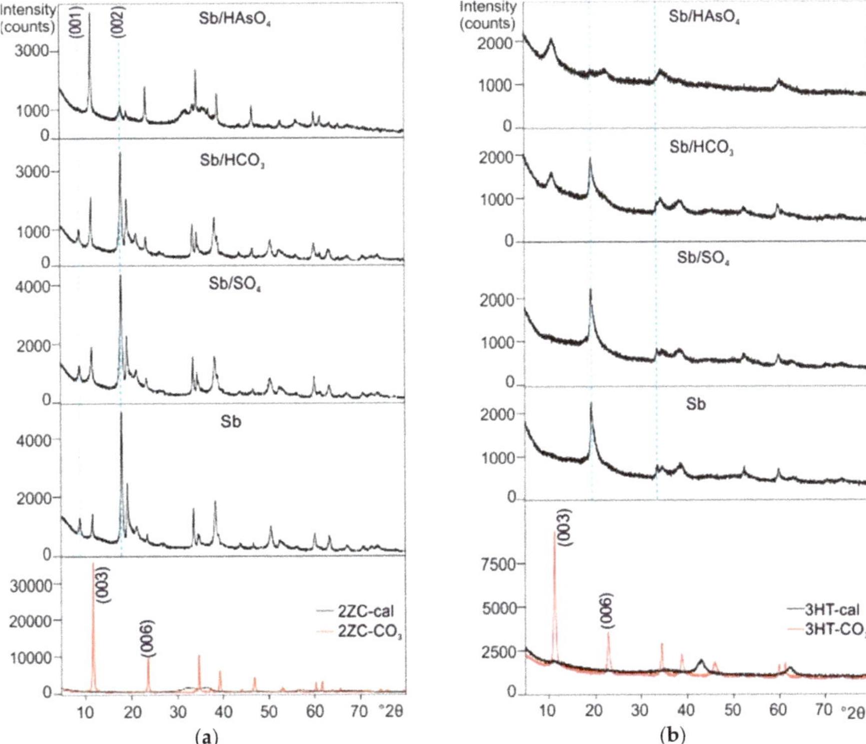

Figure 3. XRD patterns of (**a**) 2ZC-CO_3 and its calcined product 2ZC-cal and of (**b**) 3HT-CO_3 and its calcined product 3HT-cal, and XRD patterns of sorbents recovered after the sorption experiment without competitors (Sb), and with coexistent SO_4^{2-} (Sb/SO_4), HCO_3^- (Sb/HCO_3) or $HAsO_4^{2-}$ (Sb/$HAsO_4$). The blue dashed lines indicate the peaks of the (**a**) zincalstibite-like and (**b**) brandholzite-like compounds formed after the sorption experiments.

After calcination, in the XRD pattern of 2ZC-cal, the absence of LDH basal reflections and the presence of broad peaks at 32.1° and 36.4° 2θ, ascribable to ZnO, indicated the collapse of the lamellar LDH structure and the formation of a disordered ZnO, with Al probably dispersed in its structure [41]. Peaks of undesired phases were not detected. The XRD patterns of solids recovered after the sorption experiments showed two different LDH phases: the peaks at angular position 11.6° and 23.5° 2θ corresponded to the basal reflections (003) and (006) of 2ZC-CO_3; instead, the peaks at 5° and 18° 2θ were attributable, respectively, to the (001) and (002) reflections of a zincalstibite-like compound [36,42].

The characteristic hydrotalcite-like compound basal reflections (003) and (006), visible in the XRD patterns of 3HT-CO_3 (Figure 3b), were no longer detectable in the calcined phase (3HT-cal) which showed peaks at about 42° and 62° 2θ ascribable to a disordered MgO with the trivalent metals Fe and Al probably dispersed in its structure [26]. After the sorption experiments, the solids showed the characteristic peak of a brandholzite-like compound at 19.2° 2θ and a further brandholzite peak at 33.6° 2θ [36,43], except for the experiment with coexistent $HAsO_4^{2-}$ where the characteristic brandholzite peak was barely visible and additional peaks at low angle compatible with the basal reflections of 3HT-CO_3 were clearly recognizable.

3.2. Sorption Experiments with Su Suergiu Mine Drainage (SU1)

3.2.1. Solutions

For convenience in this section the concentrations of ions in solution will be expressed as mg/L and μg/L.

The results of the chemical analysis of SU1 slag drainage showed a Ca-SO_4 dominant chemical composition and high concentrations of Sb (9900 μg/L) and arsenic (As = 3390 μg/L) (Table 5).

Table 5. The pH values, electrical conductivity (EC) and chemical composition of SU1 slag drainage before (SU1 column) and after the sorption experiments performed with different amounts (0.1, 0.25, 0.5, 1g) of 2ZC-cal and 3HT-cal (SU1 + 2ZC-cal and SU1 + 3HT-cal columns).

		SU1	SU1 + 2ZC-cal				SU1 + 3HT-cal			
			Weight of Sorbent (g)				Weight of Sorbent (g)			
			0.1	0.25	0.5	1	0.1	0.25	0.5	1
EC	mS/cm	2.40	2.33	2.24	2.14	2.04	2.24	2.12	1.99	1.75
pH		8.2	8.1	8.1	8.0	7.9	8.4	9.2	9.4	9.7
SO_4	mg/L	1000	1080	1050	1000	960	1070	1050	960	770
HCO_3	mg/L	485	252	169	125	80	127	48	29	18
Cl	mg/L	59	64	65	63	63	64	64	63	63
NO_3	mg/L	0.6	4.4	0.7	1.3	0.6	1.4	1.9	3.5	6.3
F	mg/L	1.7	1.6	1.1	0.5	0.3	0.8	0.4	0.1	0.1
Ca	mg/L	362	278	264	234	244	131	68	80	129
Mg	mg/L	63	62	56	45	25	120	140	109	59
Na	mg/L	166	168	161	167	163	166	166	166	162
K	mg/L	4.7	4.4	4.4	4.8	4.3	4.7	4.6	4.8	4.6
Zn	μg/L	30	1490	174	260	242	<20	<20	<20	<20
Al	μg/L	<30	78	105	67	<30	700	<30	<30	<30
Fe	μg/L	<20	<20	<20	<20	<20	<20	<20	<20	<20
Sb	μg/L	9900	6190	78	34	20	9830	170	430	1000
As	μg/L	3390	92	35	8.3	4.5	150	25	6.5	<0.5

The high value of EC (electrical conductivity) is related to the high contents of Ca^{2+} and $SO_4{}^{2-}$, with $SO_4{}^{2-}$ mainly deriving from the oxidation of sulfides; moreover, the high concentration of $HCO_3{}^-$ avoids the decrease of pH, which resulted in slightly alkaline values (Table 5). The high concentration of both Sb and As is the consequence of the water interaction with the foundry slags [6,37]. It has been reported that in the Su Suergiu mine water, the Sb(III), when detected, results <2% of total dissolved Sb (water fraction < 0.45 μm) and that all the Sb(V) occurs as $Sb(OH)_6{}^-$ [6,44]. The concentration of Sb(III) determined in the SU1 sampled for sorption experiments with 3HT-cal and 2ZC-cal resulted to be 147 μg/L, therefore, in the present work, the $Sb(OH)_6{}^-$ is considered the only Sb form involved in the Sb removal processes. Excluding the experiments performed with 0.1 g of sorbents, at the end of reaction time up to 90–99% of Sb was removed from the solution, with 2ZC-cal slightly more effective than 3HT-cal (Figure 4a), whereas As was effectively removed in all experiments (Figure 4b). In almost all experiments, the Sb and As concentrations decreased close to, or below, the limits established for drinking water (Figure 4c,d).

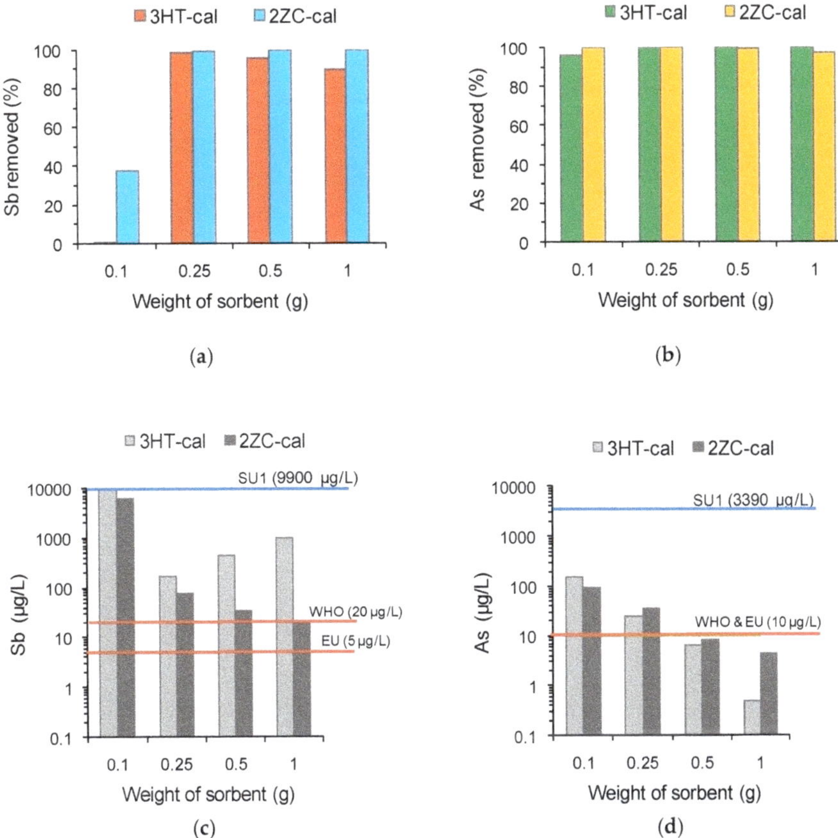

Figure 4. Results of batch sorption experiments performed with 3HT-cal and 2ZC-cal with the Su Suergiu mine drainage (SU1). Percentages of (**a**) Sb and (**b**) As removed at the end of the sorption experiments, and concentrations of residual (**c**) Sb and (**d**) As in solution at the end of the experiments for different amounts of the sorbent. In the plots (**c**) and (**d**) the blue lines indicate the starting Sb or As concentrations; the red lines indicate the limits of Sb and As set for drinking water by the World Health Organization (WHO) [2] and the European Community (EU) [3].

3.2.2. Sorbents

The XRD patterns of solids recovered after all experiments showed peaks at low angles compatible with the carbonate bearing LDHs, indicating the reconstruction of the typical lamellar LDH structure (Figures 5 and 6). In the range of pH values of the experiments, Sb and As prevail, respectively, as $Sb(OH)_6^-$ and $HAsO_4^{2-}$, but peaks attributable to a brandholzite-like compound were not visible in the solids recovered after the experiments performed with 3HT-cal (Figure 5), and only after the experiment with 0.1 g of 2ZC-cal the characteristic peak of a zincalstibite-like compound was clearly recognizable (Figure 6). Therefore, as a consequence of the complexity of the chemical composition of SU1, the $Sb(OH)_6^-$ removal did not occur through the formation of $Sb(OH)_6^-$ bearing phases, but rather, it is reasonable to suppose that $Sb(OH)_6^-$ was incorporated in the interlayer region together with other anions, i.e., CO_3^{2-} and $HAsO_4^{2-}$. It is also noticeable that all samples contained additional well defined peaks at about 30° 2θ ascribable to calcite (Figures 5 and 6), and after the experiment with 0.1 g of 2ZC-cal, the peaks attributable to monohydrocalcite were also present (Figure 6).

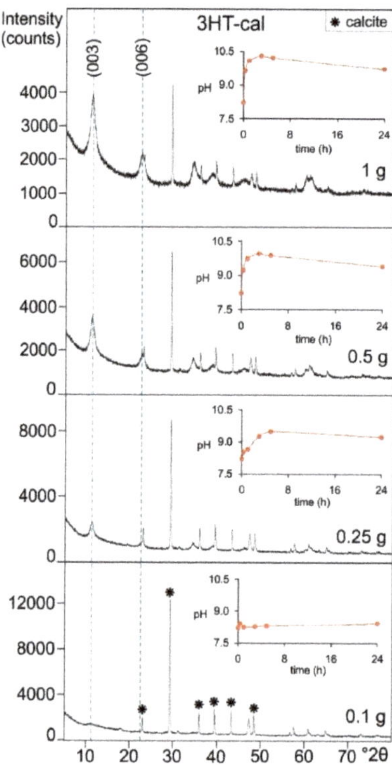

Figure 5. XRD patterns of 3HT-cal recovered after the sorption experiments with the slag drainage SU1 performed with different amounts of 3HT-cal (0.1, 0.25, 0.5, 1 g). On the upper right side of each XRD pattern the pH of solution as a function of time during each experiment is reported. The blue dashed lines indicate the basal reflection of layered double hydroxides (LDHs) formed after the sorption experiments.

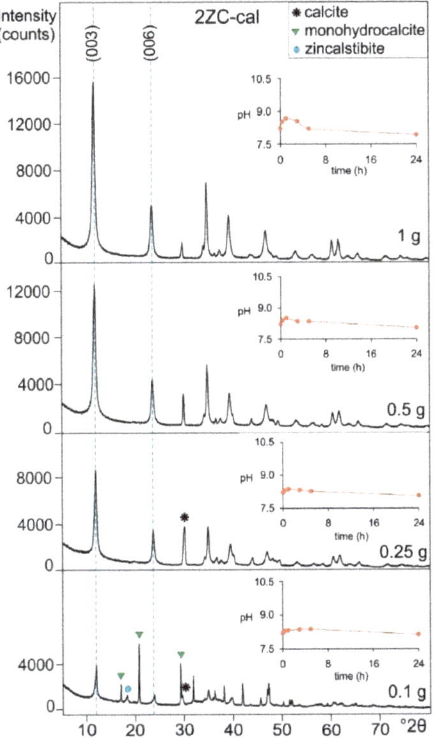

Figure 6. XRD patterns of 2ZC-cal recovered after the sorption experiments with the slag drainage SU1 performed with different amounts of 2ZC-cal (0.1, 0.25, 0.5, 1 g). On the upper right side of each XRD pattern the pH of solution as a function of time during each experiment is reported. The blue dashed lines indicate the basal reflection of LDHs formed after the sorption experiments.

4. Discussion

4.1. Effect of Coexistent Anions

The results of mineralogical characterizations and chemical analysis, in agreement with previous works, showed that 2ZC-cal removed $Sb(OH)_6^-$ from solution through the reconstruction of a zincalstibite-like compound [36], and the 3HT-cal removed $Sb(OH)_6^-$ through the formation of a low ordered brandholzite-like compound [30,36]. The zincalstibite is an $Sb(OH)_6^-$ bearing LDH (with general formula $Zn_2Al(OH)_6[Sb(OH)_6]$ [25,42], whereas the brandholzite is a non-LDH phase with general formula $Mg[Sb(OH)_6]_2{\cdot}6H_2O$, whose the layered structure is characterized by the presence of two layers, $\{[Sb(OH)_6]_9\}^{9-}$ and $\{[Sb(OH)_6]_3[Mg(H_2O)_6]_6\}^{9+}$, alternatively stacked along the *c* axis [43].

In agreement with the results of chemical analysis, the diffraction peaks of the $Sb(OH)_6^-$ bearing phases were more intense and well defined in the XRD patterns of solids recovered after the experiments wherein the greatest amounts of $Sb(OH)_6^-$ were removed from solution. In the experiments performed with 3HT-cal and HCO_3^- and $HAsO_4^{2-}$ as coexistent anions, the decrease in $Sb(OH)_6^-$ removal was linked to the appearance of the (003) and (006) basal reflections of LDHs (Figure 3b). In particular, in the experiment with coexistent HCO_3^- the competition effect was improved by the increase of solution pH values, up to 10.6 (Figure 2b), that favors the prevalence of CO_3^{2-} in solution that has a high affinity for the LDH interlayer [23,30]. Also $HAsO_4^{2-}$ has a high affinity for hydrotalcite-like compounds [27,45], but previous authors have observed that the removal capacity of calcined MgAl-LDHs is higher for $Sb(OH)_6^-$ than for $HAsO_4^{2-}$, suggesting that the uptake of $Sb(OH)_6^-$ through the selective crystallization of a brandholzite-like compound is more favorable than the sequestration from solution by the intercalation in the interlayer [46]. The results of our work showed that, when coexisting in solution, $HAsO_4^{2-}$ strongly competes with $Sb(OH)_6^-$ and is preferentially removed, probably due to its higher specific ionic charge.

The low amounts of Mg determined at the end of experiments with 3HT-cal indicated a slight dissolution of sorbent (6–15%) (Table 3). Because the concentrations of dissolved Fe and Al were always below the corresponding detection limits and no Al and/or Fe secondary phases were observed in the XRD patterns of solids recovered at the end of experiments (Figure 3b), the low amount of Al and Fe released in the solution might precipitate as amorphous solids. The Fe and Al of undissolved phase that removed the $Sb(OH)_6^-$ through the formation of brandholzite-like compound, probably remained in undetermined sites of the low ordered brandholzite structure.

The XRD pattern of the experiment performed with 2ZC-cal and $HAsO_4^{2-}$ as coexistent anion showed well defined basal reflections compatible with the original 2ZC-CO_3 LDH, while the characteristic zincalstibite-like compound peaks were scarcely visible (Figure 3a). Moreover, two broad undefined humps in the angular ranges 30–33° and 35–37° 2θ indicated that, at the end of the experiment, part of the 2ZC-cal did not react and explained the low $Sb(OH)_6^-$ removal that, unlike the experiment performed with 3HT-cal, cannot be attributable to the preferential uptake of $HAsO_4^{2-}$ (Table 2). Previous authors reported the high affinity of As(V) for sulfate bearing ZnAl-LDHs [47,48] but, at the best of our knowledge, experiments on the As(V) removal by calcined ZnAl-LDHs is lacking and needs further study.

4.2. Sorption Experiments with Su Suergiu Mine Drainage (SU1)

The results suggested that both 2ZC-cal and 3HT-cal are suitable for the Sb (and also As) removal from SU1; however, to assess their practical use, the overall quality of treated water must also be considered. The decrease of EC value after the experiments can be attributable to the decrease of Ca^{2+} and HCO_3^-. The contents of Na^+, K^+ and Cl^- did not vary significantly, instead sensible variations of SO_4^{2-}, NO_3^- and F^- were observed in a few cases. The partial dissolution of sorbents explained the irregular increase of dissolved Zn or Mg and Al. The limit of Al in drinking water is established at 0.2 mg/L by both the WHO and EU, whereas the concentration of Zn in drinking water is not regulated by the WHO and EU, but rather the Italian Legislation set 2 mg/L [49]; therefore, the concentration of Al exceeded the limit only in one experiment and the Zn limit was never reached (Table 5).

In the experiment performed with 0.1 g of 3HT-cal the amount of Sb removed was dramatically lower with respect to the other experiments with higher amounts of 3HT-cal. In this case it was possible to observe that the Sb removal did not increase with the weight of 3HT-cal used. The Ca^{2+} and HCO_3^- decreases did not show correlation (Supplementary material Figure S4). In particular, the Ca^{2+} concentration suggested a lower $CaCO_3$ precipitation at the end of experiments with 0.1 and 1 g of 3HT-cal with respect to the other ones. In the first case (i.e., 0.1 g of 3HT-cal) the $CaCO_3$ precipitation should be limited by the low increase of pH (8.2–8.4), while in the experiment with 1 g of 3HT-cal, where the pH values increase up to 10.3 (Figure 5) and CO_3^{2-} prevails in solution, the $CaCO_3$ precipitation should be hindered by the uptake of CO_3^{2-} in the interlayer during the 3HT-cal rehydration (Figure 5). This could also explain the low Sb removal in spite of the high amount of 3HT-cal. In fact, at high pH the Sb uptake was hindered by CO_3^{2-} that strongly competes for the entry in the interlayer region of the reconstructing lamellar LDH structure. The experiments with 0.5 and 0.25 g of 3HT-cal seemed the best compromise to reach the most favorable conditions for the highest Sb and As removal from 400 mL of SU1.

As observed above, at the end of the experiments with 0.25, 0.5 and 1 g of 2ZC-cal, the amount of Sb removed from solution was markedly higher with respect to 0.1 g of sorbent. Moreover, it is possible to observe that the dissolved concentrations of Sb and As slightly decreased as the amount of 2ZC-cal used increased. The decrease of Ca^{2+} and HCO_3^- observed at the end of the experiments did not show clear correlation (Supplementary material Figure S4). The sequestration of Ca^{2+} is attributable to the precipitation of calcium carbonates, which can also explain the slight decrease of Mg^{2+} in some cases (Table 5). The slight differences in residual Ca^{2+} concentration indicated the precipitation of nearly equivalent amounts of calcium carbonates in the different experiments (Table 5). Differently, the HCO_3^- concentration decreased as the amount of 2ZC-cal increased because its uptake from the solution occurred by both the precipitation of calcium carbonates and the entry in the interlayer during the LDHs reconstruction (Supplementary material Figure S4). It is possible to note that only in the experiments with 0.5 and 1 g of 2ZC-cal the pH of solution slightly increased after the addition of sorbents (up to 8.7) and successively decreased to 7.9–8.0, whereas in the experiments with 0.1 and 0.25 g, the variation of pH values was negligible and remained in the range of 8.1–8.4 (Figure 6). These limited pH variations among the experiments with 2ZC-cal can explain the precipitation of nearly equivalent amounts of calcium carbonates. The relatively low pH values can also explain the limited precipitation of calcium carbonates, deducible from the concentrations of Ca removed from solution (Supplementary material Figure S4), with respect to that observed in the experiments performed with 3HT-cal. At these pH values around 8, HCO_3^- prevails among the dissolved carbonate species, therefore it is possible that the reconstruction of LDH by rehydration of 2ZC-cal occurred via the intercalation of HCO_3^-, as well as CO_3^{2-}.

In our previous work we performed sorption experiments, carried out with ultrapure water containing only $Sb(OH)_6^-$, to determine the maximum theoretical $Sb(OH)_6^-$ sorption capacity (q_{max}) of 2ZC-cal and 3HT-cal through the Langmuir isotherm [36]. The values of q_{max} were 4.37 mmol/g (i.e., 532 mg/g) for 3HT-cal and 4.54 mmol/g (i.e., 553 mg/g) for 2ZC-cal [36]. In the present work we have observed that the coexistence of other anions in solution can affect the $Sb(OH)_6^-$ removal capacity of sorbents tested. In order to construct the isotherm, being the starting Sb concentration of sorption experiments that of SU1, the solid/liquid ratio has been changed. However, the sorption data did not fit for the calculation of the isotherm because, due to the complexity of the SU1 chemical composition, also other processes, like the precipitation of calcium carbonates, occurred during the interaction between water and calcined LDHs having an effect on the Sb removal.

It is worth mentioning that the LDHs exhibit the possibility of being reused by regeneration, operated through calcination or anion exchange, for consecutive sorption–regeneration–sorption cycles [50–54]. This is a very important characteristic for their practical use in water treatment because it can reduce the amount of post-treatment waste materials. In this regard, as far as we know, there is no data about the regeneration of ZnAl-LDH after Sb(V) adsorption; whereas it has been reported that, because brandholzite is a non-LDH mineral, the brandholzite formation connected with the Sb(V) removal by calcined MgAl-LDHs can negatively affect the MgAl-LDH regeneration capacity [46,55]. In this work we have observed that in the experiments performed with SU1, the Sb(V) removal by 3HT-cal occurred by intercalation in the LDH interlayer. It is reasonable to suppose that this removal mechanism may positively influence the effectiveness of LDHs regeneration. Therefore, further studies should be performed in order to assess the potential use of both 3HT-cal and 2ZC-cal regenerated after Sb(V) removal from real polluted water.

5. Conclusions

In this work the Sb(V) (in the $Sb(OH)_6^-$ form) removal capacity of calcined hydrotalcite-like (3HT-cal) and zaccagnaite-like (2ZC-cal) compounds has been studied in order to assess their potential for the practical use with real Sb(V) polluted water. For this purpose, first the effect of other anions on the $Sb(OH)_6^-$ removal capacity of 2ZC-cal and 3HT-cal were tested through batch sorption experiments with coexistent anions in solution. Successively batch sorption experiments were carried out with the slag drainage (SU1) sampled at the abandoned Su Suergiu mine (Sardinia, Italy) affected by relevant Sb pollution.

In agreement with previous studies, the results of our experiments with coexistent anions showed that 3HT-cal and 2ZC-cal removed $Sb(OH)_6^-$ through the formation of brandholzite-like and zincalstibite-like compounds, respectively. Among the anions tested, the competition effect on

$Sb(OH)_6^-$ removal resulted to be $HAsO_4^{2-}$ >> HCO_3^- ≥ SO_4^{2-} for 2ZC-cal, and $HAsO_4^{2-}$ >> HCO_3^- >> SO_4^{2-} for 3HT-cal.

The results of the sorption experiments showed that both 3HT-cal and 2ZC-cal effectively removed Sb and As from SU1 and, thus, these phases might have potential use for practical application with real polluted water. In the case under study, due the high concentration of carbonate species in SU1, Sb was mainly removed by intercalation in the interlayer of carbonate bearing LDHs rather than by the formation of zincalstibite-like and brandholzite-like compounds, as instead observed in the competition experiments performed with synthetic solutions. The results also indicated that the precipitation of calcium carbonates (i.e., calcite and monohydrocalcite) may favor the Sb removal, subtracting CO_3^{2-} from the solution as a possible strong competitor of Sb for the interlayer region of LDHs. Therefore, especially in the experiments with 3HT-cal, the Sb removal capacity was markedly influenced by the liquid/solid ratio that determines the solution pH, the correlated precipitation of calcium carbonates and the competition effect of carbonate species in solution. The Sb removal by intercalation in the interlayer of carbonate bearing LDHs rather than the formation of antimonate bearing phases may represent an advantage for the LDHs regeneration, therefore further studies should be addressed in that direction.

Supplementary Materials: The following are available online at http://www.mdpi.com/2073-4352/9/8/410/s1, Figure S1: The abandoned mine area of Su Suergiu (Cidu et al [34], modified); Figure S2: The linear kinetic plots of the pseudo-first order rate equation of $Sb(OH)_6^-$ sorption experiments performed without competitors and with coexistent anions performed with (**a**) 2ZC-cal and (**b**) 3HT-cal; Figure S3: The linear kinetic plots of the pseudo-second order rate equation of $Sb(OH)_6^-$ sorption experiments without competitors and with coexistent SO_4^{2-} and HCO_3^- performed with (**a**) 2ZC-cal and (**b**) 3HT-cal, and (**c**) performed with $HAsO_4^{2-}$; Figure S4: (**a**) Plot of Ca removed vs HCO_3 removed, and concentration of (**b**) Ca and (**c**) HCO_3 removed at the end of the experiments performed with 3HT-cal or 2ZC-cal and SU1. The horizontal green lines in the plots (**b**) and (**c**) indicate the starting concentrations of Ca or HCO_3, respectively.

Author Contributions: Conceptualization, E.D., F.F. and R.C.; Data curation, E.D.; Formal analysis, E.D.; Funding acquisition, F.F. and R.C.; Investigation, E.D.; Methodology, E.D.; Supervision, F.F. and R.C.; Validation, F.F. and R.C.; Writing—original draft, E.D.; Writing—review and editing, E.D., F.F. and R.C.

Funding: Authors thank the financial support from the Ministero Università Ricerca Scientifica Tecnologica through the research projects PRIN 2009 (Coordinator R. Cidu) and PRIN 2010–2011 (Coordinator P. Lattanzi).

Acknowledgments: Authors thank the Paola Meloni and Ombretta Cocco of "Research and Didactic Laboratory for the Conservation of the artistic heritage of the Monumental Cemetery of Bonaria" (Department of Mechanical, Chemical and Materials Engineering (DIMCM), University of Cagliari) for the assistance with the XRD measurements.

Conflicts of Interest: The authors declare no conflict of interest.

References

1. Filella, M.; Belzile, N.; Chen, Y.W. Antimony in the environment: A review focused on natural waters I. Occurrence. *Earth Sci. Rev.* **2002**, *57*, 125–176. [CrossRef]
2. WHO. *World Health Organization: Guidelines for Drinking Water Quality*, 4th ed.; WHO: Geneva, Switzerland, 2011.
3. Council of the European Union. Council Directive 98/83/EC of 3 November 1998 on the quality of water intended for human consumption. *Off. J. Eur. Communities* **1998**, *330*, 35–54.
4. Filella, M.; Belzile, N.; Chen, Y.W.; Quentel, F. Antimony in the environment: A review focused on natural waters II. Relevant solution chemistry. *Earth Sci. Rev.* **2002**, *59*, 265–285. [CrossRef]
5. Reimann, C.; Matschullat, J.; Birke, M.; Salminen, R. Antimony in the environment: Lessons from geochemical mapping. *Appl. Geochem.* **2010**, *25*, 175–198. [CrossRef]
6. Cidu, R.; Dore, E.; Biddau, R.; Nordstrom, D.K. Fate of Antimony and Arsenic in Contaminated Waters at the Abandoned Su Suergiu Mine (Sardinia, Italy). *Mine Water Environ.* **2018**, *37*, 151–165. [CrossRef]
7. He, M.; Wang, X.; Wu, F.; Fu, Z. Antimony pollution in China. *Sci. Total. Environ.* **2012**, *421*, 41–50. [CrossRef]
8. Okkenhaug, G.; Gebhardt, K.H.G.; Amstaetter, K.; Bue, H.L.; Herzel, H.; Mariussen, E.; Almas, Å.R.; Cornelissen, G.; Breedveld, G.; Rasmussen, G.; et al. Antimony (Sb) and lead (Pb) in contaminated shooting range soils: Sb and Pb mobility and immobilization by iron based sorbents, a field study. *J. Hazard. Mater.* **2016**, *307*, 336–343. [CrossRef] [PubMed]

9. Li, J.; Zheng, B.; He, Y.; Zhou, Y.; Chen, X.; Ruan, S.; Yang, Y.; Dai, C.; Tang, L. Antimony contamination, consequences and removal techniques: A review. *Ecotox. Environ. Saf.* **2018**, *156*, 125–134. [CrossRef]
10. Guo, X.; Wu, Z.; He, M. Removal of antimony (V) and antimony (III) from drinking water by coagulation-flocculation-sedimentation (CFS). *Water Res.* **2009**, *43*, 4327–4335. [CrossRef]
11. Bergmann, M.E.H.; Koparal, A.S. Electrochemical antimony removal from accumulator acid: Results from removal trials in laboratory cells. *J. Hazard. Mater.* **2011**, *196*, 59–65. [CrossRef]
12. Zhu, J.; Wu, F.; Pan, X.; Guo, J.; Wen, D. Removal of antimony from antimony mine flotation wastewater by electrocoagulation with aluminum electrodes. *J. Environ. Sci.* **2011**, *23*, 1066–1071. [CrossRef]
13. Ma, B.; Wang, X.; Liu, R.; Jefferson, W.A.; Lan, H.; Liu, H.; Qu, J. Synergistic process using Fe hydrolytic flocs and ultrafiltration membrane for enhanced antimony (V) removal. *J. Membr. Sci.* **2017**, *537*, 93–100. [CrossRef]
14. Saito, T.; Tsuneda, S.; Hirata, A.; Nishiyama, S.; Saito, K.; Saito, K.; Sugita, K.; Uezu, K.; Tamada, M.; Sugo, T. Removal of antimony (III) using polyol-ligand-containing porous hollow-fiber membranes. *Sep. Sci. Technol.* **2010**, *39*, 3011–3022. [CrossRef]
15. Leuz, A.K.; Mönch, H.; Johnson, C.A. Sorption of Sb(III) and Sb(V) to goethite: Influence on Sb(III) oxidation and mobilization. *Environ. Sci. Technol.* **2006**, *40*, 7277–7282. [CrossRef] [PubMed]
16. Rakshit, S.; Sarkar, D.; Punamiya, P.; Datta, R. Antimony sorption at gibbsite-water interface. *Chemosphere* **2011**, *84*, 480–483. [CrossRef] [PubMed]
17. Thanabalasingam, P.; Pickering, W.F. Specific sorption of antimony (III) by the hydrous oxides of Mn, Fe and Al. *Water Air Soil Poll.* **1990**, *49*, 175–185. [CrossRef]
18. Watkins, R.; Weiss, D.; Dubbin, W.; Peel, K.; Coles, B.; Arnold, T. Investigation into the kinetics and thermodynamics of Sb(III) adsorption on goethite (α-FeOOH). *J. Colloid Interf. Sci.* **2006**, *303*, 639–646. [CrossRef]
19. Liu, Y.; Wu, P.; Liu, F.; Li, F.; An, X.; Liu, J.; Wang, Z.; Shen, C.; Sand, W. Electroactive modified carbon nanotube filter for simultaneous detoxification and sequestration of Sb(III). *Environ. Sci. Technol.* **2019**, *53*, 1527–1535. [CrossRef]
20. Luo, J.; Luo, X.; Critteden, J.; Qu, J.; Bai, Y.; Peng, Y.; Li, J. Removal of antimonite (Sb(III)) and antimonate (Sb(V)) from aqueous solution using carbon nanofibers that are decorated with zirconium oxide (ZrO_2). *Environ. Sci. Technol.* **2015**, *49*, 11115–11124. [CrossRef]
21. Wang, L.; Wang, J.; Wang, Z.; He, C.; Lyu, W.; Yan, W.; Yang, L. Enhanced antimonate (Sb(V)) removal from aqueous solution by La-doped magnetic biochars. *Chem. Eng. J.* **2018**, *354*, 623–632. [CrossRef]
22. Wang, L.; Wang, J.; Wang, Z.; Feng, J.; Li, S.; Yan, W. Synthesis of Ce-doped magnetic biochar for effective Sb(V) removal: Performance and mechanism. *Powder Technol.* **2019**, *345*, 501–508. [CrossRef]
23. Goh, K.H.; Lim, T.T.; Dong, Z. Application of layered double hydroxides for removal of oxyanions: A review. *Water Res.* **2008**, *42*, 1343–1368. [CrossRef] [PubMed]
24. Cavani, F.; Trifirò, F.; Vaccari, A. Hydrotalcite-type anionic clays: Preparation, properties and applications. *Catal. Today* **1991**, *11*, 173–301. [CrossRef]
25. Mills, S.J.; Christy, A.G.; Génin, J.-M.R.; Kameda, T.; Colombo, F. Nomenclature of the hydrotalcite supergroup: Natural layered double hydroxides. *Mineral. Mag.* **2012**, *76*, 1289–1336. [CrossRef]
26. Carriazo, D.; del Arco, M.; Martín, C.; Rives, V. A comparative study between chloride and calcined carbonate hydrotalcites as adsorbent for Cr(VI). *Appl. Clay Sci.* **2007**, *37*, 231–239. [CrossRef]
27. Wang, S.L.; Cheng, H.L.; Ming, K.W.; Chuang, Y.H.; Chiang, P.N. Arsenate adsorption by Mg/Al-NO_3 layered double hydroxides with varying the Mg/Al ratio. *Appl. Clay Sci.* **2009**, *43*, 79–85. [CrossRef]
28. You, Y.W.; Vance, G.F.; Zhao, H.T. Selenium adsorption on Mg-Al and Zn-Al layered double hydroxides. *Appl. Clay Sci.* **2001**, *20*, 13–25. [CrossRef]
29. Iftekhar, S.; Küçük, M.E.; Srivastava, V.; Repo, E.; Sillanpää, M. Application of zinc-aluminium layered double hydroxides for adsorptive removal of phosphate and sulfate: Equilibrium, kinetic and thermodynamic. *Chemosphere* **2018**, *209*, 470–479. [CrossRef]
30. Kameda, T.; Honda, M.; Yoshioka, T. Removal of antimonate ions and simultaneous formation of a brandholzite-like compound from magnesium-aluminum oxide. *Sep. Purif. Technol.* **2011**, *80*, 235–239. [CrossRef]

31. Kameda, T.; Nakamura, M.; Yoshioka, T. Removal of antimonate ions from an aqueous solution by anion exchange with magnesium-aluminum layered double hydroxides and the formation of a brandholzite-like structure. *J. Environ. Sci. Heal. A* **2012**, *47*, 1146–1151. [CrossRef]
32. Kameda, T.; Kondo, E.; Yoshioka, T. Equilibrium and kinetics studies on As(V) and Sb(V) removal by Fe^{2+}-doped Mg-Al layered double hydroxides. *J. Environ. Manag.* **2015**, *151*, 303–309. [CrossRef] [PubMed]
33. Ardau, C.; Frau, F.; Lattanzi, P. Antimony removal from aqueous solutions by the use of Zn-Al sulphate Layered Double Hydroxide. *Water Air Soil Poll.* **2016**, *227*, 344. [CrossRef]
34. Lu, H.; Zhu, Z.; Zhang, H.; Zhu, J.; Qiu, Y. Simultaneous removal of arsenate and antimonate in simulated and practical water samples by adsorption onto Zn/Fe layered double hydroxide. *Chem. Eng. J.* **2015**, *276*, 365–375. [CrossRef]
35. Cao, D.; Zeng, H.; Yang, B.; Zhao, X. Mn assisted electrochemical generation of two-dimensional Fe-Mn layered double hydroxides for efficient Sb(V) removal. *J. Hazard. Mater.* **2017**, *336*, 33–40. [CrossRef] [PubMed]
36. Dore, E.; Frau, F. Antimonate uptake by calcined and uncalcined layered double hydroxides: Effect of cationic composition and M^{2+}/M^{3+} molar ratio. *Environ. Sci. Pollut. Res.* **2018**, *25*, 916–929. [CrossRef] [PubMed]
37. Cidu, R.; Biddau, R.; Dore, E.; Vacca, A.; Marini, L. Antimony in the soil-water-plant system at the Su Suergiu abandoned mine (Sardinia, Italy): Strategies to mitigate contamination. *Sci. Total. Environ.* **2014**, *497*, 319–331. [CrossRef] [PubMed]
38. Lagergren, S. About the theory of so-called adsorption of soluble substances. *K. Sven. Vetensk. Handl.* **1898**, *24*, 1–39.
39. Ho, Y.S.; McKay, G. A comparison of chemisorption kinetic models applied to pollutant removal on various sorbents. *Process. Saf. Environ.* **1998**, *76*, 332–340. [CrossRef]
40. Lozano, R.P.; Rossi, C.; La Iglesia, Á.; Matesanz, E. Zaccagnaite-3R, a new Zn-Al polytype from El Soplao cave (Cantabria, Spain). *Am. Minerol.* **2012**, *97*, 513–523. [CrossRef]
41. Zhang, Y.; Li, X. Preparation of Zn-Al CLDH to remove bromate from drinking water. *J. Environ. Eng. Asce* **2014**, *140*, 04014018. [CrossRef]
42. Bonaccorsi, E.; Merlino, S.; Orlandi, P. Zincalstibite, a new mineral, and cualstibite: Crystal chemical and structural relationships. *Am. Minerol.* **2007**, *92*, 198–203. [CrossRef]
43. Friedrich, A.; Wildner, M.; Tillmanns, E.; Merz, P.L. Cristal chemistry of the new mineral brandholzite, $Mg(H_2O)_6[Sb(OH)_6]_2$, and of the synthetic analogues $M^{2+}(H_2O)_6[Sb(OH)_6]_2$ (M^{2+} = Mg, Co). *Am. Minerol.* **2000**, *85*, 593–599. [CrossRef]
44. Cidu, R.; Biddau, R.; Dore, E. Determination of traces of Sb(III) using ASV in Sb-rich water samples affected by mining. *Anal. Chim. Acta* **2015**, *854*, 34–39. [CrossRef] [PubMed]
45. Türk, T.; Alp, I.; Deveci, H. Adsorption of As(V) from water using Mg–Fe-based hydrotalcite (FeHT). *J. Hazard. Mater.* **2009**, *17*, 1665–1670. [CrossRef] [PubMed]
46. Lee, S.H.; Tanaka, M.; Takahashi, Y.; Kim, K.W. Enhanced adsorption of arsenate and antimonate by calcined Mg/Al layered double hydroxide: Investigation of comparative adsorption mechanism by surface characterization. *Chemosphere* **2018**, *211*, 903–911. [CrossRef] [PubMed]
47. Ardau, C.; Frau, F.; Lattanzi, P. New data on arsenic sorption properties of Zn-Al sulphate layered double hydroxides: Influence of competition with other anions. *App. Clay Sci.* **2013**, *80*, 1–9. [CrossRef]
48. Ardau, C.; Frau, F.; Ricci, P.C.; Lattanzi, P. Sulphate-arsenate exchange properties of Zn-Al layered double hydroxides: Preliminary data. *Period. Minerol.* **2011**, *80*, 339–349.
49. GURI. *Decreto Legislativo 3 Aprile 2006, n. 152, Norme in Materia Ambientale;* Gazzetta Ufficiale della Repubblica Italiana: Roma, Italia, 2006. Suppl. ord. n. 96. (In Italian)
50. Bhaumik, A.; Samanta, S.; Mal, N.K. Efficient removal of arsenic from polluted ground water by using a layered double hydroxide exchanger. *Indian J. Chem. A* **2005**, *44*, 1406–1409.
51. Dessalegne, M.; Zewge, F.; Pfenninger, N.; Johnson, C.A.; Diaz, I. Layered double hydroxide and its calcined product for fluoride removal from groundwater of Ethiopian Rift Valley. *Water Air Soil Pollut.* **2016**, *227*, 381. [CrossRef]
52. Dore, E.; Frau, F. Calcined and uncalcined carbonate layered double hydroxides for possible water defluoridation in rural communities of the East African Rift Valley. *J. Water Process. Eng.* **2019**, *31*, 100855. [CrossRef]
53. Kuzawa, K.; Jung, Y.J.; Kiso, Y.; Yamada, T.; Nagai, M.; Lee, T.G. Phosphate removal and recovery with a synthetic hydrotalcite as an adsorbent. *Chemosphere* **2006**, *62*, 45–52. [CrossRef] [PubMed]

54. Lazaridis, N.K.; Pandi, T.A.; Matis, K.A. Chromium(VI) removal from aqueous solutions by Mg–Al–CO_3 hydrotalcite: Sorption–desorption kinetic and equilibrium studies. *Ind. Eng. Chem. Res.* **2004**, *43*, 2209–2215. [CrossRef]
55. Lee, S.H.; Choi, H.; Kim, K.W. Removal of As(V) and Sb(V) in water using magnetic nanoparticle-supported layered double hydroxide nanocomposites. *J. Geochem. Explor.* **2018b**, *184*, 247–254. [CrossRef]

Article

Mg-Fe Layered Double Hydroxides Enhance Surfactin Production in Bacterial Cells

Pei-Hsin Chang [1], Si-Yu Li [1,2], Tzong-Yuan Juang [3,*] and Yung-Chuan Liu [1,*]

1 Department of Chemical Engineering, National Chung Hsing University, 145 Xingda Road, South District, Taichung 40227, Taiwan
2 Innovation and Development Center of Sustainable Agriculture, NCHU, Taichung 40227, Taiwan
3 Department of Cosmeceutics, China Medical University, Taichung 40402, Taiwan
* Correspondence: tyjuang@mail.cmu.edu.tw (T.-Y.J.); ycliu@dragon.nchu.edu.tw (Y.-C.L.)

Received: 20 June 2019; Accepted: 11 July 2019; Published: 12 July 2019

Abstract: In this study, four additives—montmorillonite, activated carbon, and the layered double hydroxides (LDHs), Mg_2Fe–LDH and Mg_2Al–LDH—were tested for their ability to promote surfactin production in a *Bacillus subtilis* ATCC 21332 culture. Among these tested materials, the addition of 4 g/L of the Mg-Fe LDH, which featured an Mg/Fe molar ratio of 2:1, produced the highest surfactin yield of 5280 mg/L. During the time course of *B. subtilis* cultivation with the added LDH, two phases of cell growth were evident: Growth and decay. In the growth phase, the cells grew slowly and secreted a high amount of surfactin; in the decay phase, the cells degraded rapidly. The production in the presence of the Mg_2Fe–LDH had three characteristics: (i) High surfactin production at low biomass, indicating a high specific surfactin yield of 3.19 g/g DCW; (ii) rapid surfactin production within 24 h, inferring remarkably high productivity (4660 mg/L/d); and (iii) a lower carbon source flux to biomass, suggesting an efficient carbon flux to surfactin, giving a high carbon yield of 52.8%. The addition of Mg_2Fe–LDH is an effective means of enhancing surfactin production, with many potential applications and future industrial scale-up.

Keywords: *Bacillus subtilis*; surfactin; quantitative analysis; fermentation; growth phase; layered double hydroxides

---

1. Introduction

Biosurfactants are amphipathic molecules produced by microorganisms [1–3] with the capability of decreasing surface and interfacial tension [4]. Depending on their chemical composition and their producing organism, biosurfactants can possess high biodegradability, low toxicity, ecological acceptability, and high efficiency. Accordingly, they have been investigated as possible alternatives to chemical surfactants [5,6]. *Bacillus* spp., bacterial strains of complicated physiological diversity, can be used to produce many bioactive peptides with potential biotechnological and biopharmaceutical applications. Among these peptides, the lipopeptides that feature an alkyl group and a circular peptide group are the most popular biosurfactants [7]; these materials include surfactins [8–10], iturins [11,12], and fengycins [13].

The surfactin produced by *B. subtilis* is one of the strongest biosurfactants available [7]. Its chemical composition is that of a cyclic lipopeptide (comprising seven amino acids) with a 12 to 19-carbon atom hydrophobic fatty acid chain [14]. Surfactin can lower the surface tension of water to 27 mN/m even when its concentration is as low as 0.005% [7,10,15,16], suggesting its great potential applicability. Nevertheless, the high expense and low yield of surfactin production have limited its commercial use. Yeh et al. found that limiting the concentration of the carbon source (glucose) affected the surfactin production mediated by *B. subtilis* [17]. Davis et al. observed the highest production of surfactin when ammonium nitrate was the nitrogen source during *B. subtilis* cultivation in a defined medium [18].

Sen et al. noted that the ratio of Mn and Fe mineral salts in the medium was a factor affecting the production of surfactin [19]. Wei and Chu found that the yield of surfactin increased dramatically, over those obtained using genetic strains, when employing 0.01 mM Mn^{2+} [20]. Furthermore, Wei et al. employed an iron-enriched (4 mM Fe^{2+}) minimal salt medium to produce 3000 mg/L of surfactin [21]. Moreover, some of these studies revealed that the addition of solid additives (e.g., activated carbon (AC) or expanded clay) could increase surfactin production significantly. For example, Yeh et al. added AC and increased the yield of surfactin to 3600 mg/L [17].

Layered double hydroxides (LDHs)—also known as anionic clays—comprise cationic brucite-like layers with exchangeable interlayer anions [22]. Because a positive ionic charge appears on the surface layer, many types of molecules can be intercalated into LDHs [23–27]. Several methods have been developed to widen the layered gallery, with globular macromolecules as intercalating agents [28,29]. Conterosito et al. intercalated various pharmaceutics drugs and cosmetic sunscreen into Mg-Al_LDH and Zn-Al_LDH. They revealed that different bioactive molecules could interact with inorganic LDH and demonstrated the relationship between the molecular length and an enlarged interlayer spacing [30]. Toson et al. showed the intercalation of organic molecules into the LDH interlayer by the liquid-assisted grinding method. The intercalation mechanism for layer widening with intercalated organic molecules was investigated [31]. Choy et al. employed supramolecular inorganic species (e.g., nanoscale Mg-Al LDH) as biomolecule reservoirs that could be used for gene and drug delivery [23,32]. Nevertheless, few studies have focused on using LDHs as additives for microbial cultivation. Kan et al. prepared Mg_2Al–LDH and investigated its effect as an additive on surfactin production and surfactin intercalation [33]. When considering the application of the surfactin-intercalated LDH as a slow release bio-pesticide, however, this aluminum salt was prohibited from field tests [34]. For agricultural applications, iron salts are generally considered less toxic. Therefore, in this present study, we prepared several Mg-Fe LDH derivatives with potentially greater practicality. We tested the effects of their addition on the production of surfactin from a *B. subtilis* culture. To our surprise, replacing the additive to Mg_2Fe–LDH had an extraordinary effect on the surfactin production. Accordingly, we examined various Mg_nFe–LDH (n = 1, 2, 3) compositions and concentrations to determine the optimal conditions for surfactin production. In addition, we examined the time course of the production in the optimal culture. An extra low biomass of cells yielded the highest surfactin production. This result was quite different from that obtained after the addition of Mg_2Al–LDH. Furthermore, we compared the effects of the LDHs with those of other additives (e.g., montmorillonite (MMT), AC), and determined the conditions for the highest production of surfactin through quantitative analysis. Herein, we also suggest possible reasons for the enhancement of surfactin production mediated by LDHs.

2. Materials and Methods

2.1. Chemicals and Reagents

$Al(NO_3)_3 \cdot 9H_2O$, $Fe(NO_3)_3 \cdot 9H_2O$, and $Mg(NO_3)_2 \cdot 6H_2O$ were purchased from SHOWA, USA. MMT was obtained from Alfa Aesar, USA. The AC was obtained from China Activated Carbon (Taipei, Taiwan); it had a diameter of 3 to 4 mm, a height of 9 mm, and a specific surface area of 1200 m^2/g, and was prepared from bituminous coal with an iodine number of 1150 mg/g. Surfactin (≥98%, Sigma–Aldrich, Missouri, MO, USA) was used as the standard. All solvents and other chemicals were of analytical grade.

2.2. Microorganisms and Culture Conditions

The strain, *B. subtilis* ATCC 21332, was obtained from Professor Wei Yu-Hong of Yuan Ze University. This strain was kept on a nutrient-agar plate at 30 °C. For cultivation, its seed medium comprised 1% glucose, 0.5% yeast extract, 1% peptone, and 1% NaCl. The seed culture was performed in Erlenmeyer flasks (500 mL) containing the seed medium (100 mL) inoculated with two loops of cells. The cultivation was conducted at 200 rpm and 30 °C for 12 h. The main shake-flask culture was conducted in an Erlenmeyer flask (500 mL) containing the main medium (100 mL) comprising 10 g/L sucrose, 5 g/L $(NH_4)_2SO_4$, 5.67 g/L Na_2HPO_4, 4.08 g/L KH_2PO_4, 0.2 g/L $MgSO_4{\cdot}7H_2O$, and 0.57 g/L $FeSO_4{\cdot}7H_2O$. The media were sterilized (121 °C, 20 min); the carbon source was autoclaved separately. The medium (90 mL) was inoculated with the seed broth (10 mL). The flasks were incubated on a rotary shaker (200 rpm, 30 °C, 5 days). When testing additives, MMT, AC, and the prepared LDHs were added (2 g/L) to the culture medium at the beginning of the culture process.

2.3. Mg_2Al–LDH and Mg_2Fe–LDH

Mg_2Al-NO_3–LDH and Mg_2Fe-NO_3–LDH were prepared through co-precipitation, as described previously [27]. To prepare the Mg_2Al–LDH sample, $Mg(NO_3)_2{\cdot}6H_2O$ (120 g) and $Al(NO_3)_3{\cdot}9H_2O$ (90 g) were dissolved in deionized H_2O (1 L). To prepare the Mg_2Fe–LDH sample, $Mg(NO_3)_2{\cdot}6H_2O$ (169 g) and $Fe(NO_3)_3{\cdot}9H_2O$ (134 g) were mixed in deionized H_2O (1 L). To prepare samples with Mg/Fe molar ratios of 1.0 and 3.0, appropriate amounts of $Mg(NO_3)_2{\cdot}6H_2O$ and $Fe(NO_3)_3{\cdot}9H_2O$ were used. Each aqueous solution was stirred vigorously at 80 °C while purging with nitrogen gas. When preparing the Mg_2Al–LDH sample, the pH was maintained at 10 ± 0.2 by adding 4 N NaOH in portions. For the Mg_2Fe-NO_3–LDH sample, the pH was adjusted to 9.5 ± 0.2 by using a mixture of NaOH and $K_3[Fe(CN)_6]$, prepared based on the following compositions: $[OH^-]/([Mg^{2+}] + [Fe^{3+}]) = 1.6$ and $[[Fe(CN)_6]^{3-}]/[Fe^{3+}] = 3$. The suspension that formed was stirred at 80 °C for 24 h. The obtained precipitates—white Mg_2Al–LDH and dark-red Mg_2Fe–LDH—were filtered off and washed (deionized H_2O). The filtered cakes were lyophilized (freeze-drying). The dried LDHs were characterized using x-ray diffraction (XRD; PANalytical, X'Pert PRO MRD, Almelo, Netherlands) and attenuated total reflectance Fourier transform infrared (ATR-FTIR) spectroscopy (Thermo Scientific, Nicolet iS50 FTIR, Madison, WI, USA).

2.4. Quantitative Analysis

To study the effects of LDH addition, the following quantitative terms are defined. The surfactin yield (mg/L) is expressed by the volumetric concentration. The carbon source yield is defined as:

$$Y_{\frac{P}{S}} = \frac{\Delta P}{\Delta S} \tag{1}$$

The productivity is defined as:

$$\text{Productivity} = \frac{\Delta P}{\Delta t} \tag{2}$$

The specific yield is defined as:

$$Y_{\frac{P}{X}} = \frac{\Delta P}{\Delta X} \tag{3}$$

where P represents the surfactin concentration, S is the carbon source concentration, X is the concentration of biomass dried cells, and t is the duration of cultivation. All concentrations are expressed herein on a volumetric basis.

2.5. Assays

The surfactin concentration was measured using a modified approach called salt-assisted homogeneous liquid-liquid extraction via high-performance liquid chromatography (HPLC) [33,35]. The culture broth (1 mL) was subjected to a centrifugation (3200 g, 3 min, 4 °C) to remove the solid pellets. The supernatant was mixed with MeCN (0.5 mL) and ammonium sulfate (0.8 g) and subjected to vigorous stirring for 1 min, and then centrifuged (3200 g, 3 min). The supernatant was filtered (0.22 μm) to obtain the sample for injection. HPLC analysis was performed under the following conditions: A reversed-phase C-18A column (5 mm, 18 mm × 100 mm BDS-Hypersil, Thermo Fisher Scientific, Waltham, MA, USA); a mobile phase comprising CF_3CO_2H, MeCN, and deionized H_2O (0.1:400:100); a flow rate of 1.0 mL/min; an injection sample volume of 20 μL; and a UV–Vis detector

(JASCO, Tokyo, Japan) operated at 220 nm. A standard curve was constructed using a freshly prepared solution of surfactin (Sigma). The chromatogram of the standard (supplemental Figure S1) revealed various ratios of the surfactin isoforms A–F. The surfactin produced using *B. subtilis* ATCC 21332 featured the same surfactin isoforms A–F at various ratios. In the surfactin assay, the whole isoforms were measured and added up for quantitative calculation. To analyze the cells' dried weight (CDW), 5 mL of the broth sample was subjected to centrifugation (12,000 *g*, 10 min) to obtain a pellet. Distilled H_2O (5 mL) was added to the pellet; after adjusting to pH 2.0, the sample was vigorously stirred (1 min). The mixture was centrifuged (12,000 *g*, 5 min). The pellet obtained was dissolved in distilled H_2O (5 mL); the pH was adjusted to 7.0 and the mixture was again subjected to centrifugation. The obtained pellet was washed with distilled H_2O (2 × 5 mL), dried (80 °C, 12 h), and then weighed. The basal spacing of the LDH was determined using a Shimadzu SD-D1 X-ray diffractometer with a Cu target (scanning rate: 1°/min). The basal spacing was estimated using the Bragg equation ($n\lambda = 2d\sin\theta$).

2.6. Statistical Analysis

Multiple flasks were run concurrently. Three flasks were employed each time for daily sampling. Each data point is expressed as a mean plus standard deviation. The Tukey test was applied for the comparison of results ($p \leq 0.05$).

3. Results and Discussion

3.1. Preparation of MgFe–LDH

The addition of a small quantity of a solid carrier (AC or expanded clay) has been claimed as an effective approach toward increasing surfactin production [17]. LDHs are layered anionic exchange substances that have been intercalated with various macromolecules for the purpose of their slow release [36,37]. The addition of Mg_2Al–LDH LDH to a surfactin production fermentation system involving *B. subtilis* incubation revealed that surfactin could indeed intercalate into the LDH layer gallery to form a surfactin–LDH complex; this phenomenon occurred with a significant increase in the production of surfactin [33]. In consideration of a slow-release composite for agricultural use, Mg_2Al–LDH would be inappropriate for field trials. For this study, therefore, we prepared Mg_2Fe–LDH instead. We examined the effect of adding this iron salt LDH to *B. subtilis* cultivation to study whether it, too, would promote surfactin production. The prepared Mg_2Fe–LDH was subjected to XRD and ATR-FTIR spectroscopic analysis. These analyses revealed an Mg_2Fe–LDH layer spacing of 7.8 Å at a value of 2θ of 11.3°, derived from the calculation of Bragg's equation (Figure 1a), and a typical adsorption peak (1381 cm^{-1}) for NO_3^- anions within the prepared LDH (data not shown). In addition, to confirm the interaction between LDH and bacterial cells, the LDH after the cultivation was collected and subjected to XRD analysis. The result in Figure 1b shows that the collected LDH did vary its 2θ from the original 11.3° to 8.3°, indicating a d-spacing of 10.8 Å. The original XRD peak with a d-spacing of 7.8 Å completely disappeared. The enlarged spacing was likely due to the LDH interaction with surfactin molecules. The isoelectric point (IEP) of surfactin is around pH 5, and the fermentation process while applying LDH to the cultivation was around pH 7.4. The pH higher than the IEP would allow the surfactin to possess a negative charge, giving the chance of anion exchange for LDH intercalation. Besides, the interlayer spacing expansion of LDH might be ascribed not only to the surface interaction of surfactin intercalation but also the combination of water and other anion molecules in the culture medium into the Mg_2Fe–LDH interlayer.

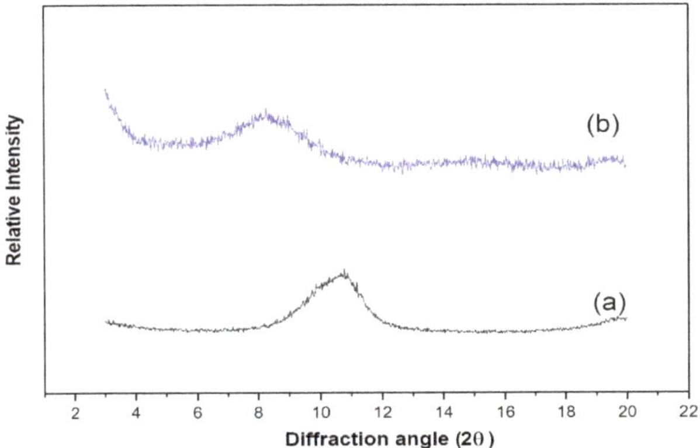

Figure 1. XRD pattern of (**a**) pristine Mg_2Fe–LDH and (**b**) Mg_2Fe–LDH collected after fermentation.

3.2. Effect of Solid Additives on Surfactin Production

As reported previously, the addition of some solid additives can enhance surfactin production [17]. For this present study, four solid additives—MMT, AC, and two LDHs—were prepared and added respectively to the *B. subtilis* culture medium; the medium prepared without any additives was used as the control during the five-day fermentation. The surfactin production increased when the culture medium contained each of these solid additives, relative to the control. The addition of MMT, AC, and the two LDHs (2 g/L) resulted in surfactin yields that had increased by 2.0-, 3.0-, 3.8-, and 4.5-fold, respectively, when compared with the control (Figure 2). It is noteworthy that the AC with the alkaline characterization might lead to surfactin linearization and surfactin binding on the AC surface, which may be an underestimation of the actual production. Thus, the LDHs were the most effective carriers for enhanced surfactin production in a culture of *B. subtilis* ATCC 21332. Furthermore, the amount of surfactin produced in the presence of Mg_2Fe–LDH was more than that produced in the presence of Mg_2Al–LDH. Indeed, Mg_2Fe–LDH had an extraordinary stimulatory effect on promoting surfactin production.

3.3. Effect of MgFe–LDH Composition on Surfactin Production

To study the effect of the Mg/Fe molar ratio on surfactin production, LDHs were prepared with Mg:Fe molar ratios of 1:1, 2:1, and 3:1 and added into *B. subtilis* cultivation. The concentrations of the additive ranged from 1 to 6 g/L in the fermentation medium. The cultivation was performed for 5 days. Figure 3 reveals that the LDH prepared with a Mg:Fe molar ratio of 2:1 had the greatest effect at promoting surfactin production. In general, LDHs possessing different ratios of divalent and trivalent metal ions possess different types of positively charged sheets and different layer dimensions in their resulting layered structures [38–41]. In the brucite-like layers of an LDH, a fraction of the divalent metal ions is replaced by trivalent metal ions, with the molar ratio of M^{3+}:(M^{3+} + M^{2+}) (x) normally positioned between 0.2 and 0.4 [24,42]. In this present study, an Mg:Fe ratio of 2:1 (x = 0.33) had the best effect on improving surfactin production. Thus, it appears that the layer size associated with the positively charged sheets of the Mg_2Fe–LDH structure had the strongest stimulatory effect on the cells.

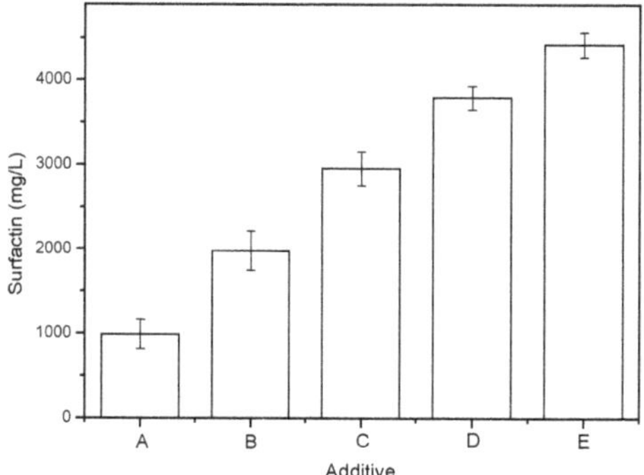

Figure 2. Effects of solid additives on surfactin yield in *B. subtilis* ATCC 21332 cultivation in a 5-day fermentation: (**A**) none; (**B**) MMT; (**C**) AC; (**D**) Mg_2Al–LDH; (**E**) Mg_2Fe–LDH. The surfactin level in the supernatant of the broth was determined. Error bars indicate the standard deviations from three tests.

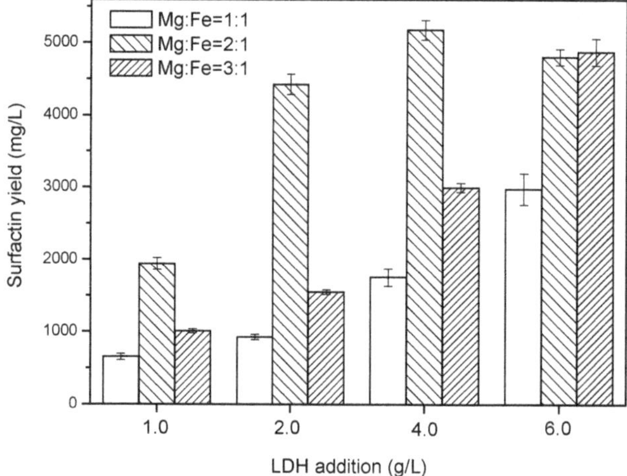

Figure 3. Effect of Mg_2Fe–LDH concentration (1, 2, 4, and 6 g/L) on the specific surfactin yield, where bars with different patterns represent various Mg^{2+}:Fe^{3+} ratios: blank, 1:1; back slash, 2:1; slash, 3:1. Error bars indicate standard deviations from three tests.

3.4. Time Course of Cultivation with LDH Addition

To study the cell growth after adding LDH, the time courses of the cultivation events performed with and without added LDH were recorded (Figure 4). Although the addition of Mg_2Fe–LDH promoted surfactin production, relative to that of the control, it was interesting to observe that the cell growth ended on the first day, where the amount of surfactin reached 4.8 g/L. In terms of product formation kinetics, this behavior was a clear growth-associated pattern: The cells grew and surfactin was produced. After day 1, the cells began to degrade in a decay phase, with the surfactin production decelerating. In contrast, the growth of cells was very rapid in the culture medium prepared without LDH, but the level of surfactin production was very low. Thus, a slight inhibition of cell growth appeared to trigger the cells to secrete more surfactin. We suspect that the surfactin secreted by the cells performed a role as a protecting agent that kept the cells from coming into direct contact with the LDH.

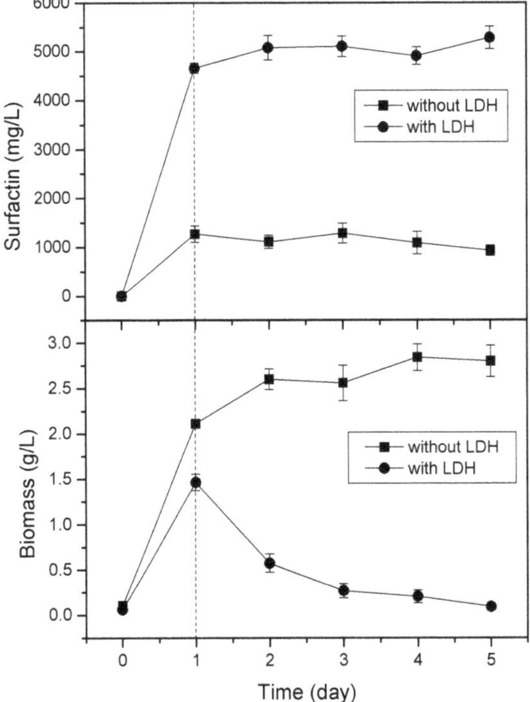

Figure 4. Time courses of surfactin and biomass production in the presence and absence of Mg_2Fe–LDH (4 g/L). Error bars indicate the standard deviations from three tests.

Figure 5A,B present microscopy images of the morphologies of the cells grown in the presence of the LDH. In the culture medium lacking the LDH, the cells had a short and rod-like morphology from day 1 to day 3 of culturing. By the fifth day, some cells became slenderer than the original short-rod cells. In contrast, in the culture medium incorporating the LDH, the cells grew in a short-rod shape on the first day, but, by the third day, most of the cells had decayed and shrunk, with many endospores present. By the fifth day, almost none of the cells were evident in the broth, with only some spores remaining in the culture. This observation is consistent with the amounts of cells measured in the study. Therefore, the addition of LDH did inhibit the growth of cells during the cell growth phase, but it also enhanced the production of surfactin. Accordingly, in addition to the high surfactin yield of the culture incorporating the LDH, an extremely high productivity also ensued. Because of the lower number of cells, not only was the specific production elevated, the carbon source conversion to surfactin was also enhanced and provided a high carbon yield.

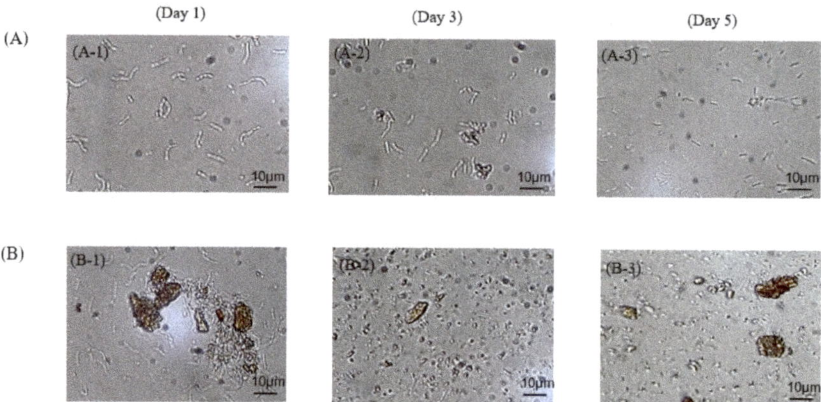

Figure 5. Optical microscopy images of *B. subtilis* ATCC 21332 growth, taken after various numbers of cultivation days; (**A**) in the absence of any LDH; (**B**) in the presence of Mg_2Fe–LDH.

3.5. Comparison of Surfactin Production

Table 1 compares the surfactin production in this present study with those reported previously in the literature. Four factors characterize surfactin production in these bioprocesses in terms of their efficiency for fermentation on industrial scale: the surfactin yield, the carbon yield, the productivity, and the specific production. Due to the variation on surfactin quantification, the surfactin assays, such as HPLC, surface tensions, and acid precipitation, were also listed. As evident in Table 1, the addition of Mg_2Fe–LDH had a unique effect in promoting surfactin production. Historically, surfactin production has improved gradually from an original yield of less than 1000 mg/L two decades ago to approximately 2000 to 3000 mg/L recently. When using this present approach, the yield of surfactin after the addition of Mg_2Fe–LDH was enhanced significantly, to greater than 5000 mg/L. Furthermore, the addition of Mg_2Fe–LDH ensured that the carbon source mostly flowed to surfactin production. Indeed, the carbon source yield was approximately 52.8%. This high carbon yield characterizes a surfactin production process with a highly efficient use of the raw material. In addition, the presence of Mg_2Fe–LDH caused the surfactin yield to reach 4660 mg/L after one day of culturing; that is, the productivity was 4660 mg/L, a remarkably high value as compared in the literature. In addition, because the number of cells decreased in the presence of Mg_2Fe–LDH, the smaller amount of biomass and the higher surfactin yield led to a specific yield of 3.19 g/g DCW. The addition of Mg_2Fe–LDH in *B. subtilis* submerged cultivation provided a high carbon yield, high productivity, and high specific production of surfactin; such a high efficiency appears well suited to industrial applications.

At the beginning of our approach, the change of Mg_2Al–LDH to Mg_2Fe–LDH was due to the practical need in agricultural applications, where the aluminum salt is prohibited from field tests. However, to our surprise, the replacement of additive to Mg_2Fe–LDH did give an extraordinarily high surfactin production. Due to this effect, the three critical characteristics affecting surfactin production were evaluated. It was found that a high specific surfactin yield, a high productivity, and a high carbon yield could be obtained in the presence of the Mg_2Fe–LDH. To explain the difference between Mg_2Fe–LDH and Mg_2Al–LDH additions, the effect of the Mg_2Fe–LDH addition with the leaking iron trace element in the culture was the possible reason for this highly efficient surfactin production. To decipher the cause of the extraordinarily high stimulatory effect of Mg_2Fe–LDH, the following considerations might be taken into account. In some previous studies, ferric ions have been found to serve as trace element stimulators, with an excellent ability to promote surfactin production [21,43,44]. In addition, the use of pristine Mg_2Al–LDH has been claimed to enhance surfactin production as a result of its toxicity toward the cells [33]. Accordingly, the presence of Mg_2Fe–LDH was expected to not only inhibit cell growth and promote surfactin production (similar to the behavior of Mg_2Al–LDH) but also to slowly release some iron salts to serve as trace elements in the medium, thereby also improving the surfactin production. The higher production obtained using Mg_2Fe–LDH, compared with that of Mg_2Al–LDH, might be due to the synergistic effect of the Mg_2Fe–LDH crystalline structure and the trace iron salts in the medium, with both combining to promote surfactin production to such a high level.

Table 1. Various approaches used for surfactin production.

Approach	Yield [a] (mg/L)	Carbon Yield (g/g carbon source)	Productivity [b] (mg/L/day)	Specific production [c] (g/g DCW)	Surfactin Quantification	Ref.
Iron addition & product in foam	800	0.02	-	0.615	Acid precipitate	[45]
Peat hydrolysate medium	160	0.004	160	0.208	HPLC	[46]
Aqueous two phase	350	0.07	350	–	Surface tensions	[47]
Optimized medium	760	0.021		–	Surface tensions	[19]
Optimized nitrogen source	439	0.015	219	0.075	HPLC	[18]
Recombinant strain	350	0035	350	–	HPLC	[48]
Strain mutation and product in foam	562	0.014	562	0.323	Surface tensions	[16]
Iron-enriched medium	3500	0.088	–	–	HPLC	[43]
Optimized iron supplement	3000	–	–	0.162	HPLC	[21]
Activated carbon addition	3600	0.09	1200	–	HPLC	[17]
Optimized trace element	3340	0.084	–	–	HPLC	[49]
Mn^{2+} addition	2600	0.065		0.289	HPLC	[20]
Mg_2Al LDH addition	3789	0.379	–	–	HPLC	This study
Mg_2Fe LDH addition	5280	0.528	4660	3.19	HPLC	This study

[a] Maximum yield in whole culture. [b] Maximum productivity in whole culture. [c] Specific production when reaching maximum yield.

4. Conclusions

We investigated the effects of LDHs on the production of biomass and surfactin in a *B. subtilis* ATCC 21332 culture. The highest yield of surfactin (5280 mg/L) was obtained after 5 days of cultivation in the presence of 4 g/L Mg_2Fe–LDH. This study demonstrated that LDHs have potential for use as additives to enhance the production of surfactin in *B. subtilis* ATCC 21332. Furthermore, microscopy revealed the inhibition of cell growth in the presence of the LDH, suggesting an efficient process for the production of surfactin through greater conversion of the carbon source.

Supplementary Materials: The following are available online at http://www.mdpi.com/2073-4352/9/7/355/s1, Figure S1: Typical surfactin standard chromatogram in HPLC showing surfactin isoform A–F.

Author Contributions: Data curation and methodology, P.-H.C. and S.-Y.L., conceptualization, writing—original draft preparation; writing—review and editing, project administration, T.-Y.J. and Y.-C.L.

Funding: This study was supported by research grants from the National Science Council of Taiwan, R.O.C. (grant nos. NSC 101-2221-E-005-061 and 106-2113-M-039-007), and China Medical University (grant no. CMU 107-N-22).

Acknowledgments: We thank Wei Yu-Hong of Yuan Ze University for sharing the strain *B. subtilis* ATCC 21332.

Conflicts of Interest: The authors declare no conflict of interest.

References

1. Maier, R.M.; Soberon-Chavez, G. Pseudomonas aeruginosa rhamnolipids: Biosynthesis and potential applications. *Appl. Microbiol. Biotechnol.* **2000**, *54*, 625–633. [CrossRef] [PubMed]
2. Kim, H.S.; Yoon, B.D.; Choung, D.H.; Oh, H.M.; Katsuragi, T.; Tani, Y. Characterization of a biosurfactant, mannosylerythritol lipid produced from Candida sp. SY16. *Appl. Microbiol. Biotechnol.* **1999**, *52*, 713–721. [CrossRef] [PubMed]
3. Banat, I.M. Biosurfactants production and possible uses in microbial enhanced oil recovery and oil pollution remediation: A review. *Bioresour. Technol.* **1995**, *51*, 1–12. [CrossRef]
4. Desai, J.D.; Banat, I.M. Microbial production of surfactants and their commercial potential. *Microbiol. Mol. Biol. Rev.* **1997**, *61*, 47–64. [PubMed]
5. Mukherjee, A.K.; Das, K. Microbial surfactants and their potential applications: An overview. *Adv. Exp. Med. Biol.* **2010**, *672*, 54–64. [PubMed]

6. Mulligan, C.N. Environmental applications for biosurfactants. *Environ. Pollut.* **2005**, *133*, 183–198. [CrossRef] [PubMed]
7. Shaligram, N.S.; Singhal, R.S. Surfactin—A review on biosynthesis, fermentation, purification and applications. *Food Technol. Biotechnol.* **2010**, *48*, 119–134.
8. Hosono, K.; Suzuki, H. Acylpeptides, the inhibitors of cyclic adenosine 3′,5′-monophosphate phosphodiesterase. I. Purification, physicochemical properties and structures of fatty acid residues. *J. Antibiot.* **1983**, *36*, 667–673. [CrossRef] [PubMed]
9. Kakinuma, A.; Hori, M.; Isono, M.; Tamura, G.; Arima, K. Determination of amino acid aequence in surfactin a crystalline peptidelipid surfactant produced by Bacillus subtilis. *Agric. Biol. Chem.* **1969**, *33*, 971–972. [CrossRef]
10. Arima, K.; Kakinuma, A.; Tamura, G. Surfactin, a crystalline peptide lipid surfactant produced by Bacillus subtilis: Isolation, characterization and its inhibition of fibrin clot formation. *Biochem. Biophys. Res. Commun.* **1968**, *31*, 488–494. [CrossRef]
11. Besson, F.; Peypoux, F.; Michel, G.; Delcambe, L. Identification of antibiotics of iturin group in various strains of Bacillus subtilis. *J. Antibiot.* **1978**, *31*, 284–288. [CrossRef] [PubMed]
12. Peypoux, F.; Guinand, M.; Michel, G.; Delcambe, L.; Das, B.C.; Lederer, E. Structure of iturine A, a peptidolipid antibiotic from Bacillus subtilis. *Biochemistry* **1978**, *17*, 3992–3996. [CrossRef]
13. Vanittanakom, N.; Loeffler, W.; Koch, U.; Jung, G. Fengycin—A novel antifungal lipopeptide antibiotic produced by *Bacillus subtilis* F-29-3. *J. Antibiot.* **1986**, *39*, 888–901. [CrossRef]
14. Liao, J.H.; Chen, P.Y.; Yang, Y.L.; Kan, S.C.; Hsieh, F.C.; Liu, Y.C. Clarification of the antagonistic effect of the lipopeptides produced by Bacillus amyloliquefaciens BPD1 against Pyricularia oryzae via in situ MALDI-TOF IMS analysis. *Molecules* **2016**, *21*, 1670. [CrossRef] [PubMed]
15. Abdel-Mawgoud, A.M.; Aboulwafa, M.M.; Hassouna, N.A. Characterization of surfactin produced by *Bacillus subtilis* isolate BS5. *Appl. Biochem. Biotechnol.* **2008**, *150*, 289–303. [CrossRef] [PubMed]
16. Mulligan, C.N.; Chow, T.Y.-K.; Gibbs, B.F. Enhanced biosurfactant production by a mutant *Bacillus subtilis* strain. *Appl. Microbiol. Biotechnol.* **1989**, *31*, 486–489. [CrossRef]
17. Yeh, M.S.; Wei, Y.H.; Chang, J.S. Enhanced production of surfactin from *Bacillus subtilis* by addition of solid carriers. *Biotechnol. Prog.* **2005**, *21*, 1329–1334. [CrossRef]
18. Davis, D.A.; Lynch, H.C.; Varley, J. The production of surfactin in batch culture by *Bacillus subtilis* ATCC 21332 is strongly influenced by the conditions of nitrogen metabolism. *Enzyme Microb. Technol.* **1999**, *25*, 322–329. [CrossRef]
19. Sen, R. Response surface optimization of the critical media components for the production of surfactin. *J. Chem. Technol. Biotechnol.* **1997**, *68*, 263–270. [CrossRef]
20. Wei, Y.-H.; Chu, I.-M. Mn^{2+} improves surfactin production by Bacillus subtilis. *Biotechnol. Lett.* **2002**, *24*, 479–482. [CrossRef]
21. Wei, Y.-H.; Wang, L.-F.; Chang, J.-S. Optimizing iron supplement strategies for enhanced surfactin production with *Bacillus subtilis*. *Biotechnol. Prog.* **2004**, *20*, 979–983. [CrossRef] [PubMed]
22. Cavani, F.; Trifirò, F.; Vaccari, A. Hydrotalcite-type anionic clays: Preparation, properties and applications. *Catal. Today* **1991**, *11*, 173–301. [CrossRef]
23. Choy, J.H.; Kwak, S.Y.; Jeong, Y.J.; Park, J.S. Inorganic layered double hydroxides as nonviral vector. *Angew. Chem. Int. Ed.* **2000**, *39*, 4041–4045. [CrossRef]
24. Wang, Q.; O'Hare, D. Recent advances in the synthesis and application of layered double hydroxide (LDH) nanosheets. *Chem. Rev.* **2012**, *112*, 4124–4155. [CrossRef] [PubMed]
25. Williams, G.R.; O'Hare, D. Towards understanding, control and application of layered double hydroxide chemistry. *J. Mater. Chem.* **2006**, *16*, 3065–3074. [CrossRef]
26. Evans, D.G.; Duan, X. Preparation of layered double hydroxides and their applications as additives in polymers, as precursors to magnetic materials and in biology and medicine. *Chem. Commun.* **2006**, *5*, 485–496. [CrossRef] [PubMed]
27. Lin, J.-J.; Juang, T.-Y. Intercalation of layered double hydroxides by poly(oxyalkylene)-amidocarboxylates: Tailoring layered basal spacing. *Polymer* **2004**, *45*, 7887–7893. [CrossRef]
28. Shau, S.-M.; Juang, T.-Y.; Lin, H.-S.; Huang, C.-L.; Hsieh, C.-F.; Wu, J.-Y.; Jeng, R.-J. Individual graphene oxide platelets through direct molecular exfoliation with globular amphiphilic hyperbranched polymers. *Polym. Chem.* **2012**, *3*, 1249–1259. [CrossRef]

29. Juang, T.-Y.; Chen, Y.-C.; Tsai, C.-C.; Dai, S.A.; Wu, T.-M.; Jeng, R.-J. Nanoscale organic/inorganic hybrids based on self-organized dendritic macromolecules on montmorillonites. *Appl. Clay Sci.* **2010**, *48*, 103–110. [CrossRef]
30. Conterosito, E.; Croce, G.; Palin, L.; Pagano, C.; Perioli, L.; Viterbo, D.; Boccaleri, E.; Paul, G.; Milanesio, M. Structural characterization and thermal and chemical stability of bioactive molecule-hydrotalcite (LDH) nanocomposites. *Phys. Chem. Chem. Phys.* **2013**, *15*, 13418–13433. [CrossRef]
31. Toson, V.; Conterosito, E.; Palin, L.; Boccaleri, E.; Milanesio, M.; Gianotti, V. Facile intercalation of organic molecules into hydrotalcites by liquid-assisted grinding: Yield optimization by a chemometric approach. *Cryst. Growth Des.* **2015**, *15*, 5368–5374. [CrossRef]
32. Choy, J.-H.; Choi, S.-J.; Oh, J.-M.; Park, T. Clay minerals and layered double hydroxides for novel biological applications. *Appl. Clay Sci.* **2007**, *36*, 122–132. [CrossRef]
33. Kan, S.-C.; Lee, C.-C.; Hsu, Y.-C.; Peng, Y.-H.; Chen, C.-C.; Huang, J.-J.; Huang, J.-W.; Shieh, C.-J.; Juang, T.-Y.; Liu, Y.-C. Enhanced surfactin production via the addition of layered double hydroxides. *J. Taiwan Inst. Chem. Eng.* **2017**, *80*, 10–15. [CrossRef]
34. Delhaize, E.; Ryan, P.R. Aluminum toxicity and tolerance in plants. *Plant Physiol.* **1995**, *107*, 315–321. [CrossRef] [PubMed]
35. Pourhossein, A.; Alizadeh, K. Salt-assisted liquid-liquid extraction followed by high performance liquid chromatography for determination of carvedilol in human plasma. *J. Rep. Pharm. Sci.* **2018**, *7*, 79–87.
36. Kuthati, Y.; Kankala, R.K.; Lee, C.-H. Layered double hydroxide nanoparticles for biomedical applications: Current status and recent prospects. *Appl. Clay Sci.* **2015**, *112–113*, 100–116. [CrossRef]
37. Mishra, G.; Dash, B.; Pandey, S. Layered double hydroxides: A brief review from fundamentals to application as evolving biomaterials. *Appl. Clay Sci.* **2018**, *153*, 172–186. [CrossRef]
38. Aisawa, S.; Takahashi, S.; Ogasawara, W.; Umetsu, Y.; Narita, E. Direct intercalation of amino acids into layered double hydroxides by coprecipitation. *J. Solid State Chem.* **2001**, *162*, 52–62. [CrossRef]
39. Khan, A.I.; O'Hare, D. Intercalation chemistry of layered double hydroxides: Recent developments and applications. *J. Mater. Chem.* **2002**, *12*, 3191–3198. [CrossRef]
40. Rives, V. Characterisation of layered double hydroxides and their decomposition products. *Mater. Chem. Phys.* **2002**, *75*, 19–25. [CrossRef]
41. Rives, V.; Ulibarri, M.A. Layered double hydroxides (LDH) intercalated with metal coordination compounds and oxometalates. *Coord. Chem. Rev.* **1999**, *181*, 61–120. [CrossRef]
42. Long, X.; Wang, Z.; Xiao, S.; An, Y.; Yang, S. Transition metal based layered double hydroxides tailored for energy conversion and storage. *Mater. Today* **2016**, *19*, 213–226. [CrossRef]
43. Wei, Y.-H.; Chu, I.-M. Enhancement of surfactin production in iron-enriched media by *Bacillus subtilis* ATCC 21332. *Enzyme Microb. Technol.* **1998**, *22*, 724–728. [CrossRef]
44. Wei, Y.-H.; Wang, L.-F.; Chang, J.-S.; Kung, S.-S. Identification of induced acidification in iron-enriched cultures of *Bacillus subtilis* during biosurfactant fermentation. *J. Biosci. Bioeng.* **2003**, *96*, 174–178. [CrossRef]
45. Cooper, D.G.; Macdonald, C.R.; Duff, S.J.; Kosaric, N. Enhanced production of surfactin from *Bacillus subtilis* by continuous product removal and metal cation additions. *Appl. Environ. Microbiol.* **1981**, *42*, 408–412. [PubMed]
46. Sheppard, J.D.; Mulligan, C.N. The production of surfactin by *Bacillus subtilis* grown on peat hydrolysate. *Appl. Microbiol. Biotechnol.* **1987**, *27*, 110–116. [CrossRef]
47. Drouin, C.M.; Cooper, D.G. Biosurfactants and aqueous two-phase fermentation. *Biotechnol. Bioeng.* **1992**, *40*, 86–90. [CrossRef] [PubMed]
48. Ohno, A.; Ano, T.; Shoda, M. Production of a lipopeptide antibiotic surfactin with recombinant Bacillus subtilis. *Biotechnol. Lett.* **1992**, *14*, 1165–1168. [CrossRef]
49. Wei, Y.-H.; Lai, C.-C.; Chang, J.-S. Using Taguchi experimental design methods to optimize trace element composition for enhanced surfactin production by *Bacillus subtilis* ATCC 21332. *Process Biochem.* **2007**, *42*, 40–45. [CrossRef]

Article

Rapid Removal and Efficient Recovery of Tetracycline Antibiotics in Aqueous Solution Using Layered Double Hydroxide Components in an In Situ-Adsorption Process

Kwanjira Panplado [1], Maliwan Subsadsana [2], Supalax Srijaranai [1] and Sira Sansuk [1,*]

1 Materials Chemistry Research Center, Department of Chemistry and Center of Excellence for Innovation in Chemistry, Faculty of Science, Khon Kaen University, Khon Kaen 40002, Thailand

2 Program of Chemistry, Faculty of Science and Technology Nakhon Ratchasima Rajabhat University, Nakhon Ratchasima 30000, Thailand

* Correspondence: sirisan@kku.ac.th

Received: 9 June 2019; Accepted: 1 July 2019; Published: 4 July 2019

Abstract: This work demonstrates a simple approach for the efficient removal of tetracycline (TC) antibiotic from an aqueous solution. The in situ-adsorption removal method involved instant precipitation formation of mixed metal hydroxides (MMHs), which could immediately act as a sorbent for capturing TC from an aqueous solution, by employing layered double hydroxide (LDH) components including magnesium and aluminum ions in alkaline conditions. By using this approach, 100% removal of TC can be accomplished within 4 min under optimized conditions. The fast removal possibly resulted from an instantaneous adsorption of TC molecules onto the charged surface of MMHs via hydrogen bonding and electrostatically induced attraction. The results revealed that our removal technique was superior to the use of LDH as a sorbent in terms of both removal kinetics and efficiency. Moreover, the recovery of captured TC was tested under the influence of various common anions. It was found that 98% recovery could be simply achieved by using phosphate, possibly due to its highly charged density. Furthermore, this method was successful for efficient removal of TC in real environmental water samples.

Keywords: tetracycline; metal hydroxides; layered double hydroxides; removal; water sample

1. Introduction

The contamination of antibiotics in environmental water is a global concern, as they are potentially toxic and harmful to both the ecosystem and human life [1]. One of the most commonly found antibiotics is tetracycline (TC), widely used in a variety of animal livestocks in order to promote growth and kill bacteria [2]. It also plays a significant role in human therapy. As a result of its misuse, TC is commonly released as an agricultural or community effluent into the environment [3,4]. Furthermore, TC can be transferred from the community to a water source such as a lake. Therefore, the purification of wastewater by the removal of TC prior to release into environmental water is still necessary.

Even though treatment methods have been developed in wastewater management, most of them such as coagulation [5], ion exchange [6], and photocatalytic degradation [7] involve intensive energy and a high cost of operation, complex procedures, and production of possible toxic products. Above all reported methods, adsorption is the most reliable and effective means, with various sorbent materials being employed. For example, multi-walled carbon nanotubes (MWCNTs) as an adsorbent provided 99.8% removal of TC within 20 min [8]. High adsorption efficiency resulted from abundant π-π interactions found from the π systems on the MWCNT surface, and benzene rings and double bonds, both C=C and C=O, in TC molecules. Similarly, it was reported that graphene oxide functionalized with magnetic particles gave 98% removal efficiency within 10 min as a consequence of strong π-π interactions between sorbent and target molecules [9]. Furthermore, zeolitic imidazolate metal organic framework ZIF-8 nanoparticles were used as an adsorbent and gave a removal efficiency of 90% as a result of a weak electrostatic interaction [10]. Furthermore, the removal of TC using Fe/Ni nanoparticle sorbent in an aqueous solution reached 97.4% removal within 2 h. [11]. In this sorbent-based adsorption technique, the removal efficiency of TC is chiefly dependent on the applied sorbents as well as their properties.

Among all these adsorbents, layered double hydroxide (LDH) has been considered as environmentally benign, and an effective adsorbent due to its high surface area and excellent anion exchangeability [12,13]. This material fundamentally consists of layers of mixed metal hydroxides linked with water molecules located in the interlayer space. The formula of LDH is generally symbolized as $[M_{1-x}{}^{2+}M_x{}^{3+}(OH)_2]^{x+}(A^{n-})_{x/n}{\cdot}mH_2O$, where M^{2+} and M^{3+} are divalent and trivalent cations, respectively. x is a molar ratio of $M^{3+}/(M^{2+} + M^{3+})$ and A is an anion located at interlayers with the negative charge n. LDH has been employed as a sorbent or functionalized for the removal of various pollutants [14–21]. However, the synthesis and characterization of LDH sorbents are commonly required. Unavoidably, these steps are commonly time- and energy-consuming.

This study presents an in situ method for removal of TC antibiotics from an aqueous solution by employing the LDH components to generate instantly formed mixed metal hydroxides (MMHs), which simultaneously act as a sorbent to capture TC molecules from an aqueous solution during their precipitation formation. Unlike other sorbent-based removal techniques, our method was simple, efficient, rapid, and eco-friendly, as the synthesis of sorbent and its characterization can be avoided, thus saving time, energy, and cost. The parameters affecting the removal efficiency were investigated thoroughly. For comparison, a kinetic removal of this contaminant with LDH used as sorbent in a conventional route was also studied. The recovery of captured TC through an ionic interfering effect was demonstrated. Additionally, the removal of TC was carried out in real natural water samples.

2. Materials and Methods

2.1. Chemicals and Reagents

Tetracycline hydrochloride ($C_{22}H_{24}N_2O_8{\cdot}HCl$) was obtained from Sigma-Aldrich (Hong Kong, China). Aluminium chloride hexahydrate ($AlCl_3{\cdot}6H_2O$) and magnesium chloride hexahydrate ($MgCl_2{\cdot}6H_2O$) were purchased from Sigma-Aldrich (St Louis, Missouri, USA). Sodium hydroxide (NaOH), sodium acetate (CH3COONa), sodium nitrate (NaNO3), and sodium sulphate (Na_2SO_4) were obtained from Carlo Erba (France). Sodium chloride (NaCl) and sodium carbonate (Na_2CO_3) were purchased from Ajax Finechem (Australia). Di-potassium hydrogen phosphate (HK_2PO_4) was obtained from BDH Prolabo (England). Methanol (CH_3OH) was purchased from LiChrosolv®(Darmstadt, Germany). All chemicals and reagents were of at least analytical grade and used as received without further purification. Deionized (DI) water was used throughout. A 1000 mg L^{-1} stock solution of TC was freshly prepared by dissolving 0.1000 g of $C_{22}H_{24}N_2O_8{\cdot}HCl$ in 100 mL of a mixed methanol:water solution (30:70% v/v). The solutions of mixed metal solutions (Mg^{2+} and Al^{3+}) with varied mole ratios were prepared by a dissolution of an appropriate amount of $MgCl_2{\cdot}6H_2O$ and $AlCl_3{\cdot}6H_2O$ in DI water.

2.2. Synthesis of LDH

As a comparative study, the removal of TC from an aqueous solution with LDH sorbent was investigated. First, LDH with a 3:1 mole ratio of Mg:Al was synthesized by co-precipitation as explained elsewhere [22]. Typically, a solution containing 0.0681 mol of $MgCl_2{\cdot}6H_2O$ (13.84 g) and 0.0227 mol of $AlCl_3{\cdot}6H_2O$ (5.48 g) in 200 mL DI was prepared. Then, 150 mL of an aqueous solution of 4.45 g NaCl was added slowly under stirring at room temperature and then 3 M NaOH was added to adjust the pH to 12. The mixture was then transferred into the Teflon coated stainless steel autoclave for hydrothermal treatment at 120 °C for 48 h. Next, the precipitates were filtered, washed thoroughly with DI water, and dried at 90 °C for 24 h.

2.3. Measurement and Characterization

To evaluate the removal efficiency, the absorption spectra of TC residuals were recorded by an Agilent 8453 UV-vis spectrophotometer (Agilent Technologies, Waldbronn, Germany). A centrifuge (H-11n, Kokusan, Tokyo) with 15 mL calibrated tubes was used for the phase separation. Power X-ray diffraction (XRD, D8 Advance, Bruker, Bremen, Germany) and attenuated total reflection Fourier transform infrared spectroscope (ATR-FTIR; Tensor 27, Bruker, Germany) were used. Scanning electron microscopy (SEM) was completed using a SNE-4500M microscope (SEC Co., Ltd, Seoul, South Korea). In addition, the surface charge of MMHs was assured by Zeta potential measurement using the Zetasizer Nano S (Malvern Instruments Ltd, Malvern, UK).

2.4. Removal Procedure

It can be noted that other divalent and trivalent cations such as Mn^{2+}, Co^{2+}, Ni^{2+}, Zn^{2+}, and Fe^{3+} can be employed [23–25]. However, the rate of precipitation of metal hydroxides should be considered as it can result in the kinetic removal of TC. From the cost and toxicity point of view, Mg^{2+} and Al^{3+} ions were selected in this study. The removal of TC molecules from an aqueous solution was studied in a batch experiment. First, a simulated water sample containing 30 mg L^{-1} TC was prepared and then conditioned with 1M NaOH. To initiate the removal of TC, a solution of Mg^{2+}/Al^{3+} ions with a fixed mole ratio of 3:1 was then added into 10 mL of TC solution. After a certain contact time, the TC-MMH particles were separated from the solution by centrifuging at 6000 rpm for 2 min. The residual TC in the solution was monitored by UV-vis measurement. The absorbance at 384 nm was used to evaluate the removal efficiency (%R) based on the following expression:

$$\%R = [(A_0 - A)/A_0] \times 100$$

where A_0 and A are the absorbance of TC before and after the removal at any time, respectively. The effects of all factors on the removal efficiency including the volume of the mixed metal solution, the volume of NaOH, mole ratio of Mg^{2+}/Al^{3+}, and the removal time were studied.

For a comparative study, the kinetic removal of TC by the LDH sorbent was investigated. The experiment was performed in a batch system with a total volume of 10 mL of 30 mg L^{-1} of TC and 0.0131 g of LDH sorbent. After a certain contact time from 15 s to 150 min, 1 mL of the solution was withdrawn for UV-vis measurement and 1 mL of DI water was introduced into the studied system. The removal efficiency was calculated by using the above expression.

3. Results and Discussion

3.1. Removal of TC

In this study, the in situ-adsorption removal strategy was based on the usage of LDH components in order to generate the mixed metal hydroxides (MMHs), which can simultaneously capture the TC molecules during their precipitation formation. First, the removal of TC from an aqueous solution was simply tested by introducing a metal solution containing both Mg^{2+} and Al^{3+} ions into an alkaline TC solution. In general, regarding its pK_a values (3.3, 7.7, and 9.7), TC is present in the aqueous solution in various forms depending on the pH of solution [26]. TC is in the TCH_2 neutral form when the pH falls in the range of 3.3–7.7. However, it is protonated and in a positively-charged $TCH_3{}^+$ form in a solution with a pH less than 3.3 and it is negatively-charged as the TCH^- form in alkaline conditions (pH > 10). As our strategy was performed in an alkaline solution at a pH of about 12, a removal of TC from an aqueous solution was accomplished through an interaction with instantly formed MMH acting as a sorbent. The removal resulted from an adsorption of TC molecules on the surface of MMHs as a consequence of an electrostatic interaction and H-bonding [27]. The UV-vis spectra of 30 mg L^{-1} TC before and after the removal for 4 min of contact time is presented in Figure 1. It was found that the absorbance of TC was reduced tremendously after the removal. This indicated the potential of the present approach for TC removal.

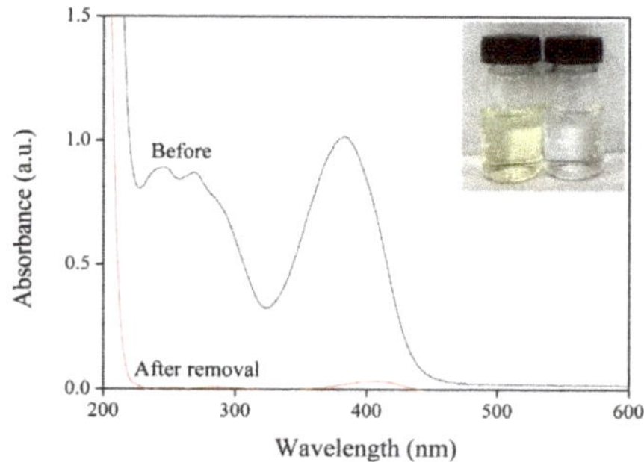

Figure 1. UV-vis absorption spectra of TC in aqueous solution at an initial concentration of 30 mg L^{-1}; before and after removal for 4 min contact time, with corresponding photograph images of the solution (inset).

3.2. *Optimization of the Removal of TC*

This method involved the use of LDH components to initiate the removal of TC from an aqueous solution. Accordingly, the main parameters including the amount of hydroxyl and metal ions were studied. These precursors could have an impact on the precipitation formation of MMHs, thereby affecting the removal efficiency. First, the impact of OH^- on the removal of 30 mg L^{-1} TC was investigated by varying the amount of OH^-, while other factors were fixed. Figure 2a shows that the removal efficiency of TC was enhanced rapidly, when increasing the amount of OH^- up to 200 mmol (200 mL). Then, the removal slightly improved until an addition of 275 mmol (275 mL) OH^-. After that, a further addition of OH^- did not benefit the removal of TC molecules. Thus, the amount of 275 mmol (275 mL) of OH^- was optimal.

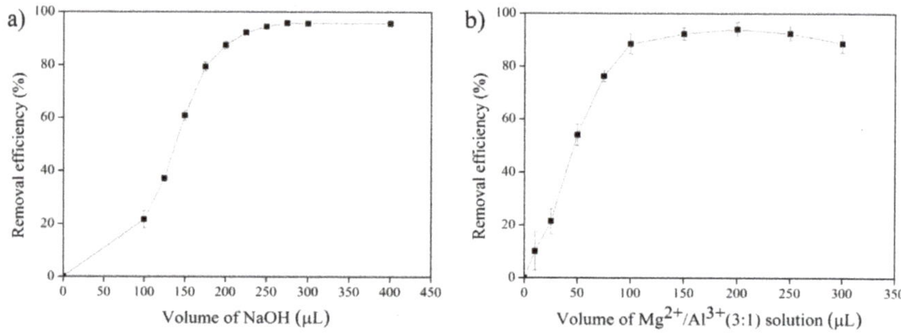

Figure 2. The effect of (**a**) the volume of 1 M NaOH and (**b**) the volume of mixed metal solution on the removal efficiency of TC in aqueous solution at an initial concentration of 30 mg L^{-1}.

The effect of both Mg^{2+} and Al^{3+} ions on the removal of TC was also examined. The mole ratio of Mg^{2+}:Al^{3+} was fixed at 3:1 as an optimal ratio, while the volume of these ions was varied and other parameters were kept constant. As presented in Figure 2b, the removal efficiency of TC was improved with increasing the volume of the mixed metal solution up to 200 μL. After that, the removal of TC became slightly lower. Similar to OH^-, an increase of Mg^{2+} and Al^{3+} content enhanced the precipitation formation of MMHs, thereby improving the interaction with TC in the solution. Thus, the removal efficiency increased when the components used to produce MMH sorbent increased as the removal is commonly dependent on the content of sorbent.

The mole ratio of Mg^{2+}:Al^{3+} was also investigated. In general, LDHs are synthesized by the hydrothermal method at a 2:1 and 3:1 mole ratio of Mg^{2+}:Al^{3+}. In this study, we simulated the formation of LDHs and employed their components to generate the simultaneous removal of TC from the solution. Under our studied conditions, the mole ratios of these ions varied from 1:1 to 4:1, while other parameters were kept constant. As can be seen in Figure 3a, the TC removal efficiency was sharply boosted by extending the mole ratio of Mg^{2+}:Al^{3+} ions up to 2:1. It was noticed that the strong precipitation of MMHs probably depended on the content of Mg^{2+} more than Al^{3+}. This resulted in an increase in the removal of TC when the content of Mg^{2+} ions was increased. However, the removal of TC reached the maximum value with the Mg^{2+}:Al^{3+} mole ratio of 3:1. Therefore, Mg^{2+}:Al^{3+} with a mole ratio of 3:1 was appropriate for the removal of TC by this approach.

The kinetic removal of TC by the proposed method was studied. The removal experiments were carried out at various contact times. The contact time was described as the time required for TC removal after the introduction of a mixed metal solution into the studied system. The experiments were tested under an optimization condition including 275 μL of NaOH, 200 μL of Mg^{2+}:Al^{3+} at a 3:1 mole ratio and an initial TC concentration of 30 mg L^{-1}. Figure 3b shows the kinetic removal of TC by our method. It was observed that the removal efficiency of TC increased rapidly and slightly improved after 15 s. However, almost complete removal of TC was achieved after 4 min of contact time. The comparison of the removal of TC by the proposed method with other methods is summarized in Table 1. The results indicated that the proposed technique was more efficient and rapid than other removal methods. In addition, our method required no synthetic step of sorbents or catalysts, thus saving time, chemicals and energy.

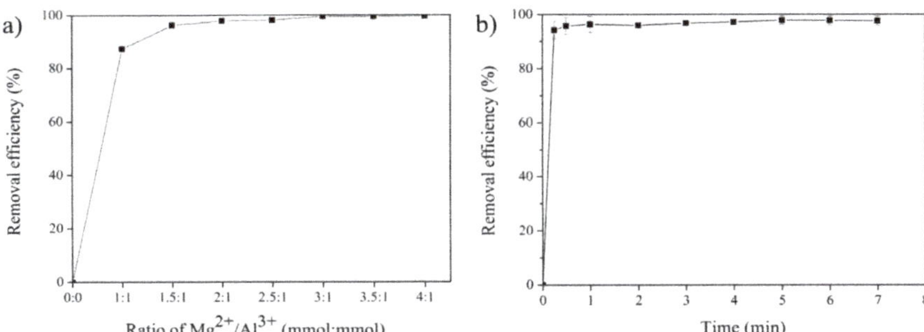

Figure 3. The effect of (**a**) mixed metal concentrations and (**b**) contact time on the removal efficiency of TC in aqueous solution at an initial concentration of 30 mg L^{-1}.

Table 1. Comparison of the removal of TC by the proposed method with other methods.

Material	Removal Method	Removal Time (min)	Removal Efficiency (%)	Ref.
GO/AC hybrid membrane	filtration	19	98.9	[28]
g-C_3N_4/Bi_3TaO_7	photocatalytic	90	89.2	[29]
Carbon doped g-C_3N_4	photocatalytic	90	90	[30]
ZnO@montmorillonite	photocatalytic	75	94	[31]
MWCNTs	adsorption	20	99.8	[8]
GO-MPs	adsorption	10	98	[9]
Fe/Ni BNPs	adsorption	120	97.4	[11]
MMHs	*in situ*-adsorption	4	99.8	This study

3.3. Characterization of MMH Sorbent and Confirmation for TC Removal

To confirm the removal of TC by our strategy, an instantly formed MMH sorbent was collected and then characterized. First, the surface charge of MMH was confirmed by the Zeta potential measurement. The result indicated that the surface of MMH particles was positively-charged with +19 mV. In the presence of TC, it was observed that the surface charge of MMH particles was reduced to −0.1 mV. The reduction of surface charge possibly resulted from the adsorption of TC molecules, present as a negative form (TCH^-) in the studied alkaline conditions, on the surface of MMH. The interaction of MMH with TC was also investigated by FT-IR measurement. As presented in Figure 4, when compared with the spectra of TC and MMH sorbent (without TC removal), a decrease in the intensity of the sharp peak at approximately 3700 cm^{-1}, corresponding to the -OH group in the MMH sorbent after removal of TC, was observed. This possibly implied a surface interaction of MMH with TC molecules via hydrogen bonding. It was also found that after the removal of TC, a characteristic peak at 1524 cm^{-1}, which was attributed to the vibration of the NH_2 group of TC molecules, shifted to 1597 cm^{-1}. These results confirmed the interaction between TC and MMH particles instantly generated in the system. Moreover, the presence of TC on the surface of MMH was also investigated by SEM measurement. As displayed in Figure 5, it was revealed that TC appeared on the surface of the MMH sorbent, confirming the interaction of TC molecules with MMH sorbent.

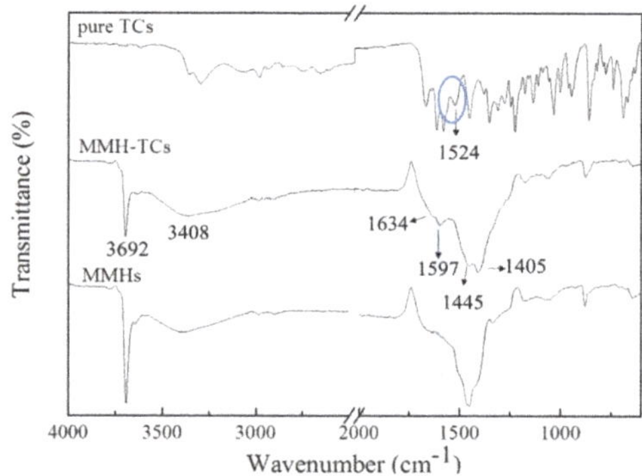

Figure 4. FT-IR spectra of pure TC, MMHs, and TC-MMHs.

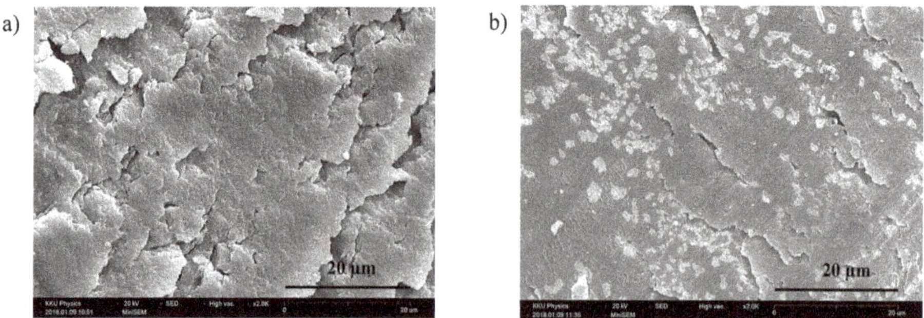

Figure 5. SEM images of (**a**) MMHs and (**b**) TC-MMHs.

3.4. Comparison on TC Removal with LDH Sorbent

We also investigated the removal of TC by using an LDH sorbent. LDH was prepared by a hydrothermal method, as explained in Section 2.2. Figure 6a represents the SEM image of LDH. It was seen that typical plate-like particles were observed. In addition, the XRD pattern confirmed the characteristics of LDH (see Figure S1 in Supplementary Materials). The experiments were conducted in a traditional way as a batch system at 30 mg L^{-1} of TC solution, as described in Section 2.4. An amount of LDH sorbent was calculated based on the optimized removal condition of the proposed method. LDH (0.0131 g) was obtained and further employed as a solid sorbent. The removal kinetics of LDH sorbent, in comparison with the proposed strategy, are displayed in Figure 6b. It was clearly seen that the removal of TC by LDH sorbent was quite slow. The removal reached an equilibrium after 60 min with a removal efficiency of 75%. This result indicated an incomplete removal of 30 mg L^{-1} of TC by the LDH sorbent. It is noted that the removal efficiency can be improved with increasing the amount of sorbent. However, this removal kinetic result implied that the present method was superior to the traditional sorbent-based removal method in terms of rapidness, efficiency, and cost-effectiveness as it eliminated the synthesis step of sorbent, thereby saving time, chemicals, and energy.

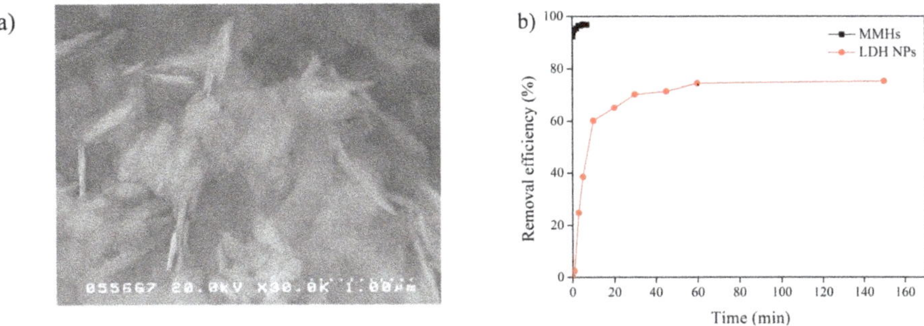

Figure 6. (**a**) SEM image of LDH sorbent and (**b**) its kinetic removal of TC in aqueous solution at an initial concentration of 30 mg L^{-1}, compared with MMHs.

3.5. Recovery of Captured TC

The ability of the proposed removal approach to recover TC from an aqueous solution was also tested. After the removal process, the precipitates containing TC were dispersed for 1 min in a solution containing an excess of various common anions (sodium salts) including Cl^-, NO_3^-, CH_3COO^-, SO_4^{2-}, CO_3^{2-}, and PO_4^{3-}. After centrifugation for phase separation, the solution was then taken for analysis of TC content. These anions were employed as they could interfere with the adsorption ability of TC molecules on the surface of MMH particles. Thus, recovery of TC could be achieved. Figure 7 represents the UV-vis spectra of TC collected after the recovery step and the recovery efficiency regarding different anions used. It was found that the recoveries of TC were ordered with the use of PO_4^{3-}, CO_3^{2-}, SO_4^{2-}, NO_3^-, CH_3COO^-, and Cl^-. The effect of these anions on both the precipitation formation of MMH particles and adsorption of TC possibly resulted from their charge density. Hence, 98% of recovery was obtained in the case of PO_4^{3-} used. In addition, this exhibited the advantage of the present removal strategy as the recovery of TC was simply obtained, regarding the loosely formed MMH sorbent.

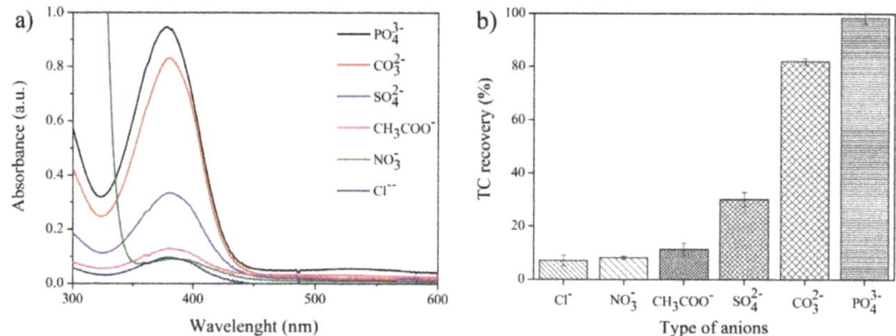

Figure 7. (**a**) UV-vis absorption spectra of TC and (**b**) the recovery efficiency after recovery of TC with various anions.

3.6. Removal of TC in Real Environmental Water Samples

The applicability of the proposed method was further assessed by the removal of TC in environmental water samples. The water samples were collected from various water sources including Nam Pong River, Ubonlratana Dam, Nong Kot Lake, Si Than Lake, Kaennakorn Lake, and the wastewater treatment pond located at Srinagarind Hospital, Khon Kaen, Thailand. It was found from an initial analysis that these water samples had a pH in the range of 6.6–7.5 without the presence of TC. Accordingly, to evaluate the applicability of the proposed method, all water samples were first filtrated and then spiked with 30 mg L^{-1} of TC. After removal using our strategy, the removal efficiency was reported, as presented in Figure 8. The results revealed that the removal of TC performed in every simulated water sample was 92%. The result indicated the applicability of our approach for the removal of TC in real water samples. High removal efficiencies were obtained even with the presence of other ionic species or contaminants, commonly found in natural water. Therefore, the interference of other compounds or ions is negligible. It can also be noted that other ionic species could facilitate the precipitation formation of MMHs, increasing the adsorption ability of TC. This led to an enhanced removal efficiency of TC from the aqueous solution.

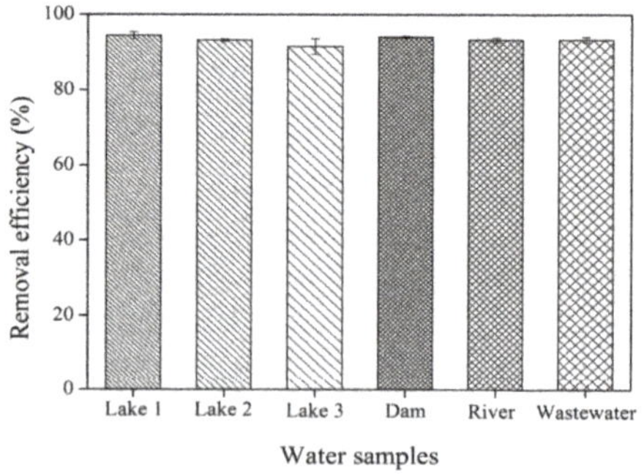

Figure 8. Removal of TC added in various environmental water samples at an initial concentration of 30 mg L^{-1}.

4. Conclusions

This study demonstrates a simple strategy for the efficient removal of tetracycline (TC) antibiotic from an aqueous solution. The in situ-adsorption method involves the utilization of layered double hydroxide (LDH) components to initiate precipitation of mixed metal hydroxides (MMHs), concurrently acting as a sorbent for instant adsorption of TC molecules from an alkaline solution. Mg^{2+} and Al^{3+}, present as chloride forms, were used. Under optimized conditions, 99.5% removal efficiency can be obtained within 4 min due to a strong electrostatic interaction and hydrogen bonding between TC molecules and the positively-charged surface of MMHs. When compared with the removal adsorption by the LDH sorbent, our method is much better and faster. With our removal strategy, 98% recovery of TC captured by MMHs can be simply achieved by dispersion in a phosphate solution. Moreover, almost complete removal of TC from simulated environmental water samples including a dam, river, three lakes and wastewater plant can be obtained.

Supplementary Materials: The following are available online at http://www.mdpi.com/2073-4352/9/7/342/s1, Figure S1: XRD pattern of LDH

Author Contributions: Conceptualization, S.S. (Sira Sansuk); investigation, K.P. and M.S.; methodology, K.P. and M.S.; supervision, S.S. (Supalax Srijaranai); writing—original draft preparation, K.P.; writing—review and editing, S.S. (Sira Sansuk).

Funding: This research received no external funding.

Acknowledgments: We gratefully acknowledge the Science Achievement Scholarship of Thailand (SAST).

Conflicts of Interest: The authors declare no conflict of interest.

References

1. Tan, H.; Ma, C.; Song, Y.; Xu, F.; Chen, S.; Wang, L. Determination of tetracycline in milk by using nucleotide/lanthanide coordination polymer-based ternary complex. *Biosens. Bioelectron.* **2013**, *50*, 447–452. [CrossRef] [PubMed]
2. Schwarz, S.; Kehrenberg, C.; Walsh, T.R. Use of antimicrobial agents in veterinary medicine and food animal production. *Int. J. Antimicrob. Agents* **2001**, *17*, 431–437. [CrossRef]
3. Li, D.; Yang, M.; Hu, J.; Ren, L.; Zhang, Y.; Li, K. Determination and fate of oxytetracycline and related compounds in oxytetracycline production wastewater and the receiving river. *Environ. Toxicol. Chem.* **2008**, *27*, 80–86. [CrossRef] [PubMed]
4. Chee-Sanford, J.C.; Aminov, R.I.; Krapac, I.J.; Garrigues-Jeanjean, N.; Mackie, R.I. Occurrence and diversity of tetracycline resistance genes in lagoons and groundwater underlying two swine production facilities. *Appl. Environ. Microbiol.* **2001**, *67*, 1494–1502. [CrossRef] [PubMed]
5. Canizares, P.; Martinez, F.; Jimenez, C.; Lobato, J.; Rodrigo, M. Coagulation and electrocoagulation of wastes polluted with dyes. *Environ. Sci. Technol.* **2006**, *40*, 6418–6424. [CrossRef] [PubMed]
6. Fu, L.; Shuang, C.; Liu, F.; Li, A.; Li, Y.; Zhou, Y.; Song, H. Rapid removal of copper with magnetic poly-acrylic weak acid resin: Quantitative role of bead radius on ion exchange. *J. Hazard. Mater.* **2014**, *272*, 102–111. [CrossRef] [PubMed]
7. Peik-See, T.; Pandikumar, A.; Ngee, L.H.; Ming, H.N.; Hua, C.C. Magnetically separable reduced graphene oxide/iron oxide nanocomposite materials for environmental remediation. *Catal. Sci. Technol.* **2014**, *4*, 4396–4405. [CrossRef]
8. Zhang, L.; Song, X.; Liu, X.; Yang, Y.; Pan, F.; Lv, J. Studies on the removal of tetracycline by multi-walled carbon nanotubes. *Chem. Eng. J.* **2011**, *178*, 26–33. [CrossRef]
9. Lin, Y.; Xu, S.; Jia, L. Fast and highly efficient tetracyclines removal from environmental waters by graphene oxide functionalized magnetic particles. *Chem. Eng. J.* **2013**, *225*, 679–685. [CrossRef]
10. Wu, C.; Xiong, Z.; Lia, C.; Zhang, J. Zeolitic imidazolate metal organic framework ZIF-8 with ultra-high adsorption capacity bound tetracycline in aqueous solution. *RSC Adv.* **2015**, *5*, 82127–82137. [CrossRef]
11. Dong, H.; Jiang, Z.; Zhang, C.; Deng, J.; Hou, K.; Cheng, Y.; Zhang, L. Removal of tetracycline by Fe/Ni bimetallic nanoparticles in aqueous solution. *J. Colloid Interface Sci.* **2018**, *513*, 117–125. [CrossRef] [PubMed]
12. Ai, L.; Zhang, C.; Meng, L. Adsorption of methyl orange from aqueous solution on hydrothermal synthesized Mg-Al layered double hydroxide. *J. Chem. Eng. Data* **2011**, *56*, 4217–4225. [CrossRef]
13. Cho, S.; Jung, S.; Jeong, S.; Bang, J.; Park, J.; Park, Y.; Kim, S. Strategy for synthesizing quantum dot-layered double hydroxide nanocomposites and their enhanced photoluminescence and photostability. *Langmuir* **2013**, *29*, 441–447. [CrossRef] [PubMed]

14. Zhou, J.Z.; Wu, Y.Y.; Liu, C.; Orpe, A.; Liu, Q.; Xu, Z.P.; Qian, G.R.; Qiao, S.Z. Effective self−purification of polynary metal electroplating wastewaters through formation of layered double hydroxides. *Environ. Sci. Technol.* **2010**, *44*, 8884–8890. [CrossRef] [PubMed]
15. Wen, T.; Wu, X.L.; Tan, X.L.; Wang, X.K.; Xu, A.W. One-pot Synthesis of water-swellable Mg−Al layered double hydroxides and graphene oxide nanocomposites for efficient removal of As (V) from aqueous solutions. *ACS Appl. Mater. Interfaces* **2013**, *5*, 3304–3311. [CrossRef] [PubMed]
16. Zaghouane-Boudiaf, H.; Boutahala, M.; Arab, L. Removal of methyl orange from aqueous solution by uncalcined and calcined MgNiAl layered double hydroxides (LDHs). *Chem. Eng. J.* **2012**, *187*, 142–149. [CrossRef]
17. Rojas, R. Copper, lead and cadmium removal by Ca Al layered double hydroxides. *Appl. Clay Sci.* **2014**, *87*, 254–259. [CrossRef]
18. Goh, K.; Lim, T.; Dong, Z. Application of layered double hydroxides for removal of oxyanions: A review. *Water Res.* **2008**, *42*, 1343–1368. [CrossRef]
19. Berner, S.; Araya, P.; Govan, J.; Palza, H. Cu/Al and Cu/Cr based layered double hydroxide nanoparticles as adsorption materials for water treatment. *J. Ind. Eng. Chem.* **2018**, *59*, 134–140. [CrossRef]
20. Abdelkader, N.; Bentouami, A.; Derriche, Z.; Bettahar, N.; De Menorval, L.C. Synthesis and characterization of Mg−Fe layer double hydroxides and its application on adsorption of orange G from aqueous solution. *Chem. Eng. J.* **2011**, *169*, 231–238. [CrossRef]
21. Ahmed, I.M.; Gasser, M.S. Adsorption study of anionic reactive dye from aqueous solution to Mg−Fe−CO_3 layered double hydroxide (LDH). *Appl. Surf. Sci.* **2012**, *259*, 650–656. [CrossRef]
22. Wang, Q.; O'Hare, D. Recent advances in the synthesis and application of layered double hydroxide (LDH) nanosheets. *Chem. Rev.* **2012**, *112*, 4125–4155. [CrossRef]
23. Tian, H.; Bao, W.T.; Jiang, Y.; Wang, L.; Zhang, L.; Sha, O.; Wu, C.; Gao, F. Fabrication of Ni-Al LDH/nitramine-N-doped graphene hybrid composites via a novel self-assembly process for hybrid supercapacitors. *Chem. Eng. J.* **2018**, *354*, 1132–1140. [CrossRef]
24. Liang, T.; Xuan, H.; Xu, Y.; Gao, J.; Han, X.; Yang, J.; Han, P.; Wang, D.; Du, Y. Rational assembly of coal-layered double hydroxide on reduced graphene oxide with enhanced electrochemical performance for energy storage. *ChemElectroChem* **2018**, *5*, 2424–2434. [CrossRef]
25. Xing, J.; Du, J.; Zhang, X.; Shao, Y.; Zhang, T.; Xu, C.A. Ni-P@NiCo LDH core-shell nanorod-decorated nickel foam with enhanced areal specific capacitance for high-performance supercapacitors. *Dalton Trans.* **2017**, *46*, 10064–10072. [CrossRef] [PubMed]
26. Wessels, J.M.; Ford, W.E.; Szymezak, W.; Schneider, S. The complexation of tetracycline and anhydrotetracycline with Mg^{2+} and Ca^{2+}: A spectroscopic study. *J. Phys. Chem. B* **1998**, *102*, 9323–9331. [CrossRef]
27. Sansuk, S.; Srijaranai, S.; Srijaranai, S. A new approach for removing anionic organic dyes from wastewater based on electrostatically driven assembly. *Environ. Sci. Technol.* **2016**, *50*, 6477–6484. [CrossRef] [PubMed]
28. Liu, M.; Liu, Y.; Bao, D.; Zhu, G.; Yang, G.; Geng, J.; Li, H. Effective removal of tetracycline antibiotics from water using hybrid carbon membranes. *Sci. Rep.* **2017**, *7*, 43717. [CrossRef] [PubMed]
29. Luo, B.; Chen, M.; Zhang, Z.; Xu, J.; Li, D.; Xu, D.; Shi, W. Highly efficient visible-light-driven photocatalytic degradation of tetracycline by a Z-scheme g-C_3N_4/Bi_3TaO_7 nanocomposite photocatalyst. *Dalton Trans.* **2013**, *46*, 8431–8438. [CrossRef]
30. Penneri, S.; Ganguly, P.; Mohan, M.; Nair, B.N.; Mohamed, A.; Warrier, K.G.; Hareesh, U.S. Photoregenerable, Bifunctional granules of carbon-doped g-C_3N_4 as adsorptive photocatalyst for the efficient removal of tetracycline antibiotic. *ACS Sustain. Chem. Eng.* **2017**, *5*, 1610–1618. [CrossRef]
31. Zyoud, A.; Jondi, W.; AlDaqqah, N.; Asaad, S.; Qamhieh, N.; Hajamohideen, A.; Helal, M.; Kwon, H.; Hilal, H. Self-sensitization of tetracycline degradation with simulated solar light catalyzed by ZnO@montmorillonite. *Solid State Sci.* **2017**, *74*, 131–143. [CrossRef]

Article

The Fabrication of Calcium Alginate Beads as a Green Sorbent for Selective Recovery of Cu(II) from Metal Mixtures

Niannian Yang [1], Runkai Wang [1,*], Pinhua Rao [1,*], Lili Yan [1], Wenqi Zhang [2], Jincheng Wang [1] and Fei Chai [1]

[1] School of Chemistry and Chemical Engineering, Shanghai University of Engineering Science, Shanghai 201620, China; m040116128@sues.edu.cn (N.Y.); liliyan@sues.edu.cn (L.Y.); wjc406@sues.edu.cn (J.W.); m040117127@sues.edu.cn (F.C.)

[2] School of Civil Engineering, Kashgar University, Xinjiang 844000, China; zhangwenqi@sues.edu.cn

* Correspondence: wangrunkai@sues.edu.cn (R.W.); raopinhua@sues.edu.cn (P.R.); Tel.: +86-021-67791217 (R.W.); +86-021-67791211 (P.R.)

Received: 3 April 2019; Accepted: 13 May 2019; Published: 17 May 2019

Abstract: Calcium alginate (CA) beads as a green sorbent were easily fabricated in this study using sodium alginate crosslinking with $CaCl_2$, and the crosslinking pathway was the exchange between the sodium ion of α-L-guluronic acid and Ca(II). The experimental study was conducted on Cu(II), Cd(II), Ni(II) and Zn(II) as the model heavy metals and the concentration was determined by inductively coupled plasma optical emission spectrometry (ICP-OES). The characterization and sorption behavior of the CA beads were analyzed in detail via using scanning electron microscopy (SEM), fourier transform infrared spectroscopy (FTIR) and X-ray photoelectron spectroscopy (XPS). The adsorption experiments demonstrated that the CA beads exhibited a high removal efficiency for the selective adsorption of Cu(II) from the tetra metallic mixture solution and an excellent adsorption capacity of the heavy metals separately. According to the isotherm studies, the maximum uptake of Cu(II) could reach 107.53 mg/g, which was significantly higher than the other three heavy metal ions in the tetra metallic mixture solution. Additionally, after five cycles of adsorption and desorption, the uptake rate of Cu(II) on CA beads was maintained at 92%. According to the properties mentioned above, this material was assumed to be applied to reduce heavy metal pollution or recover valuable metals from waste water.

Keywords: alginate beads; green sorbent; selective adsorption; heavy metals

---

1. Introduction

In recent years, agriculture and the mining, chemical fertilizer, leather, battery and paper industries have developed vigorously, and the phenomenon of heavy metal wastewater directly or indirectly being discharged into the environment has become more and more serious, particularly in developing countries [1,2]. Heavy metal ions in wastewater are mainly comprised of zinc (Zn), nickel (Ni), cadmium (Cd), copper (Cu) and so on. Among them, copper is widely used in industrial production, such as in printed circuit boards (PCBs), and it is also one of the indispensable nutrients (trace elements) in the human body [3,4]. When copper-containing wastewater is discharged into the environment beyond its self-purification range, the high toxicity and non-biodegradability of copper ions poses a serious threat to animal and human health. Hence, investigating how to effectively remove copper ions from wastewater is very important to the ecological environment. In addition, the recovery of copper from wastewater also has certain economic benefits.

To make the recovery of Cu more meaningful, the literature reported many methods. For example, Coruh et al. [5] selected vitrification as the technology to deal with industrial copper, mixing it with other inorganic wastes and materials and sintering them into glass for reuse. Printed circuit boards (PCBs) are widely used in the electronics industry [6,7], resulting in a large amount of copper-containing wastewater. Liu et al. [8] used electrolysis to recover 97% copper from waste PCBs, and Mdlovu et al. [9] reported the use of the microemulsion process to recover copper nanoparticles with diameters of 20–50 nm. In addition to the methods reported above, adsorption was also selected to treat copper-containing wastewater due to its advantages, such as low initial cost and process simplicity [10,11]. However, the instability of the adsorbents and difficulty in the separation still limit their practical applications. Overall, for the treatment of heavy metals in wastewater, availability and cost effectiveness play an important role in the synthesis of the adsorbents. This has made people pay attention to abundant, renewable and environmentally-friendly marine resources, such as bio-sorbents [12]. Torres-Caban et al. [13] proved that a calcium alginate/spent coffee grounds composite bead was an effective biological absorbent for the removal of copper ions, which first made us notice the role of alginate in the process of adsorption.

Alginate derived from brown algae is a highly popular material for the biosorption of heavy metals due to its advantages such as low cost and high affinity via gelation [14,15]. Abundant functional groups have been found in sodium alginate, such as carboxyl and hydroxyl groups, which can crosslink with cations [16,17]. The reason for this is that the carboxyl group is a negative group, and it can adsorb electrostatically with heavy metal ions and produce chelation at the same time. Sodium alginate reacts with divalent cations such as Ca(II), Ba(II) and Sr(II), to form insoluble hydrogels, which are crosslinked to form a reticular structure called the "egg box" structure, and the crosslinking pathway is the exchange between the sodium ions of α-L-guluronic acid and divalent ions [18,19]. In all types of alginate materials, alginate beads can be easily recovered from water [20,21]. Consequently, calcium alginate (CA) is a promising biomaterial for the biosorption of heavy metals [22]. In previous studies, it has been found that CA has a selective adsorption effect on some metal ions, which is extremely important for the recovery and re-utilization of metal ions. Hence, we want to investigate whether CA has selective adsorption on copper under the interference of cadmium, zinc and nickel metal ions.

In our study, the alginate beads as a green sorbent were easily fabricated with the Ca(II) crosslink, maintaining the high efficiency of the selective recovery of Cu(II) from the metal mixtures and the good adsorption of the heavy metals separately. The experimental study was conducted on Cu(II), Cd(II), Ni(II) and Zn(II) as the model heavy metals. The morphology and structure, functional groups, adsorption mechanism of CA beads were investigated by various characterization methods, such as scanning electron microscopy (SEM), fourier transform infrared spectroscopy (FTIR) and X-ray photoelectron spectroscopy (XPS). All pH measurements were adopted using a LEICI PHS-2F pH meter. Additionally, the alginate beads showed good reusability after five rounds of simple sorption–desorption procedures.

2. Materials and Methods

2.1. Materials

Sodium alginate and calcium(II) chloride ($CaCl_2$) were purchased from Adamas-beta. Calcium chloride was used for crosslinking of the alginate beads and sodium alginic acid was used to fabricate the alginate beads. Four types of heavy metals, $Cu(NO_3)_2 \cdot 3H_2O$, $Zn(NO_3)_2 \cdot 6H_2O$, $Ni(NO_3)_2 \cdot 6H_2O$ and $Cd(NO_3)_2 \cdot 4H_2O$, were purchased from Aladdin. To adjust the pH of the solution, 1 M hydrochloric acid (HCl) and 1 M sodium hydroxide (NaOH) were applied, which were acquired from Sinopharm Chemical Reagent Co., Ltd. All other chemicals used in this study were of analytical grade without purification. Distilled water (DW) with specific resistivity greater than 18 MΩ·cm was used in all experiments.

2.2. Preparation of Calcium Alginate Beads

Sodium alginate powder was added to distilled water to obtain a yellow viscous sodium alginate solution. The obtained solution remained stationary until there were no air bubbles. To synthesize the spherical bio-sorbent, sodium alginate solution was added dropwise into $CaCl_2$ (1%, w/v) solution under gentle stirring with a dropper and the bead was formed immediately. After 2 hours of curing, the sphere became compact, and a CA bead with a diameter of about 3 mm was obtained. The gel ball was rinsed with distilled water several times to remove free Ca(II) and stored in distilled water for further use.

2.3. Material Characterizations

The concentration of Cu(II), Zn(II), Ni(II) and Cd(II) used in all experiments was determined by inductively coupled plasma optical emission spectrometry (ICP-OES, Varian 700-ES, Walnut Creek, CA, USA), using 2% HNO_3 as the medium. The standard solution and the solution of heavy metal ions to be measured were acidic. Scanning electron microscopy (SEM, Hitachi SU8010, Ibaraki, Japan) was used to obtain the information of the physical structure and morphology of alginate beads. Fourier transform infrared spectroscopy (FTIR, PerkinElmer Spectrum Two, Waltham, MA, USA) was recorded in the 400–4000 cm^{-1} region on a FTIR spectrophotometer using a KBr disk method. Thermogravimetric analysis (TGA, TA Q600 SDT, New Castle, DE, USA) was carried out in a nitrogen gas flow from room temperature to 600 °C with a heating rate of 10 K/min. The point of zero charge (STARTER 2100, Parsippany, NJ, USA) was measured within the pH range from 3.0 to 10.0 by addition of 0.1 N HCl and NaOH. The samples were further analyzed by X-ray photoelectron spectroscopy (XPS, Thermo Fisher 250XI, Waltham, MA, USA) for the Cu 2p, Cd 3d, Zn 2p and Ni 2p regions. The charging shifts of the spectra were calibrated by placing the C1s peak at 284.8 eV from the adventitious carbon. Then, the results obtained from XPS were analyzed using the non-linear least squares curve fitting program (XPSPEAK4.1, software, Hong Kong, China).

2.4. Adsorption Experiments

For kinetic studies, an adsorption experiment was carried out using 200 mg/L of Cu(II), Zn(II), Ni(II) and Cd(II) in distilled water, respectively, of which the initial pH was 5.5, 6.3, 6.6 and 6.6, respectively. The saturated adsorption time was determined, and the samples were taken at a predetermined time interval. The adsorption isotherms of CA beads in a monomer solution and a tetra-metallic mixture solution were studied to evaluate their saturated adsorption capacity. The beads (1 g) were fully dispersed in 50 mL of monometallic solution and tetra-metallic mixture solution with different concentrations ranging from 50–800 mg/L at 120 r/min for 12 h at a temperature of 25 °C in a thermostabilized warm bath. In the pH effect experiment, five groups of tetra-metallic mixture solutions with a pH of 2.0, 3.0, 4.0, 5.0 and 6.0, respectively, were prepared and adjusted by adding NaOH (0.1 mol/L) or HNO_3 (0.1 mol/L).

The suspended impurities in the solution were removed by membrane filtration. The concentration of the metal ions was determined by ICP-OES and the amount of adsorption could be calculated using the following equation:

$$q = \frac{(C_0 - C)V}{M} \tag{1}$$

where q is the adsorption of the metal ions (mg/g), C_0 and C are the initial and final metal ion concentrations, respectively (mg/L), V is the total volume of suspension (L), and M is the dry mass of adsorbent (g).

2.5. Desorption and Reuse Experiments

Desorption experiments were also conducted using 1% $CaCl_2$ solutions and 0.1 M HNO_3 as the desorption reagents. In this experiment, 1 g of CA beads were added to 50 mL tetra-metallic mixture solutions with pH 5.7. After sorption, the heavy metal loaded CA beads were subsequently suspended in a 0.1 M HNO_3 eluting agent to evaluate the desorption performance. Then, the CA beads were washed with 1% $CaCl_2$ solutions to make them neutral and distilled water several times to remove the free Ca(II) ions from the beads. Four sorption–desorption cycles were conducted to assess the reusability of CA beads.

3. Results and Discussion

3.1. Characterization of CA Beads

The CA beads were synthesized through crosslinking with a diameter of 3 mm. Figure 1a shows the overall appearance of CA beads after drying; the diameter was reduced to about 1 mm. Figure 2b,c shows that there were a lot of ravines on the surface of CA beads, which increased the surface area and provided more adsorption sites. The surface of the CA bead was composed of wire-like Ca-alginate and a honeycomb network could be observed in the internal structure of CA (Figure 1d).

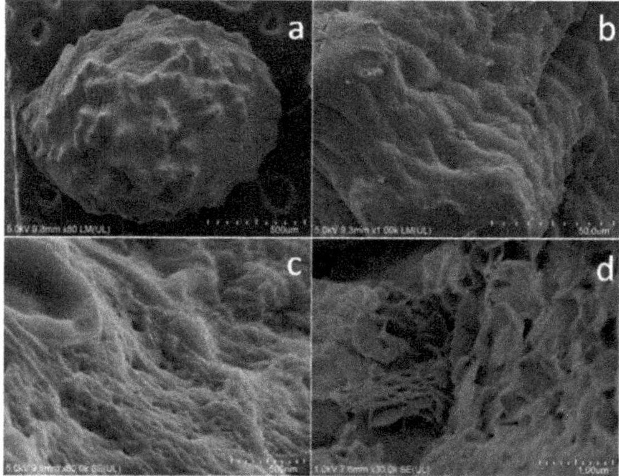

Figure 1. Scanning electron microscopy (SEM) images of the outer surface (**a–c**) and the internal structure (**d**) of the dried calcium alginate (CA) bead.

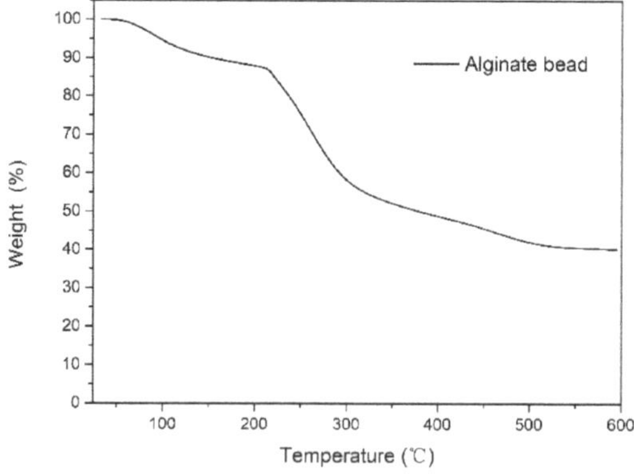

Figure 2. Thermogravimetric analysis (TGA) curves of the CA beads.

To evaluate the hydration and thermal decomposition of the CA beads, thermogravimetric analysis (TGA) was applied. As seen in Figure 2, the first mass loss was attributed to the dehydration process and a 12.6 wt% loss up to 210 °C. The second part was a thermal degradation stage, which resulted in pyrolysis from 210 °C to 510 °C with a weight loss of 45.8 wt%. The third part represented the conversion of the remaining materials to carbon residues and $CaCO_3$ and the residual weights were 40.2 wt% [23].

3.2. Adsorption Kinetics

In industrial applications, adsorption kinetics are important for process design and operation, which required us to achieve adsorption equilibrium under certain system conditions. The kinetic behavior of the metal ions removed by CA beads is presented in Figure 3. The experimental data were fitted with pseudo-first-order and pseudo-second-order kinetics [24,25].

$$\text{Pseudo} - \text{first} - \text{order model}: \ q_t = q_1(1 - \exp(-k_1 t)) \tag{2}$$

$$\text{Pseudo} - \text{second} - \text{order model}: \ q_t = \frac{q_2^2 k_2 t}{1 + q_2 k_2 t} \tag{3}$$

where both q_1 and q_2 are the amount of metal ions adsorbed at equilibrium (mg/g), q_t is the amount of metal ions adsorbed at any time t (mg/g), k_1 and k_2 are equilibrium rate constants of the pseudo-first-order and pseudo-second-order, respectively (g/mg min).

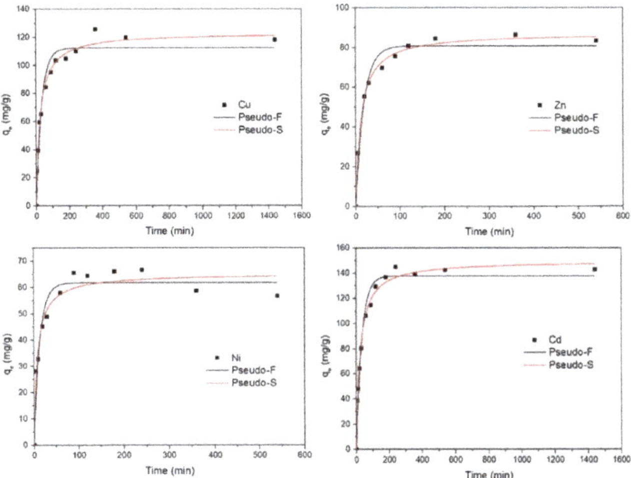

Figure 3. Kinetic adsorption plots of Cu(II), Zn(II), Ni(II) and Cd(II) by CA beads, respectively. Reaction conditions: T = 293 K, $[Cu(II)]_0 = [Zn(II)]_0 = [Ni(II)]_0 = [Cd(II)]_0 = 200$ mg/L.

The results for the kinetic model parameters are summarized in Table 1. The higher coefficient of determination (R^2) value of the pseudo-second-order model demonstrated that the adsorption rate was dominated by the rate of the chemical reaction [26]. In this experiment, the ion exchange occurred between the metal ions and Ca(II) and the adsorption rate of the CA was dependent on the exchange rate. The obtained k_2 values for Cu(II), Zn(II), Ni(II) and Cd(II) adsorption onto CA beads indicated that the adsorption rate was Ni(II) > Zn(II) > Cd(II) ≈ Cu(II).

Table 1. Parameters of pseudo-first-order and pseudo-second-order kinetic models. q_1 and q_2 = the amount of metal ions adsorbed at equilibrium; k_1 and k_2 = equilibrium rate constants of the pseudo-first-order and pseudo-second-order, R^2 = coefficient of determination, respectively.

Metal Type	Pseudo-First-Order			Pseudo-Second-Order		
	q_1 (mg/g)	k_1 (min^{-1})	R^2	q_2 (mg/g)	k_2 (g/mg min)	R^2
Cd	137.68	0.0300	0.9636	149.62	0.0436	0.9866
Cu	112.34	0.0304	0.9547	123.03	0.0422	0.9888
Zn	80.81	0.0548	0.9688	87.34	0.0837	0.9955
Ni	61.75	0.0743	0.9426	65.16	0.1243	0.9559

3.3. *Adsorption Isotherm*

In order to evaluate the maximum adsorption capacity of CA beads, the adsorption isotherms were measured in the initial metal concentrations ranging from 0–800 mg/L. The adsorption equilibrium data were fitted with Langmuir and Freundlich isotherm models [27,28].

$$\text{Langmuir model}: \ q_e = \frac{q_m K_L C_e}{1 + K_L C_e} \tag{4}$$

$$\text{Freundlich model}: \ q_e = K_F C_e^{1/n} \tag{5}$$

where q_e is the equilibrium adsorption capacity of heavy metal ions (mg/g), q_m indicates the maximum adsorption capacity for heavy metal ions (mg/g), C_e is the equilibrium concentration after adsorption (mg/L), K_L and K_F denote equilibrium constants of Langmuir (L/mg) and Freundlich (mg/g) (L/g), respectively, and n is the Freundlich exponent.

Figure 4 shows the adsorption isotherm of Cu(II), Zn(II), Ni(II) and Cd(II) separately on the CA beads and Table 2 lists the parameter values along with their correlation coefficients. The CA beads showed good adsorption towards Cu(II), Zn(II), Ni(II) and Cd(II), with a maximum adsorption of 140.55 mg/g, 174.60 mg/g, 114.69 mg/g and 216.82 mg/g, respectively. As shown in Table 3, calcium alginate beads exhibited significant advantages over other low-cost biosorption materials in terms of their maximum adsorption capacity for Cu(II).

Figure 5 shows the selective removal of Cu(II) on CA beads. With the increase of equilibrium concentration, the adsorption effect of CA beads on Cu(II) first increased rapidly and then tended to slow until equilibrium was reached, while the adsorption capacity of Zn(II), Ni(II) and Cd(II) decreased slowly. The fitting parameters of the Langmuir and Freundlich models are shown in Table 2 and the adsorption data were more consistent with the Langmuir model (R^2 = 0.9920) than the Freundlich model (R^2 = 0.8126) according to the Cu(II)* adsorption coefficient, which suggested that the adsorption of Cu(II) was the mono-layer sorption during the sorption process. Due to the competitive sorption between Cu(II) and other metals, the isotherm data of Zn(II), Ni(II) and Cd(II) were unable to be fit by the Langmuir or Freundlich models. Therefore, the CA beads showed better sorption capacity toward Cu(II) selectively than the other three heavy metals, indicating that the CA beads could be applied to the selective recovery of Cu(II) from polymetallic solutions.

Table 2. Parameters of the Langmuir and Freundlich models. q_m = the maximum adsorption capacity for heavy metal ions; K_L and K_F = equilibrium constants of Langmuir and Freundlich, respectively; n = Freundlich exponent; R^2 = coefficient of determination.

Metal	Langmuir Model			Freundlich Model		
	q_m (mg/g)	K_L (L/mg)	R^2	K_F (mg/g) $(L/g)^{1/n}$	n	R^2
Cu	140.55	0.0553	0.9605	46.9337	5.6609	0.8729
Cd	216.82	0.0177	0.8917	31.2836	3.3430	0.8939
Zn	174.60	0.0055	0.8821	10.0481	2.4589	0.8570
Ni	114.69	0.0108	0.9842	14.1111	3.2582	0.8690
Cu *	107.53	0.0639	0.9920	44.6145	7.2228	0.8126

* Parameters of Langmuir and Freundlich models for selective adsorption of Cu(II) by CA in mixed solution.

Table 3. Comparison of adsorption capacity of low-cost adsorbents for Cu(II).

Adsorbents	Maximum Adsorption Capacity of Cu(II)	References
IDA-modified cellulose	69.6 mg/g	[29]
Alkali leaching wire rope sludge	36.48 mg/g	[30]
EDTA-functionalized bamboo activated carbon	42.19 mg/g	[31]
Waste coffee	92.78 mg/g	[32]
Kapok-DTPA	101.0 mg/g	[33]
CA beads	140.55 mg/g	This study

Where, IDA, EDTA and DTPA represent iminodiacetic acid, ethylene diamine tetraacetic acid, and diethylenetriamine pentaacetic acid, respectively.

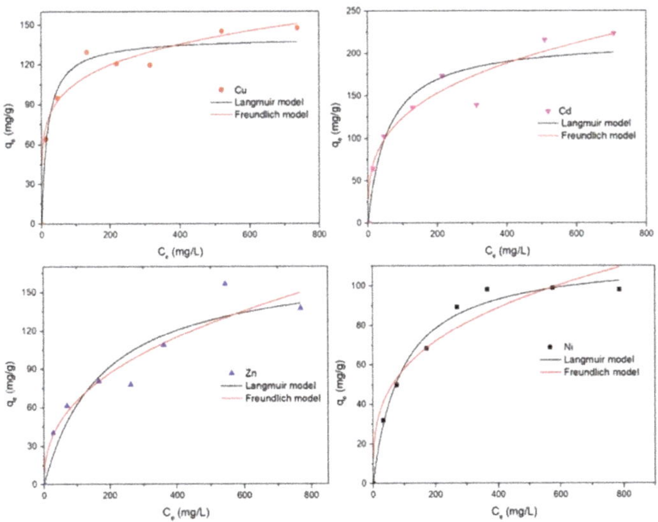

Figure 4. The adsorption isotherm by CA beads in the Cu(II), Zn(II), Ni(II) and Cd(II) heavy metal solution at different initial concentrations. Reaction conditions: T = 293 K, $[Cu(II)]_0$ = $[Zn(II)]_0$ = $[Ni(II)]_0$ = $[Cd(II)]_0$ = 0–800 mg/L, reaction time = 12 h.

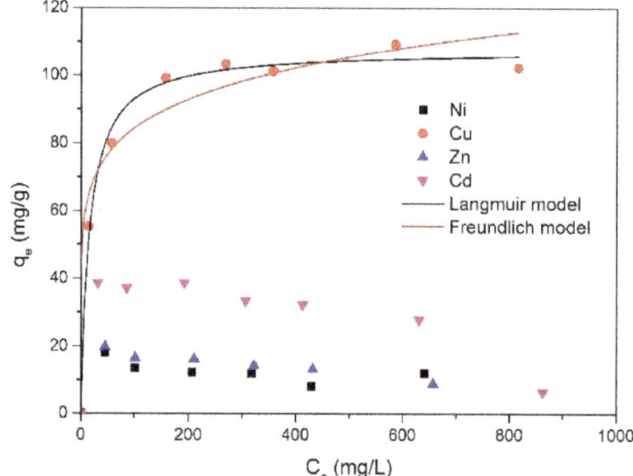

Figure 5. The selective removal and adsorption isotherm by CA beads in the mixed heavy metals solution at different initial concentrations. Reaction conditions: T = 293 K, $[Cu(II)]_0$ = $[Zn(II)]_0$ = $[Ni(II)]_0$ = $[Cd(II)]_0$ = 0–800 mg/L, reaction time = 12 h. q_e = the equilibrium adsorption capacity of heavy metal ions; C_e = the equilibrium concentration after adsorption.

A dimensionless R_L constant was introduced to reveal the essential characteristics of the Langmuir model, and the adsorption process was further evaluated. The correlation values of R_L could be calculated from the following equation.

$$R_L = \frac{1}{1 + bC_i} \tag{6}$$

where C_i is the initial concentration of Cu(II) (mg/L) and b is the dimensionless constant of Langmuir. The values of R_L indicated that the adsorption process may be an unfavorable trend ($R_L > 1$), linear ($R_L = 1$), favorable ($0 < R_L < 1$) or irreversible ($R_L = 0$) [34]. Here, the adsorption process of Cu(II) by CA beads showed that the b value was 0.0639, the calculated R_L value was within the range of 0.0175–0.2434 and just fell within the range of 0–1, indicating that CA beads were favorable to the Cu(II) adsorption process.

3.4. *Effect of pH*

The effect of pH on Cu(II) adsorption by CA beads in the mixed metal solutions was investigated with the pH value in the range of 2–6. Figure 6 shows the uptake capacities of heavy metals using the CA beads at various pH values. When the pH value was lower than 2.0, few heavy metal ions were adsorbed due to the competition of the many H^+ ions. With the increase of initial pH value from 2 to 6, the adsorption capacity of CA beads to Cu(II) increased significantly, and when the pH was 6, the maximum adsorption was observed. However, there was no obvious change in the adsorption amount of the other three kinds of metal ions, which were only slightly increased.

The point of zero charge (PZC) was employed to analyze the surface charges of the CA beads, and the measured value was 8.2. When the pH was lower than 8.2, the surface of the material was positively charged because of the introduction of calcium ions. At a pH above 8.2, there was a negative charge due to the reaction between the -OH on the surface and the OH· in the solution, leading to the formation of a negatively charged functional group, O^-. Therefore, the adsorption process of heavy metal ions by CA beads could be affected by pH, and there is a competition mechanism between H^+ and metal ions with carboxyl groups in acidic environments. Additionally, the carboxylic (-COOH) and hydroxyl (-OH) groups of the CA beads were changed to the protonated forms and the Ca(II) was released into the solution under acidic conditions [35].

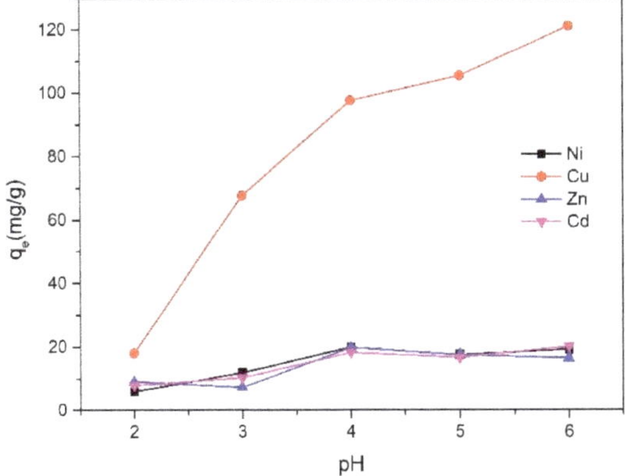

Figure 6. Effect of pH in a quaternary metal system. Reaction conditions: T = 293 K, $[Cu(II)]_0 = [Zn(II)]_0 = [Ni(II)]_0 = [Cd(II)]_0$ = 200 mg/L, reaction time = 12 h.

3.5. Adsorption Mechanism Analysis

FTIR was used to analyze the molecular structure of chemical compounds and the FTIR spectra of CA beads is shown before and after metal ion adsorption in Figure 7. For the CA beads before adsorption, the dominant peak at 3425.28 cm^{-1} was due to the vibration stretching of the O-H bond, indicating that hydroxyl (-OH) groups existed in the beads. Bands at 2926.01 cm^{-1} referred to the vibration stretching of -CH. The absorption peaks around 1610.24 cm^{-1}, 1418.32 cm^{-1} and 1034.37 cm^{-1} could be attributed to the asymmetric and symmetric stretching vibrations of -COO (carboxylate) and the stretching vibration of C-O, respectively [36]. After the sorption of metal ions, the FTIR spectra displacement slightly changed. This may be because the increased ionic volume weakened the stretching and torsional vibration of the functional groups, thus causing the displacement of adsorption peaks. The decrease of peak vibration intensity indicated that metal ions bind to -OH and -COO formed in the CA beads. In addition, no new peaks were produced, indicating that the functional groups of the adsorbents did not change, and the ion exchange process between the metal ions and the CA beads was possible.

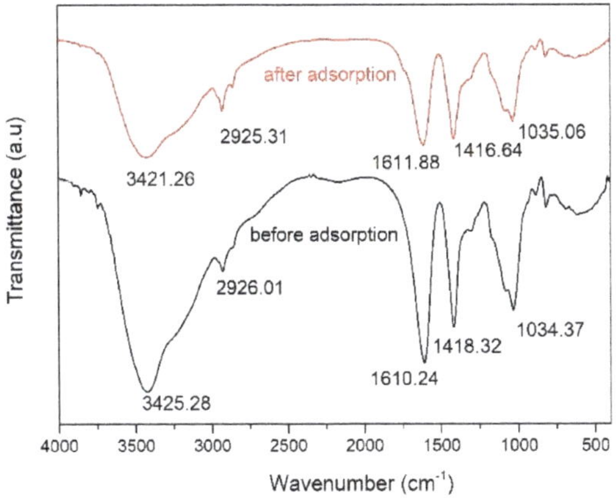

Figure 7. FTIR images of CA beads before and after the adsorption.

Figure 8 shows the XPS diagram of Cu 2p, Cd 3d, Zn 2p and Ni 2p (after sorption), further exploring the adsorption mechanism. The binding energy of Cu 2p, Cd 3d, Zn 2p and Ni 2p were

933.55/953.44 eV, 405.59/412.34 eV, 1021.97/1044.94 eV and 856.96 eV, respectively, after deconvolution, indicating that the adsorbed metal ions were in chemical states without further oxidation or reduction and the valence states were not changed during the ion exchange process [37–40].

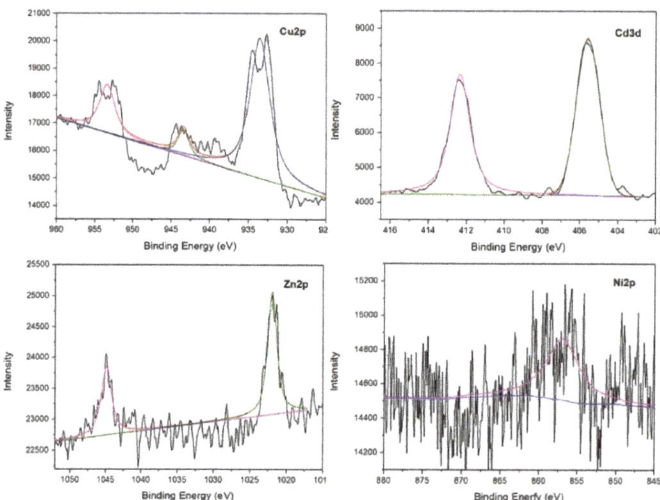

Figure 8. X-ray photoelectron spectroscopy (XPS) spectra of CA beads after the adsorption of mixed metal ions at a concentration of 500 mg/L in aqueous solution.

Cu(II) preferentially bound to the active sites of CA beads when the four ions were coexistent at the same concentration (Figure 9). The selective adsorption mechanism of copper ions on CA beads was via divalent metal ions exchanging ions with Ca(II) in the "egg box" structure [41]. The adsorption sites were mainly negative groups (COO-) that could adsorb the cation in the solution, and the stability of the bond with Cu(II) was better [42]. Therefore, compared with Zn(II), Ni(II) and Cd(II), the affinity for Cu(II) was the strongest in the ion exchange process between Ca(II) and metal ions in CA beads. This competitive relationship made the CA bead a potential material that could be applied to the recovery the copper ions from heavy metal mixtures.

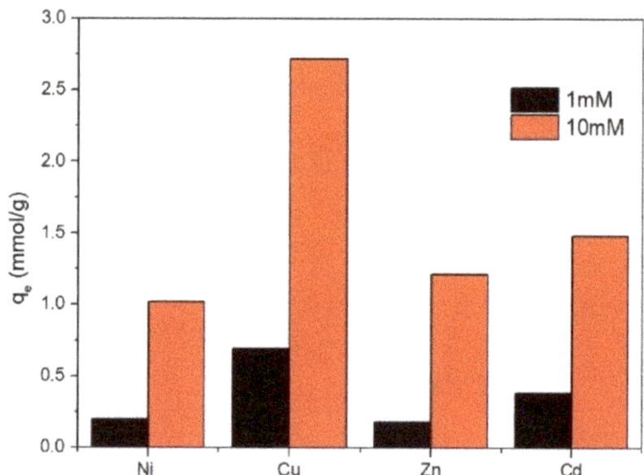

Figure 9. Effect of Cu(II) competitive adsorption in a quaternary metal system. Reaction conditions: T = 293 K, pH = 5, $[Cu(II)]_0 = [Zn(II)]_0 = [Ni(II)]_0 = [Cd(II)]_0$ = 1 mM/10 mM, reaction time = 12 h.

3.6. Desorption and Reusability Experiment

Regeneration of loaded sorbent is a key factor in water treatment processes and the desorption of the adsorbed Cu(II), Zn(II), Ni(II) and Cd(II) ions with 0.1 M HNO_3 was investigated. Figure 10 showed the reusability of CA beads with four heavy metal ions. The results revealed that after five cycles, the CA beads still had selective adsorption for Cu(II) and the adsorption capacity was more than 92%. Therefore, it can be explained that CA beads have good reusability during reuse experiments and have economic potential in wastewater treatment.

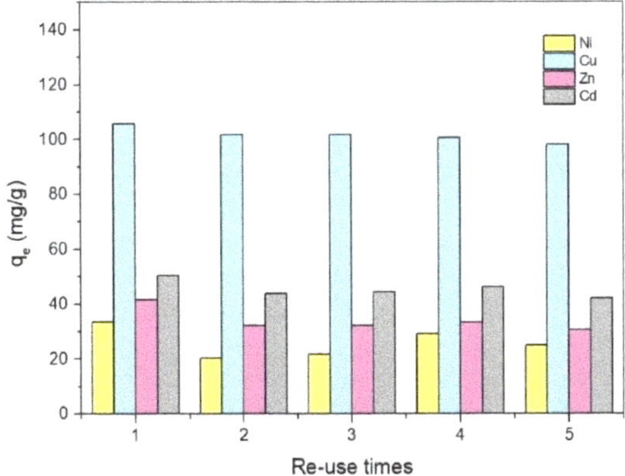

Figure 10. Effect of recycling on Cu(II), Zn(II), Ni(II) and Cd(II) ion adsorption. Reaction conditions: T = 293 K, $[Zn(II)]_0$ = $[Ni(II)]_0$ = $[Cd(II)]_0$ = 200 mg/L; reaction time = 12 h.

4. Conclusions

A spherical CA bead with a diameter of 3 mm was prepared by crosslinking the hydroxyl and carboxyl groups of sodium alginate with Ca(II) to form an insoluble hydrogel; there were a large number of active sites in the porous honeycomb structure for metal ions to attach. According to the pseudo-second-order and Langmuir isotherm model, the adsorption mechanism was explained well and the chemical reaction dominated the rate of the adsorption process. The maximum uptake of Cu(II) could reach 107.53 mg/g in the mixed heavy metal solution and the valence state of the four metal ions was not changed according to XPS analysis during the adsorption process. Cu(II) exchanged ions with Ca(II), binding with α-L-guluronic acid in the "egg box" structure. The selective adsorption was indicated through the isotherm experiments, giving this material a high potential for the continuous treatment of the selective recovery of copper from multi-metal solutions.

Author Contributions: Conceptualization, R.W. and P.R.; methodology, N.Y. and F.C.; software, L.Y. investigation, N.Y., R.W. and W.Z.; resources, R.W. and J.W.; writing—original draft preparation, N.Y.; writing—review and editing, N.Y., R.W. and P.R.; supervision, R.W. and P.R.; project administration, R.W., J.W. and P.R.; funding acquisition, R.W. and J.W.

Funding: This research was funded by Shanghai Sailing Program (Grant No. 17YF1407200), the "Capacity Building Project of Some Local Colleges and Universities in Shanghai" (Grant No. 17030501200), SUES Sino-foreign cooperative innovation center for city soil ecological technology integration (Grant No. 2017PT03) and the Project of Shanghai University Young Teacher Training Scheme (Grant No. ZZGCD16018).

Conflicts of Interest: The authors declare no conflict of interest.

References

1. Mazur, L.P.; Cechinel, M.A.P.; de Souza, S.; Boaventura, R.A.R.; Vilar, V.J.P. Brown marine macroalgae as natural cation exchangers for toxic metal removal from industrial wastewaters: A review. *J. Environ. Manag.* **2018**, *223*, 215–253. [CrossRef]
2. Huang, Y.; Wu, H.; Shao, T.; Zhao, X.; Peng, H.; Gong, Y.; Wan, H. Enhanced copper adsorption by DTPA-chitosan/alginate composite beads: Mechanism and application in simulated electroplating wastewater. *Chem. Eng. J.* **2018**, *339*, 322–333. [CrossRef]
3. Wang, F.; Lu, X.; Li, X.-Y. Selective removals of heavy metals (Pb^{2+}, Cu^{2+}, and Cd^{2+}) from wastewater by gelation with alginate for effective metal recovery. *J. Hazard. Mater.* **2016**, *308*, 75–83. [CrossRef]
4. Yi, X.; He, J.; Guo, Y.; Han, Z.; Yang, M.; Jin, J.; Gu, J.; Ou, M.; Xu, X. Encapsulating Fe_3O_4 into calcium alginate coated chitosan hydrochloride hydrogel beads for removal of Cu (II) and U (VI) from aqueous solutions. *Ecotoxicol. Environ. Saf.* **2018**, *147*, 699–707. [CrossRef] [PubMed]
5. Coruh, S.; Ergun, O.N.; Cheng, T. Treatment of copper industry waste and production of sintered glass-ceramic. *Waste Manag. Res.* **2006**, *24*, 234–241. [CrossRef]
6. Arrabito, G.; Errico, V.; Zhang, Z.; Han, W.; Falconi, C. Nanotransducers on printed circuit boards by rational design of high-density, long, thin and untapered ZnO nanowires. *Nano Energy* **2018**, *46*, 54–62. [CrossRef]
7. Errico, V.; Arrabito, G.; Plant, S.R.; Medaglia, P.G.; Palmer, R.E.; Falconi, C. Chromium inhibition and size-selected Au nanocluster catalysis for the solution growth of low-density ZnO nanowires. *Sci. Rep.* **2015**, *5*, 12336. [CrossRef]
8. Liu, X.; Tan, Q.; Li, Y.; Xu, Z.; Chen, M. Copper recovery from waste printed circuit boards concentrated metal scraps by electrolysis. *Front. Environ. Sci. Eng.* **2017**, *11*, 10. [CrossRef]
9. Mdlovu, N.V.; Chiang, C.; Lin, K.; Jeng, R. Recycling copper nanoparticles from printed circuit board waste etchants via a microemulsion process. *J. Clean. Prod.* **2018**, *185*, 781–796. [CrossRef]
10. Petrovič, A.; Simonič, M. Removal of heavy metal ions from drinking water by alginate-immobilised Chlorella sorokiniana. *Int. J. Environ. Sci. Technol.* **2016**, *13*, 1761–1780. [CrossRef]
11. Pawar, R.R.; Lalhmunsiama; Gupta, P.; Sawant, S.Y.; Shahmoradi, B.; Lee, S.M. Porous synthetic hectorite clay-alginate composite beads for effective adsorption of methylene blue dye from aqueous solution. *Int. J. Biol. Macromol.* **2018**, *114*, 1315–1324. [CrossRef]
12. Torres-Caban, R.; Vega-Olivencia, C.A.; Alamo-Nole, L.; Morales-Irizarry, D.; Roman-Velazquez, F.R.; Mina-Camilde, N. Removal of Copper from Water by Adsorption with Calcium-Alginate/Spent-Coffee-Grounds Composite Beads. *Materials* **2019**, *12*, 395. [CrossRef]
13. Cataldo, S.; Gianguzza, A.; Merli, M.; Muratore, N.; Piazzese, D.; Liveri, M.L. Experimental and robust modeling approach for lead(II) uptake by alginate gel beads: Influence of the ionic strength and medium composition. *J. Colloid Interface Sci.* **2014**, *434*, 77–88. [CrossRef] [PubMed]
14. Guo, J.; Han, Y.; Mao, Y.; Wickramaratne, M.N. Influence of alginate fixation on the adsorption capacity of hydroxyapatite nanocrystals to Cu^{2+} ions. *Colloids Surf. A* **2017**, *529*, 801–807. [CrossRef]
15. He, J.; Chen, J.P. A comprehensive review on biosorption of heavy metals by algal biomass: Materials, performances, chemistry, and modeling simulation tools. *Bioresour. Technol.* **2014**, *160*, 67–78. [CrossRef] [PubMed]
16. Feng, Y.; Wang, Y.; Wang, Y.; Zhang, X.F.; Yao, J. In-situ gelation of sodium alginate supported on melamine sponge for efficient removal of copper ions. *J. Colloid Interface Sci.* **2018**, *512*, 7–13. [CrossRef]
17. Davis, T.A.; Volesky, B.; Mucci, A. A review of the biochemistry of heavy metal biosorption by brown algae. *Water Res.* **2003**, *37*, 4311–4330. [CrossRef]
18. Nussinovitch, A.; Dagan, O. Hydrocolloid liquid-core capsules for the removal of heavy-metal cations from water. *J. Hazard. Mater.* **2015**, *299*, 122–131. [CrossRef]
19. Sánchez, M.; Vásquez-Quitral, P.; Butto, N.; Díaz-Soler, F.; Yazdani-Pedram, M.; Silva, J.; Neira-Carrillo, A. Effect of Alginate from Chilean Lessonia nigrescens and MWCNTs on $CaCO_3$ Crystallization by Classical and Non-Classical Methods. *Crystals* **2018**, *8*, 69. [CrossRef]
20. Hu, Z.H.; Omer, A.M.; Ouyang, X.K.; Yu, D. Fabrication of carboxylated cellulose nanocrystal/sodium alginate hydrogel beads for adsorption of Pb(II) from aqueous solution. *Int. J. Biol. Macromol.* **2018**, *108*, 149–157. [CrossRef]

21. Vu, H.C.; Dwivedi, A.D.; Le, T.T.; Seo, S.-H.; Kim, E.-J.; Chang, Y.-S. Magnetite graphene oxide encapsulated in alginate beads for enhanced adsorption of Cr(VI) and As(V) from aqueous solutions: Role of crosslinking metal cations in pH control. *Chem. Eng. J.* **2017**, *307*, 220–229. [CrossRef]
22. Dechojarassri, D.; Omote, S.; Nishida, K.; Omura, T.; Yamaguchi, H.; Furuike, T.; Tamura, H. Preparation of alginate fibers coagulated by calcium chloride or sulfuric acid: Application to the adsorption of Sr^{2+}. *J. Hazard. Mater.* **2018**, *355*, 154–161. [CrossRef]
23. Lai, Y.C.; Chang, Y.R.; Chen, M.L.; Lo, Y.K.; Lai, J.Y.; Lee, D.J. Poly(vinyl alcohol) and alginate cross-linked matrix with immobilized Prussian blue and ion exchange resin for cesium removal from waters. *Bioresour. Technol.* **2016**, *214*, 192–198. [CrossRef]
24. Choudhary, B.C.; Paul, D.; Borse, A.U.; Garole, D.J. Surface functionalized biomass for adsorption and recovery of gold from electronic scrap and refinery wastewater. *Sep. Purif. Technol.* **2018**, *195*, 260–270. [CrossRef]
25. Majidnia, Z.; Fulazzaky, M.A. Photoreduction of Pb(II) ions from aqueous solution by titania polyvinylalcohol–alginate beads. *J. Taiwan Inst. Chem. Eng.* **2016**, *66*, 88–96. [CrossRef]
26. Cataldo, S.; Gianguzza, A.; Milea, D.; Muratore, N.; Pettignano, A. Pb(II) adsorption by a novel activated carbon—Alginate composite material. A kinetic and equilibrium study. *Int. J. Biol. Macromol.* **2016**, *92*, 769–778. [CrossRef]
27. Arshad, F.; Selvaraj, M.; Zain, J.; Banat, F.; Haija, M.A. Polyethylenimine modified graphene oxide hydrogel composite as an efficient adsorbent for heavy metal ions. *Sep. Purif. Technol.* **2019**, *209*, 870–880. [CrossRef]
28. Srikantan, C.; Suraishkumar, G.K.; Srivastava, S. Effect of light on the kinetics and equilibrium of the textile dye (Reactive Red 120) adsorption by Helianthus annuus hairy roots. *Bioresour. Technol.* **2018**, *257*, 84. [CrossRef]
29. Barsbay, M.; Kavaklı, P.A.; Tilki, S.; Kavaklı, C.; Güven, O. Porous cellulosic adsorbent for the removal of Cd (II), Pb(II) and Cu(II) ions from aqueous media. *Radiat. Phys. Chem.* **2018**, *142*, 70–76. [CrossRef]
30. Kong, M.; Wang, L.; Chao, J.; Ji, Z.; Peng, F.; Yang, F.; Zhang, Y. Removal of Cu^{2+} and Ni^{2+} from Wastewater by Using Modified Alkali-Leaching Residual Wire Sludge as Low-Cost Adsorbent. *Water Air Soil Pollut.* **2019**, *230*, 65. [CrossRef]
31. Lv, D.; Liu, Y.; Zhou, J.; Yang, K.; Lou, Z.; Baig, S.A.; Xu, X. Application of EDTA-functionalized bamboo activated carbon (BAC) for Pb(II) and Cu(II) removal from aqueous solutions. *Appl. Surf. Sci.* **2018**, *428*, 648–658. [CrossRef]
32. Botello-González, J.; Cerino-Córdova, F.J.; Dávila-Guzmán, N.E.; Salazar-Rábago, J.J.; Soto-Regalado, E.; Gómez-González, R.; Loredo-Cancino, M. Ion Exchange Modeling of the Competitive Adsorption of Cu(II) and Pb(II) Using Chemically Modified Solid Waste Coffee. *Water Air Soil Pollut.* **2019**, *230*, 73. [CrossRef]
33. Duan, C.; Zhao, N.; Yu, X.; Zhang, X.; Xu, J. Chemically modified kapok fiber for fast adsorption of Pb^{2+}, Cd^{2+}, Cu^{2+}; from aqueous solution. *Cellulose* **2013**, *20*, 849–860. [CrossRef]
34. Zhao, Y.; Chen, Y.; Zhao, J.; Tong, Z.R.; Jin, S.H. Preparation of SA-g-(PAA-co-PDMC) polyampholytic superabsorbent polymer and its application to the anionic dye adsorption removal from effluents. *Sep. Purif. Technol.* **2017**, *188*, 329–340. [CrossRef]
35. Yan, L.G.; Yang, K.; Shan, R.R.; Yan, T.; Wei, J.; Yu, S.J.; Yu, H.Q.; Du, B. Kinetic, isotherm and thermodynamic investigations of phosphate adsorption onto core-shell Fe_3O_4@LDHs composites with easy magnetic separation assistance. *J. Colloid Interface Sci.* **2015**, *448*, 508–516. [CrossRef]
36. Algothmi, W.M.; Bandaru, N.M.; Yu, Y.; Shapter, J.G.; Ellis, A.V. Alginate-graphene oxide hybrid gel beads: An efficient copper adsorbent material. *J. Colloid Interface Sci.* **2013**, *397*, 32–38. [CrossRef]
37. Lim, S.F.; Zheng, Y.M.; Zou, S.W.; Chen, J.P. Characterization of copper adsorption onto an alginate encapsulated magnetic sorbent by a combined FT-IR, XPS, and mathematical modeling study. *Environ. Sci. Technol.* **2008**, *42*, 2551. [CrossRef]
38. Zhao, X.; Wang, Y.; Wu, H.; Fang, L.; Liang, J.; Fan, Q.; Li, P. Insights into the effect of humic acid on Ni(II) sorption mechanism on illite: Batch, XPS and EXAFS investigations. *J. Mol. Liq.* **2017**, *248*, 1030–1038. [CrossRef]
39. Duchoslav, J.; Steinberger, R.; Arndt, M.; Stifter, D. XPS study of zinc hydroxide as a potential corrosion product of zinc: Rapid X-ray induced conversion into zinc oxide. *Corros. Sci.* **2014**, *82*, 356–361. [CrossRef]

40. Lalhmunsiama; Gupta, P.L.; Jung, H.; Tiwari, D.; Kong, S.; Lee, S. Insight into the mechanism of Cd(II) and Pb(II) removal by sustainable magnetic biosorbent precursor to *Chlorella vulgaris*. *J. Taiwan Inst. Chem. Eng.* **2017**, *71*, 206–213. [CrossRef]
41. Rodrigues, J.R.; Lagoa, R. Copper ions binding in Cu-alginate gelation. *J. Carbohydr. Chem.* **2006**, *25*, 219–232. [CrossRef]
42. Cataldo, S.; Gianguzza, A.; Pettignano, A.; Piazzese, D.; Sammartano, S. Complex formation of copper(II) and cadmium(II) with pectin and polygalacturonic acid in aqueous solution. An ISE-H^+ and ISE-Me^{2+} electrochemical study. *Int. J. Electrochem. Sci.* **2012**, *7*, 6722–6737.

Article

Luminescent Layered Double Hydroxides Intercalated with an Anionic Photosensitizer via the Memory Effect

Alexandre C. Teixeira [1], Alysson F. Morais [1], Ivan G.N. Silva [1], Eric Breynaert [2] and Danilo Mustafa [1,*]

1 Instituto de Física da Universidade de São Paulo, São Paulo 05508-090, Brazil; alexandre.candido.teixeira@usp.br (A.C.T.); afmorais@if.usp.br (A.F.M.); ingsilva@if.usp.br (I.G.N.S.)

2 Center for Surface Chemistry and Catalysis (COK-KAT), KU Leuven, B-3001 Leuven, Belgium; Eric.Breynaert@biw.kuleuven.be

* Correspondence: dmustafa@if.usp.br

Received: 12 February 2019; Accepted: 9 March 2019; Published: 14 March 2019

Abstract: Layered double hydroxides (LDHs) containing Eu^{3+} activators were synthesized by coprecipitation of Zn^{2+}, Al^{3+}, and Eu^{3+} in alkaline NO_3^--rich aqueous solution. Upon calcination, these materials transform into a crystalline ZnO solid solution containing Al and Eu. For suitably low calcination temperatures, this phase can be restored to LDH by rehydration in water, a feature known as the memory effect. During rehydration of an LDH, new anionic species can be intercalated and functionalized, obtaining desired physicochemical properties. This work explores the memory effect as a route to produce luminescent LDHs intercalated with 1,3,5-benzenetricarboxylic acid (BTC), a known anionic photosensitizer for Eu^{3+}. Time-dependent hydration of calcined LDHs in a BTC-rich aqueous solution resulted in the recovery of the lamellar phase and in the intercalation with BTC. The interaction of this photosensitizer with Eu^{3+} in the recovered hydroxide layers gave rise to efficient energy transfer from the BTC antennae to the Eu^{3+} ions, providing a useful tool to monitor the rehydration process of the calcined LDHs.

Keywords: layered double hydroxide; memory effect; rare earth; europium; 1,3,5-benzenetricarboxylic acid

---

1. Introduction

Layered double hydroxides (LDHs), also called cationic clays, are a class of anion-exchange materials with a general chemical formula of $[M^{II}_{1-x}M^{III}_x(OH)_2]^{x+}[A^{n-}_{x/n}]^{x-}\cdot yH_2O$ (M: metal, A: anion). The isomorphic substitution of divalent metal cations (M^{II}) in otherwise neutral brucite-like $M^{II}(OH)_2$ sheets with trivalent cations (M^{III}) introduces positive charges in the hydroxide layers. In the overall LDH structure, these charges are compensated by the intercalation of anions (A^{n-}) in the interlamellar space, as illustrated in Figure 1.

Synthetic LDHs exhibit a wide flexibility in their composition, as a score of metal cations and polyvalent anions can be introduced in the structure, tuning their chemico-physical properties [1–3]. By changing both the interlayer and the metal components, LDHs have been tuned towards different applications, serving as heterogeneous catalysts, catalyst supports [4], water treatment agents [5], luminescent materials [6–9], etc.

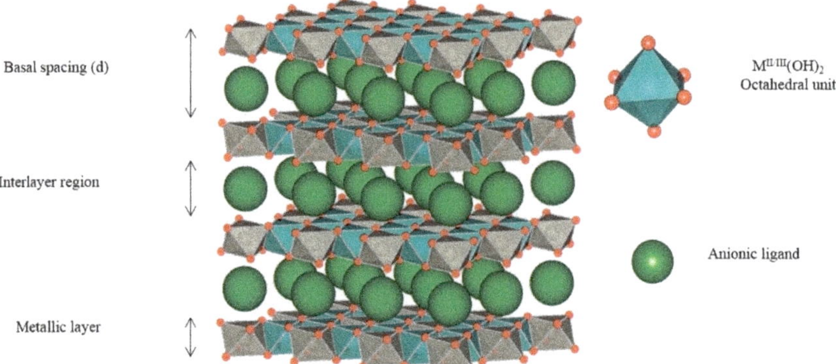

Figure 1. Layered double hydroxides (LDHs) are built up from sheets of $[M^{II}_{1-x}M^{III}_{x}(OH)_2]^{x+}$ octahedral units intercalated with anions. Each octahedron is formed from a metal cation (M^{II} or M^{III}) 6-coordinated to OH^- groups (red spheres).

The introduction of trivalent rare earth elements (RE^{3+}) in LDH matrixes has revealed a series of 2D structured materials that have promising luminescent properties [7,10]. Rare earth elements form a subgroup of column 3 in the periodic table and exhibit very low influence of the ligand field in the energy of their spectral lines. This characteristic optical property results from the shielding of the 4f sub-shell by their filled 5d and 6s most external sub-shells [11,12]. Additionally, RE^{3+} ions present very low molar extinction coefficients because their 4f intraconfigurational transitions are forbidden, as demonstrated by Laporte's rule ($\Delta\ell = \pm 1$). To increase the total observable luminescence, it is necessary to embed these elements in highly light-absorbing matrixes that serve as harvesting antennae, transferring the harvested energy to the luminescent center (antenna effect). This increases the excited state population of the RE^{3+}, thereby increasing the overall luminescence [13–15].

The antenna effect has been demonstrated in RE-containing LDHs synthesized by direct coprecipitation with anionic antenna molecules or by exploiting anion exchange to intercalate anionic sensitizer ligands in the interlamellar space. A series of anionic sensitizers, including 4-biphenylacetate [8], sulfonates, and other carboxylates [16,17], have been intercalated in LDHs to incorporate RE^{3+} in their hydroxide layers. Several authors have also reported LDHs intercalated with Eu^{3+} complexes, where not only the ligand, but also the RE^{3+} is located in the interlamellar space [6,18].

Controlled thermal decomposition of LDHs converts their structure into a mixed metal oxyhydroxide phase, before reaching a fully oxidic end product [5,9,19,20]. Starting from the oxyhydroxide phase, the lamellar LDH structure can be recovered by re-hydrating the product in aqueous solutions containing anions, a phenomenon known as the memory effect [5,9,20–23]. During this process, selective adsorption of dissolved anionic species A^{n-} can efficiently occur, as indicated by the large anion exchange capacity observed for thermally treated samples [24]. The most frequently investigated application of the memory effect in LDHs involves the adsorption and subsequent removal of anionic dyes from waste water [5,25]. A less explored application is its use for the intercalation of anionic photosensitizing molecules.

The memory effect of the layered double hydroxides has been described in the literature for a large number of combinations of metal cations [21,23,26,27]. Different applications can be explored for different chemical compositions. For instance, Wong et al. [27] have shown that the memory effect of LDHs containing Li and Al can be used to sense water uptake in organic coatings. Ni et al. [25] have explored the memory effect of Zn/Al LDHs in the removal of methyl orange from aqueous solutions. Targeting environmental remediation, Gao et al. [23] have investigated the influence of humic acid on the memory effect of Mg/Al and Zn/Al LDHs. The removal of boron species from waste water was also explored [28].

In this work, an underused strategy for generating luminescent LDHs via the memory effect was explored. The luminescent and structural properties of thermally treated $[Zn_2Al_{0.95}Eu_{0.05}(OH)_6]\cdot(NO_3^-)$ LDHs after hydration in the presence of 1,3,5-benzenetricarboxilate (BTC) were investigated. Calcination was performed at 350 and 600 °C, followed by a time-dependent rehydration process in a BTC-rich aqueous solution, equilibrating the samples over a period between 1 min and 5 days. The resulting hydrated phases were characterized by both powder X-ray diffraction and luminescence spectroscopy.

2. Materials and Methods

H_3BTC (97 mol%, Sigma-Aldrich), $Zn(NO_3)_2\cdot6H_2O$ (98 mol%, Vetec), $Al(NO_3)_3\cdot9H_2O$ (98 mol%, LabSynth), and NaOH (97 mol%, Vetec) were used without further purification. $Eu(NO_3)_3\cdot6H_2O$ was prepared by dissolving Eu_2O_3 (CSTARM, 99,99 mol%) in concentrated nitric acid, which led to the subsequent crystallization of $Eu(NO_3)_3\cdot6H_2O$ in accordance to the procedure described by Silva et al. [29].

2.1. Sample Preparation

Layered double hydroxides with nominal molar composition $[Zn_2Al_{0.95}Eu_{0.05}(OH)_6]\cdot(NO_3^-)$ (ZnAlEu-NO_3 LDH) and $[Zn_2Al_{0.95}Eu_{0.05}(OH)_6]\cdot(BTC^{3-})_{0.33}$ (ZnAlEu-BTC LDH) were synthesized by coprecipitation. A volume of 10 mL of a 1 mol L^{-1} solution containing the metal precursors $Zn(NO_3)_2\cdot6H_2O$, $Al(NO_3)_2\cdot9H_2O$, and $Eu(NO_3)_2\cdot6H_2O$ in the ratio 2:0.95:0.05 was added dropwise (~10 mL h^{-1}) to 200 mL of an alkaline (pH 8) solution containing the dissolved ligands. During the synthesis, this solution was continuously stirred, purged with $N_{2(g)}$, and pH-stated at pH 8 using an automatic titrator (Metrohm 785 DMP Titrino). The resulting suspension was equilibrated in a closed vessel at 60 °C for two days, followed by centrifugation and subsequent rinsing with distilled water. The resulting solid phase was dried in air at 60 °C for three days.

For the LDHs intercalated with NO_3^-, the nitrate anions originated from the precursor metal salts. For LDHs intercalated with BTC^{3-}, the amount of BTC (2.3×10^{-3} mol) was chosen to exceed the positive charges in the hydroxide layers by a factor of at least 2, with BTC/Al = 2:3.

Calcination of the dried ZnAlEu-NO_3 LDHs was performed by heating the LDH in a muffle furnace for four hours at, respectively, 200 °C (C-LDH-200), 350 °C (C-LDH-350), or 600 °C (C-LDH-600), ramping up the oven temperature from ambient to the final temperature at a rate of 5 °C min^{-1}.

Rehydration of the calcined samples was achieved by suspending 100 mg of the calcined LDHs in 10 mL of a 12 mmol L^{-1} BTC aqueous solution. This solution was prepared by dissolving BTC in a heated (60 °C) aqueous solution, subsequently titrated with a 1M NaOH solution (ca. 0.3 mL) to finally reach a pH between 3.5 and 4.0. Rehydration of the LDH was performed under continuous stirring (~300 rpm), varying the rehydration time from 1 min up to five days.

2.2. Characterization

A thermogravimetric analysis was performed on 13 mg of the ZnAlEu–NO_3 LDH using a TGA Q500 (TA Instruments, New Castle, DE, USA), by ramping up the temperature from room temperature to 800 °C at a rate of 5 °C min^{-1} using an air flow of 60 mL min^{-1}.

Powder X-ray diffraction (PXRD) analyses were performed using a D8 Discover (Bruker, Atibaia, Brazil) diffractometer in Bragg–Brentano geometry using Cu Kα radiation (λ = 1.5418 Å). Data were recorded from 4° to 70° 2θ in steps of 0.05° using an integration time of 1 s.

Luminescence spectra were obtained on a SPEX® Fluorolog® 3 (Horiba, São Paulo, SP, Brazil) equipped with a 450 W Xenon lamp as an excitation source.

3. Results and Discussion

X-ray diffraction patterns of the as-prepared ZnAlEu–NO_3 LDHs (Figure 2A) readily revealed the characteristic (003) and (006) basal reflections at 9.90° and 19.85° 2θ, which indicated a basal spacing (see Figure 1) of 8.96 Å, typical for NO_3^--intercalated LDHs [30]. From the partially overlapped (110) and (113) reflections around 60.45 2θ, the metal-to-metal distance within the hydroxide layers could also be estimated as 'a = $2d_{(110)}$ = 3.06 Å' [3]. The (00*l*) basal reflections of the ZnAlEu–BTC LDHs at 6.6°, 13.2°, 20.2°, and 26.2° 2θ confirmed the basal spacing of 13.4 Å previously reported for BTC-intercalated LDHs (as illustrated in Figure 2C) [7].

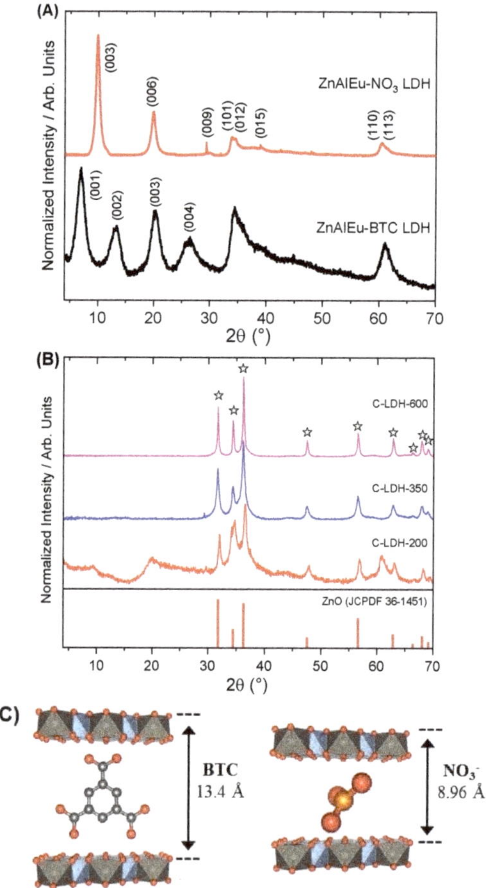

Figure 2. Powder X-ray diffraction patterns of (**A**) the as-prepared ZnAlEu LDHs and (**B**) ZnAlEu–NO_3 LDHs calcined at 200 °C (C-LDH-200), 350 °C (C-LDH-350), and 600 °C (C-LDH-600). Starred peaks are related to the formation of a ZnO-like phase. The diffractogram of ZnO is shown for comparison. (**C**) Intercalation scheme and basal spacing for LDHs with 1,3,5-benzenetricarboxilate (BTC) and NO_3^- [30].

LDHs can be dehydrated and dehydroxylated by calcination at temperatures as low as 200 °C [31]. For the ZnAlEu–NO_3 LDHs, thermogravimetric analysis showed three main mass loss events (Figure 3). The first one, centered around 90 °C, is related to the desorption of physisorbed and weakly bound surface water. A more pronounced mass loss was observed at 220 °C, related to the removal of crystal water, the dehydroxylation of vicinal OH groups within the same layer, and the subsequent condensation of OH groups from adjacent layers when the layer structure collapsed. At this time, a mixed metal oxyhydroxide phase was obtained, containing a series of oxygen sites $O^{2-}{}_{(C\text{-}LDH)}$ and oxygen vacancies resulting from the following endothermic transformation [21], which occurred above 180 °C:

$$2\,OH^-{}_{(LDH)} \rightarrow H_2O_{(g)} + O^{2-}{}_{(C\text{-}LDH)}$$

The PXRD pattern of the sample calcined at 200°C (C-LDH-200) (Figure 2B) indicated the formation of a ZnO-like phase. The disappearance of the basal reflections of the LDHs with the retention of the partially overlapped (110) and (113) reflections indicated the collapse of the layers, but with incomplete dehydroxylation (as illustrated in Figure 4).

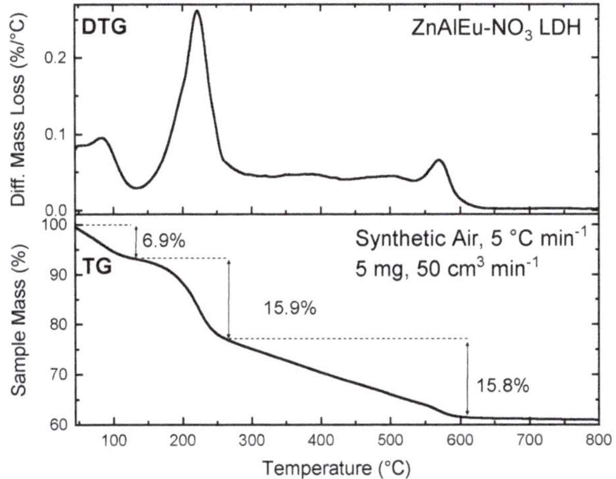

Figure 3. Thermogravimetric analysis of the ZnAlEu–NO_3 LDHs.

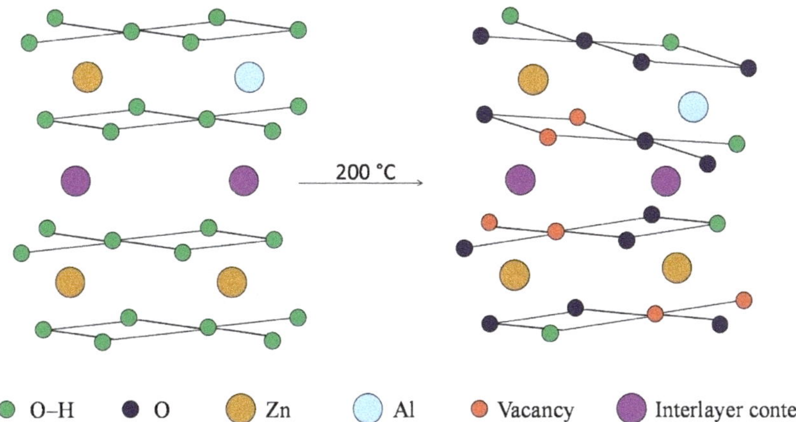

Figure 4. Schematic overview of the dehydroxylation of LDHs at low temperatures.

The final mass loss event observed in the TGA over a broad temperature range from 300 to 620 °C (Figure 3) was assigned to further dehydroxylation, continuous decomposition, and release of the intercalated NO_3^-. As described by Valente et al. [32], the extent of this last event is related to the close packing induced by the collapse of the interlayer structure, thereby hindering the release of the interlayer products. The experimental mass loss observed at 600 °C in the TG experiment is in agreement with a theoretical loss of 16 wt % that was expected upon dehydroxylation and complete removal of NO_3^- from a $[Zn_2Al_{0.95}Eu_{0.05}(OH)_6]\cdot(NO_3)\cdot(H_2O)$ LDH phase.

By calcining ZnAlEu–NO_3 LDH at 350 °C, i.e., above the second mass loss event in the thermogravimetric analysis (Figure 3), the material was converted into ZnO-like crystals, exhibiting no signs of segregation of crystalline Al_2O_3 or Eu_2O_3 [33]. Some authors attribute the absence of diffraction lines for Al_2O_3 or Eu_2O_3 to the formation of a ZnO-based solid solution embedding the trivalent ions [34–36]. Even upon calcination at 600 °C, no indication for Al_2O_3 or Eu_2O_3 could be observed.

Figure 5 shows the evolution of the X-ray diffraction patterns as a function of the rehydration time of calcined LDHs suspended BTC-rich aqueous solutions. Overall, this rehydration process can be expressed by the following equation [23]:

$$M^{II}_{1-x}M^{III}_{x}O_{1+x/2} + (x/n)A^{n-} + (m + 1 + x/2)H_2O \rightarrow M^{II}_{1-x}M^{III}_{x}(OH)_2(A^{n-})_{x/n}\cdot mH_2O + xOH^-$$

For samples calcined at 350 °C (Figure 5 left), the emergence of the reflection at 9.90° 2θ indicated a recovery of a lamellar structure after 1 min of rehydration. As described by Valente et al. [32], NO_3^- could become partially confined upon the collapse of the LDH structure, which could assist the recovery of the structure upon rehydration. After 30 min, BTC started to become intercalated in the previously recovered layered phase. The (00*l*) basal reflections of the recovered BTC-intercalated lamellar phase presented a slight shift to higher 2θ, indicating that the as-synthesized and recovered phases possessed different hydration contents. After three days of rehydration, the reflections resulting from the oxide phase broadened considerably, which indicated a breakdown of the oxide domains in the sample.

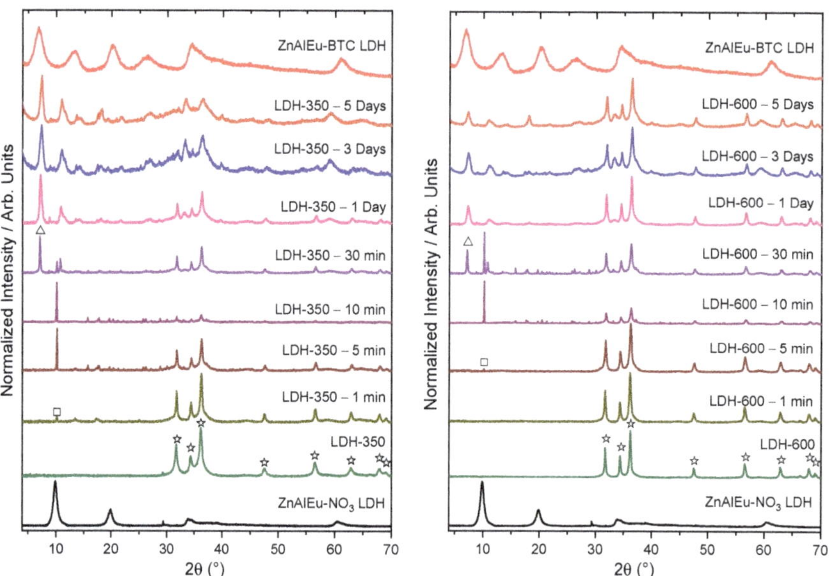

Figure 5. Powder X-ray diffraction patterns for different hydration times of ZnAlEu–NO_3 LDH samples calcined at 350 °C (**left**) and 600 °C (**right**). Starred peaks relate to the formation of a ZnO phase. Reflections associated with the reappearance of a layered phase are marked with open squares, while the basal reflections of LDH intercalated with BTC (7.0°, 12.7 Å) are marked with an open triangle. The PXRD patters of the samples ZnAlEu–NO_3 LDH and ZnAlEu–BTC LDH are shown for comparison.

For specimens treated at 600 °C (Figure 5 right), the recovery of the lamellar phase was delayed compared to that of samples treated at 350 °C. After five days, a lower degree of regeneration was observed for the LDH calcined at 600 °C (C-LDH-600).

The photoluminescence properties of the samples were investigated by measuring their excitation and emission spectra (Figure 6). The excitation spectra (Figure 6 left) of all samples were collected at room temperature, monitoring the $(Eu^{3+})^5D_0 \rightarrow {}^7F_2$ emission at 614 nm. In the calcined samples, no zinc oxide $O^{2-} \rightarrow Eu^{3+}$ Ligand-to-Metal-Charge Transfer band (LMCT) was found, in accordance to other studies of Eu^{3+}-doped ZnO [37]. In the hydrated materials, the broad excitation band in the higher energy region (235–315 nm) was composed by two bands owing to the energy transference from the BTC to Eu^{3+}. One band was centered around 265 nm and was attributed to $(BTC)O^{2-} \rightarrow Eu^{3+}$ LMCT [29,38–40]. The other band, centered at 290 nm, corresponded to the BTC singlet transition $S_0 \rightarrow S_n$ [41–43] with subsequent energy transfer to Eu^{3+}.

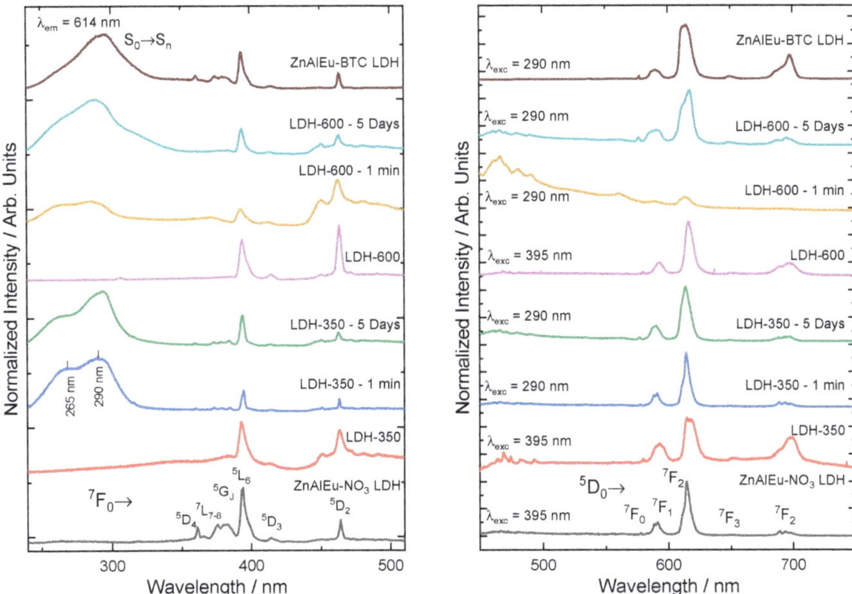

Figure 6. **(left)** Normalized excitation spectra recorded at 300 K, monitoring the $(Eu^{3+})^5D_0 \rightarrow {}^7F_2$ emission at 614 nm of the ZnAlEu–NO_3 LDH (as-prepared) together with the calcined and rehydrated samples. **(right)** Normalized emission spectra (300 K) under excitation at the BTC ligand-to-metal transference band (290 nm) for the rehydrated samples and under excitation at the $(Eu^{3+})^7F_0 \rightarrow {}^5L_6$ excitation band at 395 nm for the as-prepared and calcined samples. The labels of the 4f-4f intraconfigurational transitions of Eu^{3+} are shown. The spectra of the sample ZnAlEu–BTC LDH are shown for comparison.

The emission spectrum under direct excitation of the (Eu^{3+}) ${}^7F_0 \rightarrow {}^5L_6$ excitation band (395 nm) of the as-synthesized ZnAlEu–NO_3 LDHs (Figure 6 right) showed the characteristic emission bands of Eu^{3+}. Similar emission spectra were observed for the samples calcined at 350 and 600 °C, but with broader emission bands (especially for C-LDH-350). The broadening of the emission bands arose from slightly different sites occupied by Eu^{3+} in the matrix, which produced slightly different emission profiles [44]. This indicated that the number of sites occupied by Eu^{3+} increased after calcination. Compared to the ${}^5D_0 \rightarrow {}^7F_1$ transition, the higher intensity of the ${}^5D_0 \rightarrow {}^7F_2$ transition was enabled by forced electric dipole [45–48], which indicated the absence of inversion symmetry around the average Eu^{3+} site.

After rehydrating in a BTC-rich aqueous solution, the calcined samples started to recover their lamellar structure. At the same time, these samples also started to transfer energy from the BTC antenna to Eu^{3+}. Figure 6 shows the emission spectra under excitation of the BTC $S_0 \rightarrow S_n$ transition after time-dependent hydration of the samples C-LDH-350 and C-LDH-600. Although no intercalation with BTC was found after hydrating the sample C-LDH-350 for 1 min, the photosensitizer was still able to transfer energy to Eu^{3+}. This effect was probably due to BTC molecules adsorbed on the outer surface of the particles, which recovered to the LDH phase faster than the bulk [23]. This effect was seen in the sample C-LDH-600 only for longer hydration times, which indicated that the rehydration of this sample was less efficient than that of C-LDH-350, as was also inferred from the PXRD results.

4. Conclusions

In summary, to produce luminescent LDH phases, ZnAlEu–NO_3 LDHs calcined at up to 600 °C can only partially be restored to the LDH phase through rehydration in BTC-rich aqueous solutions. The calcination temperature appears to be an important factor for an efficient rehydration, as samples calcined at 350 °C showed a higher degree of regeneration compared to those treated at 600 °C. For samples treated at 350 °C, the recovery of the LDH structure started after 1 min of rehydration, while for the samples treated at 600 °C the regeneration only started after 5 min. After 30 min, the pH of the rehydration solution increased as a result of the dissolution of the oxides, and this facilitated the intercalation of BTC. The interaction of this photosensitizer with Eu^{3+} in the recovered hydroxide layers gave rise to efficient energy transfer from the BTC antenna molecule to the Eu^{3+} ions, which proved to be a useful tool to monitor the rehydration process of the calcined LDHs.

Author Contributions: Conceptualization, E.B. and D.M.; methodology, A.F.M., I.G.N.S., E.B. and D.M.; validation, A.C.T. and A.F.M.; formal analysis, A.C.T, A.F.M. and I.G.N.S.; investigation, A.C.T., A.F.M. and I.G.N.S.; resources, D.M.; data curation, A.C.T., A.F.M., I.G.N.S. and D.M.; writing—original draft preparation, A.F.M., I.G.N.S., E.B. and D.M.; writing—review and editing, A.F.M., I.G.N.S., E.B. and D.M.; visualization, A.F.M. and I.G.N.S.; supervision, E.B. and D.M.; project administration, D.M.; funding acquisition, D.M.

Funding: This research was funded by Fundação de Amparo à Pesquisa do Estado de São Paulo (FAPESP, 2015/19210-0), Coordenação de Aperfeiçoamento de Pessoal de Nível Superior (CAPES, 1723707, Finance Code 001), and Conselho Nacional de Desenvolvimento Científico e Tecnológico (CNPq, 142127/2014-0 and 403055/2016-4).

Acknowledgments: The authors acknowledge the Laboratory of Crystallography (IFUSP, São Paulo) for assistance with the PXRD measurements.

Conflicts of Interest: The authors declare no conflict of interest.

References

1. Nalawade, P.; Aware, B.; Kadam, V.J.; Hirlekar, R.S. Layered double hydroxides: A review. *J. Sci. Ind. Res.* **2009**, *68*, 267–272.
2. Khan, A.I.; O'Hare, D. Intercalation chemistry of layered double hydroxides: recent developments and applications. *J. Mater. Chem.* **2002**, *12*, 3191–3198. [CrossRef]
3. Duan, X.; Evans, D.G.; He, J.; Kang, Y.; Khan, A.I.; Leroux, F.; Li, B.; Li, F.; O'Hare, D.; Slade, R.C.T.; et al. *Structure and Bonding—Layered Double Hydroxides*, 1st ed.; Duan, X., Evans, D.G., Eds.; Springer: Berlin/Heidelberg, Germany, 2005; Volume 119, ISBN 3-540-28279-3.
4. Fan, G.; Li, F.; Evans, D.G.; Duan, X. Catalytic applications of layered double hydroxides: recent advances and perspectives. *Chem. Soc. Rev.* **2014**, *43*, 7040–7066. [CrossRef] [PubMed]
5. Santos, R.M.M.; Tronto, J.; Briois, V.; Santilli, C.V. Thermal decomposition and recovery properties of ZnAl–CO_3 layered double hydroxide for anionic dye adsorption: insight into the aggregative nucleation and growth mechanism of the LDH memory effect. *J. Mater. Chem. A* **2017**, *5*, 9998–10009. [CrossRef]
6. Sarakha, L.; Forano, C.; Boutinaud, P. Intercalation of luminescent Europium(III) complexes in layered double hydroxides. *Opt. Mater.* **2009**, *31*, 562–566. [CrossRef]

7. Morais, A.F.; Silva, I.G.N.; Sree, S.P.; de Melo, F.M.; Brabants, G.; Brito, H.F.; Martens, J.A.; Toma, H.E.; Kirschhock, C.E.A.; Breynaert, E.; et al. Hierarchical self-supported ZnAlEu LDH nanotubes hosting luminescent CdTe quantum dots. *Chem. Commun.* **2017**, *53*, 7341–7344. [CrossRef] [PubMed]
8. Gunawan, P.; Xu, R. Lanthanide-doped layered double hydroxides intercalated with sensitizing anions: Efficient energy transfer between host and guest layers. *J. Phys. Chem. C* **2009**, *113*, 17206–17214. [CrossRef]
9. Zhao, Y.; Li, J.-G.; Fang, F.; Chu, N.; Ma, H.; Yang, X. Structure and luminescence behaviour of as-synthesized, calcined, and restored MgAlEu-LDH with high crystallinity. *Dalt. Trans.* **2012**, *41*, 12175. [CrossRef] [PubMed]
10. Zhuravleva, N.G.; Eliseev, A.A.; Lukashin, A.V.; Kynast, U.; Tretyakov, Y.D. Energy transfer in luminescent Tb- and Eu-containing layered double hydroxides. *Mendeleev Commun.* **2004**, *14*, 176–178. [CrossRef]
11. Meggers, W.F. Electron configurations of "rare-earth" elements. *Science* **1947**, *105*, 514–516. [CrossRef] [PubMed]
12. Biggemann, D.; Mustafa, D.; Tessler, L.R. Photoluminescence of Er-doped silicon nanoparticles from sputtered SiO_x thin films. *Opt. Mater.* **2006**, *28*, 842–845. [CrossRef]
13. Binnemans, K. Lanthanide-based luminescent hybrid materials. *Chem. Rev.* **2009**, *109*, 4283–4374. [CrossRef] [PubMed]
14. Binnemans, K. Interpretation of europium(III) spectra. *Coord. Chem. Rev.* **2015**, *295*, 1–45. [CrossRef]
15. Mustafa, D.; Biggemann, D.; Martens, J.A.; Kirschhock, C.E.A.; Tessler, L.R.; Breynaert, E. Erbium enhanced formation and growth of photoluminescent Er/Si nanocrystals. *Thin Solid Films* **2013**, *536*, 196–201. [CrossRef]
16. Gao, X.; Hu, M.; Lei, L.; O'Hare, D.; Markland, C.; Sun, Y.; Faulkner, S. Enhanced luminescence of europium-doped layered double hydroxides intercalated by sensitiser anions. *Chem. Commun.* **2011**, *47*, 2104–2106. [CrossRef] [PubMed]
17. Zhang, Z.; Chen, G.; Liu, J. Tunable photoluminescence of europium-doped layered double hydroxides intercalated by coumarin-3-carboxylate. *RSC Adv.* **2014**, *4*, 7991. [CrossRef]
18. Gago, S.; Pillinger, M.; Sá Ferreira, R.A.; Carlos, L.D.; Santos, T.M.; Gonçalves, L.S. Immobilization of Lanthanide Ions in a Pillared Layered Double Hydroxide. *Chem. Mater.* **2005**, *17*, 5803–5809. [CrossRef]
19. Bharali, P.; Saikia, R.; Boruah, R.K.; Goswamee, R.L. A comparative study of thermal decomposition behaviour of Zn-Cr, Zn-Cr-Al and Zn-Al type layered double hydroxides. *J. Therm. Anal. Calorim.* **2004**, *78*, 831–838. [CrossRef]
20. Kowalik, P.; Konkol, M.; Kondracka, M.; Próchniak, W.; Bicki, R.; Wiercioch, P. Memory effect of the CuZnAl-LDH derived catalyst precursor - In situ XRD studies. *Appl. Catal. A Gen.* **2013**, *464–465*, 339–347. [CrossRef]
21. Marchi, A.J.; Apesteguía, C.R. Impregnation-induced memory effect of thermally activated layered double hydroxides. *Appl. Clay Sci.* **1998**, *13*, 35–48. [CrossRef]
22. Klemkaite, K.; Prosycevas, I.; Taraskevicius, R.; Khinsky, A.; Kareiva, A. Synthesis and characterization of layered double hydroxides with different cations (Mg, Co, Ni, Al), decomposition and reformation of mixed metal oxides to layered structures. *Cent. Eur. J. Chem.* **2011**, *9*, 275–282. [CrossRef]
23. Gao, Z.; Sasaki, K.; Qiu, X. Structural Memory Effect of Mg-Al and Zn-Al layered Double Hydroxides in the Presence of Different Natural Humic Acids: Process and Mechanism. *Langmuir* **2018**, *34*, 5386–5395. [CrossRef] [PubMed]
24. Dos Santos, R.M.M.; Gonçalves, R.G.L.; Constantino, V.R.L.; da Costa, L.M.; da Silva, L.H.M.; Tronto, J.; Pinto, F.G. Removal of Acid Green 68:1 from aqueous solutions by calcined and uncalcined layered double hydroxides. *Appl. Clay Sci.* **2013**, *80–81*, 189–195. [CrossRef]
25. Ni, Z.-M.; Xia, S.-J.; Wang, L.-G.; Xing, F.-F.; Pan, G.-X. Treatment of methyl orange by calcined layered double hydroxides in aqueous solution: Adsorption property and kinetic studies. *J. Colloid Interface Sci.* **2007**, *316*, 284–291. [CrossRef] [PubMed]
26. Hobbs, C.; Jaskaniec, S.; McCarthy, E.K.; Downing, C.; Opelt, K.; Güth, K.; Shmeliov, A.; Mourad, M.C.D.; Mandel, K.; Nicolosi, V. Structural transformation of layered double hydroxides: an in situ TEM analysis. *npj 2D Mater. Appl.* **2018**, *2*, 4. [CrossRef]

27. Wong, F.; Buchheit, R.G. Utilizing the structural memory effect of layered double hydroxides for sensing water uptake in organic coatings. *Prog. Org. Coatings* **2004**, *51*, 91–102. [CrossRef]
28. Theiss, F.L.; Ayoko, G.A.; Frost, R.L. Removal of boron species by layered double hydroxides: A review. *J. Colloid Interface Sci.* **2013**, *402*, 114–121. [CrossRef] [PubMed]
29. Silva, I.G.N.; Rodrigues, L.C.V.; Souza, E.R.; Kai, J.; Felinto, M.C.F.C.; Hölsä, J.; Brito, H.F.; Malta, O.L. Low temperature synthesis and optical properties of the R_2O_3:Eu^{3+} nanophosphors (R^{3+}: Y, Gd and Lu) using TMA complexes as precursors. *Opt. Mater.* **2015**, *40*, 41–48. [CrossRef]
30. Marappa, S.; Radha, S.; Kamath, P.V. Nitrate-intercalated layered double hydroxides - Structure model, order, and disorder. *Eur. J. Inorg. Chem.* **2013**, *2013*, 2122–2128. [CrossRef]
31. Mokhtar, M.; Inayat, A.; Ofili, J.; Schwieger, W. Thermal decomposition, gas phase hydration and liquid phase reconstruction in the system Mg/Al hydrotalcite/mixed oxide: A comparative study. *Appl. Clay Sci.* **2010**, *50*, 176–181. [CrossRef]
32. Valente, J.S.; Rodriguez-Gattorno, G.; Valle-Orta, M.; Torres-Garcia, E. Thermal decomposition kinetics of MgAl layered double hydroxides. *Mater. Chem. Phys.* **2012**, *133*, 621–629. [CrossRef]
33. Zhao, X.; Zhang, F.; Xu, S.; Evans, D.G.; Duan, X. From layered double hydroxides to ZnO-based mixed metal oxides by thermal decomposition: Transformation mechanism and UV-blocking properties of the product. *Chem. Mater.* **2010**, *22*, 3933–3942. [CrossRef]
34. Millange, F.; Walton, R.I.; O'Hare, D. Time-resolved in situ X-ray diffraction study of the liquid-phase reconstruction of Mg–Al–carbonate hydrotalcite-like compounds. *J. Mater. Chem.* **2000**, *10*, 1713–1720. [CrossRef]
35. Silva, I.G.N.; Morais, A.F.; Brito, H.F.; Mustafa, D. $Y_2O_2SO_4$:Eu^{3+} nano-luminophore obtained by low temperature thermolysis of trivalent rare earth 5-sulfoisophthalate precursors. *Ceram. Int.* **2018**, *44*, 15700–15705. [CrossRef]
36. Silva, I.G.N.; Cunha, C.S.; Morais, A.F.; Brito, H.F.; Mustafa, D. Eu^{3+} or Sm^{3+} -Doped terbium-trimesic acid MOFs: Highly efficient energy transfer anhydrous luminophors. *Opt. Mater.* **2018**, *84*, 123–129. [CrossRef]
37. Lima, S.A.M.; Sigoli, F.A.; Davolos, M.R.; Jafelicci, M. Europium(III)-containing zinc oxide from Pechini method. *J. Alloys Compd.* **2002**, *344*, 280–284. [CrossRef]
38. Kumar, M.; Seshagiri, T.K.; Mohapatra, M.; Natarajan, V.; Godbole, S.V. Synthesis, characterization and studies of radiative properties on Eu^{3+}-doped $ZnAl_2O_4$. *J. Lumin.* **2012**, *132*, 2810–2816. [CrossRef]
39. García-Hipólito, M.; Hernández-Pérez, C.; Alvarez-Fregoso, O.; Martínez, E.; Guzmán-Mendoza, J.; Falcony, C. Characterization of europium doped zinc aluminate luminescent coatings synthesized by ultrasonic spray pyrolysis process. *Opt. Mater.* **2003**, *22*, 345–351. [CrossRef]
40. Liu, K.; You, H.; Zheng, Y.; Jia, G.; Zhang, L.; Huang, Y.; Yang, M.; Song, Y.; Zhang, H. Facile shape-controlled synthesis of luminescent europium benzene-1,3,5-tricarboxylate architectures at room temperature. *CrystEngComm* **2009**, *11*, 2622–2628. [CrossRef]
41. Silva, I.G.N.; Mustafa, D.; Andreoli, B.; Felinto, M.C.F.C.; Malta, O.L.; Brito, H.F. Highly luminescent Eu^{3+}-doped benzenetricarboxylate based materials. *J. Lumin.* **2016**, *170*, 364–368. [CrossRef]
42. Silva, I.G.N.; Mustafa, D.; Felinto, M.C.F.C.; Faustino, W.M.; Teotonio, E.E.S.; Malta, O.L.; Brito, H.F. Low Temperature Synthesis of Luminescent RE_2O_3:Eu^{3+} Nanomaterials Using Trimellitic Acid Precursors. *J. Braz. Chem. Soc.* **2015**, *26*, 2629–2639. [CrossRef]
43. Karbowiak, M.; Zych, E.; H ls, J. Crystal-field analysis of Eu^{3+} in Lu_2O_3. *J. Phys. Condens. Matter* **2003**, *15*, 2169–2181. [CrossRef]
44. Brito, H.F.; Malta, O.M.L.; Felinto, M.C.F.C.; Teotonio, E.E.S. Luminescence Phenomena Involving Metal Enolates. In *PATAI'S Chemistry of Functional Groups*; John Wiley & Sons, Ltd.: Chichester, UK, 2010; pp. 131–184. ISBN 9780470061688.
45. Gollakota, P.; Dhawan, A.; Wellenius, P.; Lunardi, L.M.; Muth, J.F.; Saripalli, Y.N.; Peng, H.Y.; Everitt, H.O. Optical characterization of Eu-doped β-Ga_2O_3 thin films. *Appl. Phys. Lett.* **2006**, *88*, 221906. [CrossRef]
46. Trojan-Piegza, J.; Zych, E.; Hreniak, D.; Strek, W.; Kepinski, L. Structural and spectroscopic characterization of Lu_2O_3:Eu nanocrystalline spherical particles. *J. Phys. Condens. Matter* **2004**, *16*, 6983–6994. [CrossRef]

47. Zych, E. Concentration dependence of energy transfer between Eu^{3+} ions occupying two symmetry sites in Lu_2O_3. *J. Phys. Condens. Matter* **2002**, *14*, 5637–5650. [CrossRef]
48. Mustafa, D.; Silva, I.G.N.; Bajpe, S.R.; Martens, J.A.; Kirschhock, C.E.A.; Breynaert, E.; Brito, H.F. Eu@COK-16, a host sensitized, hybrid luminescent metal–organic framework. *Dalt. Trans.* **2014**, *43*, 13480–13484. [CrossRef] [PubMed]

Review

Layered Double Hydroxides in Bioinspired Nanotechnology

Giuseppe Arrabito [1,*], Riccardo Pezzilli [2], Giuseppe Prestopino [2,*] and Pier Gianni Medaglia [2]

[1] Dipartimento di Fisica e Chimica—Emilio Segrè, Università di Palermo, Viale delle Scienze, Edificio 17, 90128 Palermo, Italy

[2] Dipartimento di Ingegneria Industriale, Università di Roma "Tor Vergata", Via del Politecnico 1, 00133 Roma, Italy; riccardo.pezzilli@gmail.com (R.P.); medaglia@uniroma2.it (P.G.M.)

* Correspondence: giuseppedomenico.arrabito@unipa.it (G.A.); giuseppe.prestopino@uniroma2.it (G.P.)

Received: 26 May 2020; Accepted: 6 July 2020; Published: 11 July 2020

Abstract: Layered Double Hydroxides (LDHs) are a relevant class of inorganic lamellar nanomaterials that have attracted significant interest in life science-related applications, due to their highly controllable synthesis and high biocompatibility. Under a general point of view, this class of materials might have played an important role for the origin of life on planet Earth, given their ability to adsorb and concentrate life-relevant molecules in sea environments. It has been speculated that the organic–mineral interactions could have permitted to organize the adsorbed molecules, leading to an increase in their local concentration and finally to the emergence of life. Inspired by nature, material scientists, engineers and chemists have started to leverage the ability of LDHs to absorb and concentrate molecules and biomolecules within life-like compartments, allowing to realize highly-efficient bioinspired platforms, usable for bioanalysis, therapeutics, sensors and bioremediation. This review aims at summarizing the latest evolution of LDHs in this research field under an unprecedented perspective, finally providing possible challenges and directions for future research.

Keywords: origin of life; DNA; layer double hydroxide; synthetic biology; bioinspired devices; biosensors; bioanalysis

1. Introduction

Layered double hydroxides (LDHs) are an important class of two-dimensional (2D) layered materials belonging to the group of hydrotalcite-like (HT) compounds [1–4]. They are constituted by stacks of positively charged hydroxyl layers of bivalent ions (e.g., Ca^{2+}, Zn^{2+}, Mg^{2+} and Ni^{2+}) and trivalent metallic ions (e.g., Al^{3+}, Fe^{3+}, Cr^{3+} and In^{3+}). The isomorphic substitution of some of the bivalent metal ions by the trivalent metal ions forms a positive residual charge on the metal-hydroxide framework, which is in turn balanced by exchangeable interlayer anions to maintain the global electroneutrality. The general formula of LDHs is $[M_{1-x}{}^{2+}M_x{}^{3+}(OH)_2]^{x+}\cdot[A_{x/n}]^{n-}\cdot mH_2O$, where M^{2+} and M^{3+} are the divalent and trivalent metal ions, respectively; A^{n-} are inorganic or organic anions; m is the number of interlayer water; and $x = M^{3+}/(M^{2+} + M^{3+})$ is the layer charge density, or molar ratio [5]. It is worth pointing out that, in LDHs, the electric charge of the layers and of the interlayer ions is the opposite of that found in the vast majority of layered materials such as silicate clays (cationic clays), which feature negatively charged host layers and exchangeable cations in the interlayer spaces. Indeed, LDHs are also usually known as anionic clays [6]. Their unique physicochemical properties [7], i.e., biocompatibility, lamellar structure and compositional diversity, make them suitable for adsorption processes of a large variety of molecules, ranging from organic molecules to biomacromolecules. LDHs also play a prominent role in photocatalysis [8] and, more recently, in many life sciences related applications [9], especially for biosensors [10], drug delivery [11] and tissue bioengineering [12]. In the last years, the application of LDHs and their composites in biological, chemical and environmental processes have been extensively reviewed [13–16].

In particular, the role of LDHs as catalysts in relevant organic chemistry reactions, as photocatalytic centers, and their emerging application in cellular biology were extensively summarized in a previous review from our group [5]. Notably, the well-investigated chemistry- and biology-related applications of LDHs summarized in that review article highlight the conspicuous studies focused on the synthesis, characterization and applications of LDH-based systems. In fact, similar to ZnO-materials [17,18], LDHs can be synthesized under mild conditions [19] following the coprecipitation method, which consists in the addition of a basic substance to an aqueous solution containing the salts of two different metals, namely M^{2+} and M^{3+}, leading to the precipitation of the metal hydroxides, with the subsequent formation of LDHs [20–22]. Such synthetic approach can be leveraged to produce LDH materials on demand, with the desired chemical composition, aggregation state and particle size [15]. The latter can be tuned ranging from nanoclusters [23] to micron scale [24]. Notably, highly porous and highly dispersed LDHs with high surface area allow the full potential of LDH-based compounds to be efficiently exploited in many applications, such as heterogeneous catalysis and catalyst supports. Single layer LDHs are also ideal building blocks for functional assembly. However, the LDHs synthesized by conventional coprecipitation methods tend to easily aggregate, due to electrostatic interactions, interlamellar hydrogen bonding networks and their hydrophilic nature [25], thus resulting in low surface area and large particle size. To overcome these issues, many research efforts have optimized down to single- or few-layer nanosheets. Delamination of LDHs into single layers in polar solvents, such as formamide, butanol and acrylate, is a well-known process, as extensively reviewed by Wang and O'Hare [26]. In this context, it is worth mentioning that carbonate (CO_3^{2-}) contamination during both synthesis and post-treatment processes complicates LDH delamination, to such an extent that the direct exfoliation of a CO_3^{2-}-LDH is believed to be almost impossible [27]. The conventional approach to overcome this consists in the development of anion-exchange protocols to replace carbonate ions with other intercalating anions featuring weaker electrostatic interactions with the LDH structure [25,28]. Recently, a direct exfoliation method of CO_3^{2-}-LDHs, which avoids anion-exchange reactions, using liquid phase delamination via ultrasonic tip has been demonstrated [29]. Ecofriendly delamination approaches constituted by washing/stirring the LDH dispersion in decarbonated water have also been reported [30–32]. Isolation and recovery of delaminated LDHs without aggregation is a big challenge hindering practical application and commercial exploitation. In the last years, with the aim to increase LDH dispersion and organophilicity and to fabricate stable delaminated LDH dry powders, O'Hare and coworkers developed post-treatment routes for LDHs using both aqueous miscible organic (AMO-LDHs) and aqueous immiscible (AIM-LDHs) organic solvents, yielding highly porous, highly dispersed LDH powders of exfoliated nanosheets [33,34]. Some experimental protocols also permit films of interconnected LDH nanoplatelets to be grown onto aluminum surfaces as hierarchically nanostructured coatings by soaking aluminum thick foils in an aqueous solution containing a soluble zinc salt [35–39]. In this case, the trivalent ions (Al^{3+}) are provided by the sacrificial aluminum foil, whose role is both reactant and substrate, and ultimately allows the LDH film adhesion to be improved. LDH-based materials have triggered conspicuous studies in the field of organic chemistry, mainly due to their excellent ability as heterogeneous catalysts for the preparation of fine chemicals, intermediates and valuable molecules, benefiting from the simplicity of their preparation from cheap and non-harmful precursors, and the low toxicity of their possibly produced decomposition products [5]. The high reactivity of LDHs is related to their layered structure which features a high anion-exchange capability [40–44], and to their high basicity, due to either structural hydroxyl anions in hydrated material or O^{2-}-M^{n+} pairs in the case of water-free calcined material [45]. A crucial feature is also represented by the metal ratio. In fact, the incorporation of specific metals in the octahedral layer can produce LDH compounds with unexpected and novel properties, as well as modulate the existing chemical reactivity. It is worth mentioning literature where the arrangement of cations within the LDH lattice is discussed. As a matter of fact, the ordering of cations is believed to have crucial effects on many of the physicochemical properties of LDHs, affecting the charge density of the metal hydroxide sheets, and the overall bonding, mobility, orientation and reactivity of the chemical species in the interlayer space and on the surface [46–49]. In particular, the M^{2+}-M^{3+} cation order is important for the catalytic activity of LDHs [50], and the different distribution of the M^{3+} cation in the M^{2+} matrix regulates interlayer anionic charge and, in turn, intercalation reactions [51].

The intriguing catalytic properties and the high compatibility with biological systems of LDHs continuously open new research directions. Among them, up-to-date research topics of interest both for fundamental research and applications come from the presumed role of LDHs in the evolution of natural systems and, in turn, from the realization of a novel class of LDH-based life-inspired devices. As a matter of fact, the path permitting a complete understanding of the role of LDHs within the origin of life is still unclear and needs more research. However, some studies point out their possible role in this exciting research field. In fact, cationic clays and LDHs [52,53] might have played an important role for the origin of life on planet Earth, having triggered the formation of life-relevant prebiotic settings through the evolution of molecular systems, leading to the formation of polymerized molecules and finally to the emergence of life [54]. In a pre-biological Earth model, the ion exchange properties arising from the positively charged sites in LDHs allowed the latter to concentrate amino acids, and to safely promote biopolymer formation, offering also protection from the effects of UV irradiation. A mechanism of prebiotic information storage and transfer in LDH matrices based on replication and conservation of M^{2+}-M^{3+} cation arrangements was also proposed by Greenwell and Coveney [55]. In a sort of reverse evolutionary process, in the recent years, learning from nature, researchers have conceived the concepts of biomimetism and bioinspiration [56] in nanomaterials science to achieve unprecedented advanced functionalities based on designs that referred to structures in living creatures [57–61]. Under this exciting perspective, the most recent studies in the field of LDHs are striving to recapitulate life-like scenarios in the construction of hybrid LDH-based platforms, mimicking compositions or structures of natural species (e.g., biological macromolecules and microorganisms), with the aim to further improve their efficiency and their sustainability in the context of the natural environment [62,63]. Indeed, the low cost, biocompatibility and extraordinary adaptability of LDHs in many different purposes have allowed the effective application of bioinspired synthesis routes to LDH-based assembly of nanostructured materials and devices, which combine the eco-friendly nature of these materials and their versatility in the context of materials science. Another intriguing strategy to achieve bio-functionalization consists in associating biological species (e.g., amino acids, nucleosides, oligo nucleotides, DNA, proteins and enzymes) to LDH host matrices, as recently reviewed by Forano et al. [15] for LDH-biohybrids based on immobilization of bacteria and algae.

Motivated by the emerging role of bioinspired nanomaterials research, we felt the need to review the role of LDHs in this scenario, by investigating the fundamental science aspects and the possible emerging nanotechnology-related applications. At first, the present review provides an insight into the link between the role of LDHs in the origin of life, in particular focusing on their catalytic activities that promoted the formation of pre-biotic molecules and the possibility to fabricate artificial LDHs-based artificial compartments to mimic prebiotic platforms. These studies are speculative, mostly driven by the willingness to explore basic aspects of LDHs in their possible but still not completely clear role within the topic of origin of life. Secondly, this review updates and extends our previously review article [5], covering different aspects and, more in general, aiming at bridging fundamental science with new approaches that may support the design of LDH-based systems for the fabrication of life-like and life-inspired devices, especially in the fields of composites and coatings, sensors, life-inspired catalysis and bioremediation. To guide the reader, this review is divided into the following sections: the role of clays in the origin of life (Section 2), LDH interaction with biomolecules (Section 3), LDH-based compartments fabrication (Section 4), LDH-bioinspired devices assembly (Section 5) and finally the conclusions and perspectives (Section 6).

2. The Role of Clays in the Origin of Life

The origin of life (or abiogenesis) is among the most intriguing questions in science [64]. Although it is not possible to know all the details of this process, some scientific hypotheses attempt to explain the synthesis of the molecules which were necessary for the emergence of primitive forms of life in planet Earth 3.8 billion years ago [65] as deriving from a reducing atmosphere (Miller's experiment [66]), meteorites impacts [67] or from metal sulfides in deep-sea hydrothermal vents, as explained in the famous article of Russel [68]. Recent studies from Das at al. [69] leveraged computational methods to shed light on the role of hydrogen cyanide (HCN) as a source of ribonucleic acid (RNA) and protein precursors. The authors also argued that just the interaction of HCN with water would have sufficed to trigger a series of reactions leading to the life-essential precursors (nucleic acids and proteins), as reported in Figure 1.

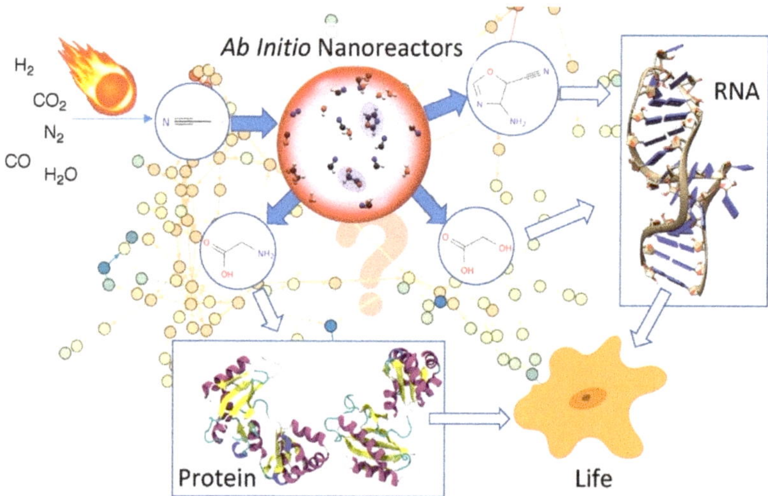

Figure 1. The synthesis of life relevant molecules (RNA, proteins) starting from inorganic materials and leading to the onset of chemical evolution. Figure reproduced from Ref. [70] https://pubs.acs.org/doi/10.1021/acscentsci.9b00832; Copyright (2019) American Chemical Society, further permissions related to the material excerpted should be directed to the ACS.

Once prebiotic molecules formed in water, cationic and anionic clays were also present on the surface of the Earth and could then interact with prebiotic molecules. In fact, whereas the probability that molecules could encounter in a 3D liquid environment is quite low, the solid mineral surfaces permitted to adsorbed molecules and permit them to interact and react [71]. According to some theories, radionuclide-induced defect sites in iron-rich minerals could have triggered the formation of prebiotic life [72]. Among all the possible hypothesis for the origin of life, one of the most relevant is that of hydrothermal vents [73], since they provided the possible evolutionary transition from geochemical to biochemical processes in absence of sun light. Hydrothermal vents are crevices in the seafloor from which geothermally heated water pours into ocean water. Among them, alkaline vents (such as the Lost City hydrothermal field) could have acted as ideal electrochemical flow reactors in the ocean during the Hadean period [74], in which the ocean waters were acidic since they were rich in CO_2. More specifically, within these hydrothermal vents, a pH gradient was formed between inorganic micropores barriers containing catalytic Fe(Ni)S minerals and the acidic ocean water, resulting in proton gradients similar to the proton-motive force required for carbon fixation in bacteria and archaea. The alkaline hydrothermal field provided serpentinization reactions that produced methane and hydrogen. Such pH energy input, in addition to the catalysis provided by the Fe(Ni)S minerals, triggered the reduction of CO_2 with H_2 and the relative production of life-essential organic molecules, such as hydrocarbons, finally leading to the first primitive cellular systems. Within this scenario, the question that can arise is then if LDHs were present in the hydrothermal vents. To this regard, Cairns-Smith and Braterman speculated that inorganic Fe^{2+}/Fe^{3+} LDHs (also defined green rust) [75] were at the sites of Archean oceanic hydrothermal vents, constituting one of the most-common components of early ocean sediments. The presence of this peculiar LDH can be ascribed to the lack of oxygen in the Archean atmosphere, permitting the presence of Fe^{2+} in aqueous environments and oxidation and precipitation as $Fe^{III}OOH$. As explained by Russell [76], the green rust might have played a fundamental role in the formation of prebiotic life. In particular, within the alkaline vent model, the green rust precipitated forming individual compartments that permitted the formation of ionic gradients and catalytic centers for endergonic reactions, aided by sulfides and trace elements acting as catalytic promoters that led to the formation of pre-biotic molecules.

In this context, as first speculated by Bernal [77], clays likely played an important role in promoting the molecular organization and the formation of polymerized biomolecules (nucleotides and amino acids), given their potentially large adsorption capacity, the ability to shield molecules against ultraviolet radiation, concentrate organic chemicals and catalyze chemical polymerization reactions. Following the seminal proposal by Cairns-Smith [78], who postulated a concept of "genetic takeover", according to which carbon-based life may have gradually evolved from organic–inorganic complexes of clays with organic molecules, some remarkable experiments demonstrated that clay minerals played an active role in the abiotic origin of life. For instance, as elegantly demonstrated by Ferris [79], montmorillonite clays could have a role in the polymerization of short RNA oligomers (up to 50 mer). Similarly, the fluctuation of temperature and humidity could have triggered the amino acids polymerization in the presence of clay minerals [80]. Finally, clays could have increased the rate of formation of vesicles from fatty acid micelles, along with the possibility to encapsulate clay particles inside the vesicles in addition to RNA molecules adsorbed on the clay surfaces, as demonstrated by Szostak et al. using montmorillonite clay [81]. Once formed, such vesicles could grow by incorporating fatty acids and divide, thus mediating vesicle replication through cycles of growth and division. In this scenario, since the first hypotheses by Kuma et al. [52] and Arrhenius [53], there are not so many studies in this field that regard the role of LDHs, even if they are positively charged clays that can interact with negatively charged molecules. LDHs might have played a role in selecting DNA over RNA for genetic information, due to its ability in enhancing the Watson–Crick hydrogen-bonding when intercalated within the LDH interlayers, in comparison to RNA [55,82]. Grégoire et al. [83] also explored the possible role of LDHs in early Earth chemistry, leading to the formation of peptides. Very recently, Vasti et al. demonstrated that LDHs strongly interact with model lipid membranes [84]; these experiments could also highlight the role of LDHs in partitioning within the membranes of primitive cellular-like compartments. In a very recent hypothesis paper, Bernhardt [85] proposed that LDH clays might have promoted the synthesis of prebiotic precursors and also played other roles in the origin of life, such as in the formation of the first lipid bilayers.

The fundamental reaction that triggered aerobic life was the photosynthetic oxygen evolution, in which two water molecules produce four electrons and four protons used for the photosynthetic processes and molecular oxygen [86]. This reaction is catalyzed by a manganese-containing cofactor contained in photosystem II known as the oxygen-evolving complex. Mimicking natural systems, studies have demonstrated that LDHs can constitute ideal low-cost catalytic materials for the oxygen evolution [87], in which both the M^{2+} and M^{3+} take part to the redox reactions, as reported in Figure 2. Some relevant applications of LDHs as catalysts for oxygen evolution reaction (OER) are discussed in Section 5.3.

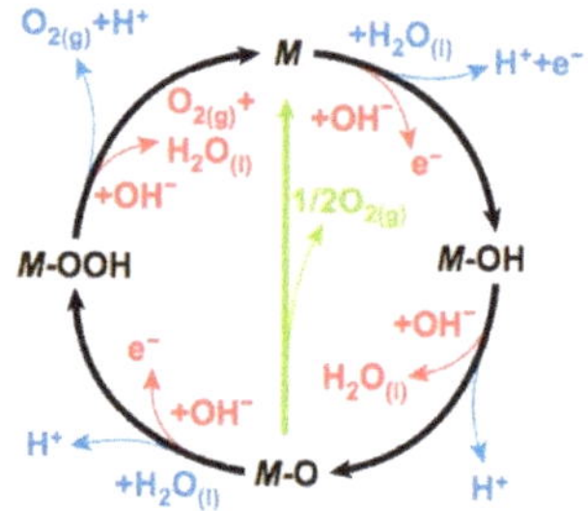

Figure 2. Simplified scheme for the oxygen evolution mechanism involving the formation of a peroxide (M–OOH) intermediate under acid (blue line) and alkaline (red line) reaction conditions. Reproduced from Ref. [87] under a Creative Commons Attribution 4.0 International License.

3. LDH Interaction with Biomolecules and Relevant Applications

3.1. Nucleic Acids

Nucleotides are the key molecules for the storage and transmission of genetic information [88]. The chemistry of their synthesis and polymerization is still a matter of debate [89]; however, based on the observation from Bernal [77], they could have had a role in concentrating and even acting as catalyst to form oligonucleotides. Moreover, they could have also been involved in protecting nucleotide against decomposition in presence of ionizing radiation [90]. Nucleic acids are composed of nitrogenous organic bases, pentose and phosphoric acids, are a typical polyanion within hydrogen-bonding groups, and can therefore enter inside the LDH interlayer, forming a nucleic acid/inorganic composite stabilized by electrostatic interactions. Inspired by the ability to protect and store nucleotides and oligomers, DNA-LDH hybrids have found remarkable applications in the field of gene delivery to cells. The reason for this was investigated by Li et al. [91], who employed dissipative particle dynamic simulations for elucidating the interaction with cellular membranes at different levels (i.e., total or partial penetration vs. anchorage—see Figure 3), as a function of the lateral size and layer number. These results shed light on the very important role of layer thickness—i.e., the lower is the thickness, the higher the probability that DNA-LDHs hybrids can be internalized, and hence serve as gene delivery tool.

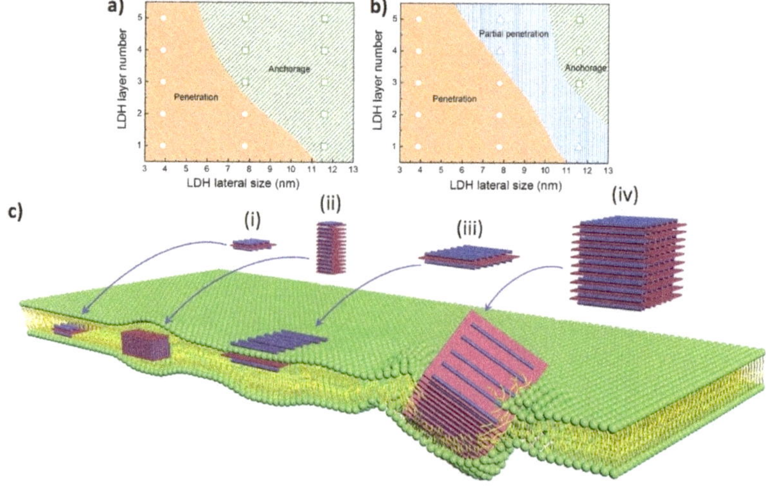

Figure 3. Interaction of DNA-LDH hybrids of different lateral layer sizes and thicknesses with model

cellular membranes. Relation between the LDH lateral size and LDH layer number at two different time points: 6.4 μs (**a**); and 32 μs (**b**). (**c**) Effect of the size of the DNA-LDH hybrids on the internalization. Reproduced from Ref. [91]. Copyright (2020), with permission from Elsevier.

3.2. Phospholipids

Cationic clays can effectively interact with living cells producing hybrid half inorganic viable structures, as demonstrated by Murase and Gonda [92] and Konnova et al. [93] in previous seminal studies on montmorillonite and hallosite clay nanotubes, respectively. Indeed, it was clarified that such clays can interact with phospholipids via electrostatic interactions [92], giving rise to apparently ordered phospholipids self-assembled layers, which can lead to *"armored cells"* if exposed to living cell membranes [93]. Notably, in the case of LDHs, the structural analysis from Itoh et al. [94] interestingly demonstrated, by means of spectroscopic and crystallographic characterizations, that stearate ions can form ordered bilayer structures within MgAl-LDH layers with a tilting angle of approximately 29° from the right angle of the layers (Figure 4).

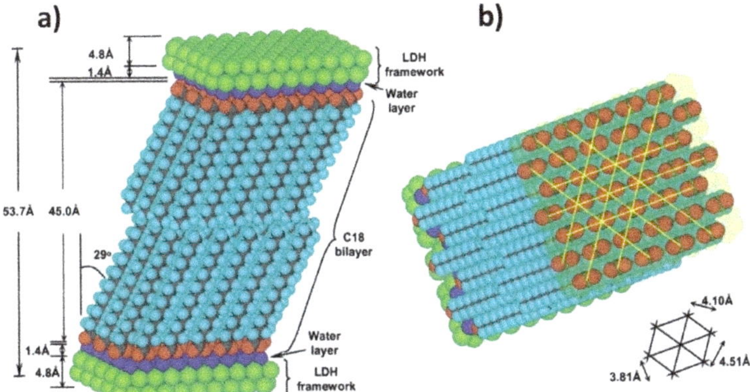

Figure 4. Scheme of stearate ions ordered layer assembly into LDH: (**a**) tilted bilayer, and (**b**) packing of stearate ions in the LDH framework. Reprinted with permission from Ref. [94] Copyright (2003) American Chemical Society.

3.3. Amino Acids

Amino acid intercalation inside LDHs is also a well-known topic [95]. An interesting application of LDHs-intercalated with amino acids is constituted by drug delivery, specifically because amino acid molecules provide the interaction sites to effectively improve drug-loading [96]. Apart from the interest in elucidating the structural features, the interaction between anionic clays and amino acids is very interesting and may offer an insight into the active role of LDHs in the polymerization of amino acids, in addition to the well-known characterization of the intercalation [95]. Differently from nucleotides, for which the prebiotic synthesis route is still speculative [89], there is a little more evidence of the possibility that amino acids were present in prebiotic systems. Accordingly, both cationic clays and LDHs might have facilitated the polymerization of amino acids, thanks to the possibility to absorb them via their negative C-termini. Interestingly, as argued by Paecht-Horowitz [97], the former can catalyze polymerization only when the amino acid is small enough to enter the interlayer space of the clay. As for the latter, the computational study from Erastova et al. [54] demonstrated the possibility that LDHs could be a functional system for amino acid adsorption and polymerization on the positively charged LDH surface. Intriguingly, all the amino acids they investigated (alanine, aspartate, histidine, leucine, lysine and tyrosine) showed an increase in adsorption upon dehydration, with the only exception of lysine. Dehydration, in turn, decreases the interlayer spacing, in accord with our experimental findings [37], favoring both the crowing of amino acids and the alignment of the N- and C-termini, and also permitting the formation of a peptide bond (see Figure 5). This reaction leads to the loss of charged group, which, in turn, allows for the introduction of a new amino acid to an adjacent site, where it then can react with the peptide's C-terminus. The growing peptide chain remains tethered at the LDH surface via the C-terminus of the latest amino acid added. Differently to nucleic acids, amino acids do not form strong association on the LDH surface. Even more, the authors observe that, within the clay layers, peptide chains should be able to undergo hydrophobic collapse, an essential mechanism of protein folding, in perfect resemblance to the ribosome systems.

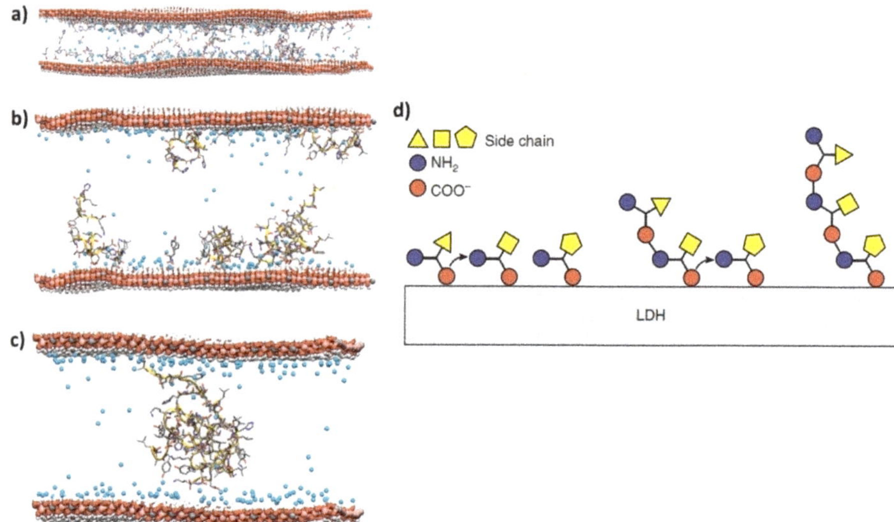

Figure 5. Amino acids on LDHs. Ensembles of (**a**) amino acids; (**b**) short; and (**c**) long peptides intercalated into LDH interlayers via their C-terminal. (**d**) Proposed mechanism for peptide bond formation inside LDH interlayers. Reproduced from Ref. [54] under the terms of the Creative Commons CC BY license.

3.4. *Carbohydrates and Cellulose*

Carbohydrates are among the most abundant biomolecules in nature. They are formed by plants through photosynthesis and serve as energy sources, structural building blocks and biorecognition elements in organisms. They can exist as oligomers and polymers; the latter are formed by linking single monomeric units with glycosidic bonds. LDHs have been leveraged as systems to prepare biomimetic composites with carbohydrates molecules showing enhanced mechanical properties, mimicking the natural structure of nacre, a biological organic–inorganic composite material produced by some mollusks and formed by the ordered stacking of calcium carbonate and crosslinked protein, resulting in very good mechanical properties (high tensile strength, about 150 MPa). To this aim, it is worth mentioning some previous studies on LDH composites with carboxymethyl cellulose (CMC) [98] and cellulose nanofibrils [99], which exhibited improved mechanical properties. As reported in Figure 6a, the interaction of CMC with LDHs is purely electrostatic, being driven by the negative electrostatic charge of the CMC that permits its insertion into MgAl-LDH galleries. In addition, alginate was also used as composite with LDHs [100], following a bioinspired approach that mimicked the mechanical properties of nacre. In particular, NiAl-LDH is prepared by coprecipitation and is subsequently exfoliated mechanically; after that, it is mixed with an alginate solution permitting to obtain alginate-coated LDH, which is subsequently crosslinked with Ca^{2+} (see Figure 6b). Importantly, the ionic crosslinking of alginate by calcium ions along with the hydrogen bonds between NiAl-LDH and alginate, allows the nacre film to be mimicked, achieving a tensile strength as high as 194 MPa, in addition to transparency higher than 70% in the visible light range.

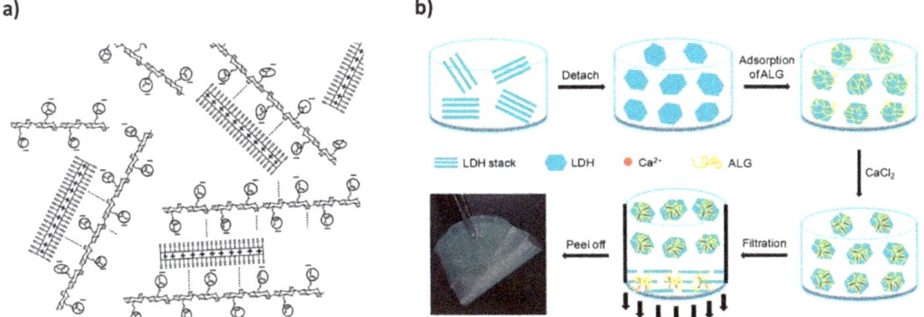

Figure 6. (**a**) Stabilization of the MgAl-LDH-CMC hybrid particle by electrostatic-driven interactions. Reprinted from Ref. [98]. Copyright (2014) with permission from Elsevier. (**b**) Production process of hybrid NiAl-LDH/Alginate-Ca^{2+} film. Reproduced from Ref. [100]—Published by The Royal Society of Chemistry.

4. Life-like Clays Based Artificial Compartments

The fabrication of artificial compartments that mimic life-like properties is a topic of large interest and has been boosted by the development of new microfluidic platforms containing droplet-based and vesicle-based artificial cells [101,102], and by the employment of printing methodologies, as shown also by our group [103,104]. The systems can be realized at solid surfaces mimicking the processes of DNA condensation [105] or in membrane-enclosed liquid compartments [106,107]. Importantly, such artificial biosystems permit to recapitulate some of the features of primitive sub-micron sized protocells, such as effect of droplet size on the molecular behavior, molecular confinement, molecular crowding, and liquid–liquid phase separation. In the context of clay-based compartments, the studies are mainly based on the fabrication of 100–10 μm sized colloidosomes or Pickering emulsions [108], showing intriguing membrane permeability [109,110], and even some life-like behavior, such as enzyme-powered motility [111]. Regarding the fabrication of synthetic compartments, printing methods are gaining increasing importance [112,113], given their rapidity, low-cost and adaptability to biomolecular, organic and inorganic inks [114] or oil droplets [115], opening up the possibility to print artificial cell-like systems with higher degree of flexibility and possibility to assemble complex predesigned droplet networks. The development of compartments containing LDHs is still at its infancy. However, the possible application of LDH-based compartments would lead to the mimicking of nanocomposites, as for instance mimicking the structure and the mechanical properties of the high strength seashell nacre. A fundamental study by Zhang and Evans [116] explored the dynamic assembly of LDH platelet-rich droplets drying onto solid surfaces, taking into consideration the interplay of capillary flows, colloidal stability and pH on the final CoAl-LDH composite deposited on a silicone coated paper. They prepared pH 3 and 10 dispersions resulting in well-dispersed and flocculated CoAl-LDH platelet suspensions, respectively. The pH-dependent colloidal stability of CoAl-LDH dispersions in the pH range from 3 to 10 was discussed by the same authors in another report [117], showing that the suspension was most stable at pH 3, due to the protonation of the OH groups resulting in repulsion forces between LDH platelets, positively charged on both basal and edge surfaces. It is worth pointing out that both stronger acid and basic conditions (i.e., pH < 3 and pH > 10) lead to LDH structural dissolution [118,119]. In the case of deposition of droplets at pH 3 (where

LDHs are well dispersed), a central hole was observed in the LDH deposit (see Figure 7a), along with a ring of higher thickness at 0.6 r from the center (where r is the radius of the droplet). The authors explained this structure by considering possible capillary effects between LDH platelets. A completely different situation occurred in the case the LDH platelets were deposited as flocculated at pH 10. In this case, the suspension produced a flatter structure due to the lack of capillary flows. As the authors state, these results could be applied to the understanding of droplet morphologies produced by inkjet printing [120], with the aim to optimize both the flow processes in a drying drop and the stacking interactions among LDHs, ultimately permitting to obtain ordered stacking and high uniformity in the LDH coverage. An elegant demonstration of LDH printing was given by Zhu et al. [121] who leveraged exfoliation processes of CoAl-LDH and NiAl-LDH nanosheets in formamide to obtain printable LDH formulations, leading to stable LDH-based solid films (see Figure 7b), in which the LDH stacking is able to mimic the structures of nacre, potentially leading to good mechanical properties. Notably, the authors also demonstrated the possibility to tune the optical and mechanical properties of the LDH films by tuning the anions that can be stacked into the galleries.

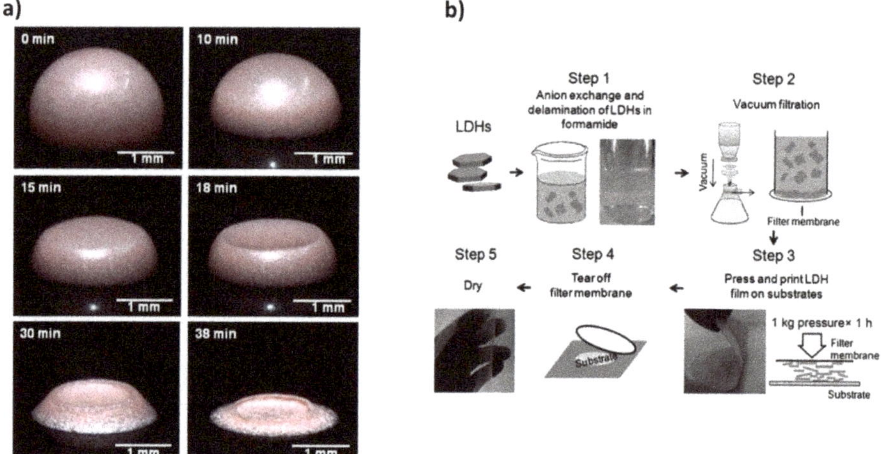

Figure 7. (**a**) Photographs of a drying drop of 2.0 vol% LDH dispersion at pH 3 as a function of drying time. Reprinted from Ref. [116]. Copyright (2013) with permission from Elsevier. (**b**) Schematic diagram of the preparation of LDH thin films and subsequent printing on substrates. Reproduced from Ref. [121] with permission from The Royal Society of Chemistry.

5. LDH-Bioinspired Devices

5.1. Applications in Composites and Coatings

Billions of years of evolution have produced optimized natural materials with outstanding properties resulting by the combination of organic and inorganic elements, which inspire scientists and engineers to design artificial materials. In recent years, the development of bioinspired composites has become an exciting direction for the fabrication of novel multifunctional materials, which exhibit a combination of outstanding mechanical, fire-shielding, optical and oxygen barrier properties. In this field, LDHs are receiving increasing attention since they offer numerous advantages over other inorganic layered materials, also in the field of nanocomposites [122,123]

Shu et al. [124] reported on a nacre-like film based on heparine/NiAl-LDH with a layered nano/microscale-hierarchical structure fabricated by the vacuum-filtration method (Figure 8a). The realized hybrid film showed a reduced elastic modulus ($Er \approx 23.4$ GP) and a hardness ($H \approx 0.27$ GPa) remarkably higher than those of other reported polymer/LDH composites. In addition to such outstanding mechanical characteristics, the LDH composite exhibited interesting UV-blocking, flame retardant, and heat-shielding properties. In a successive work, the same group [125] studied a hierarchical structure based on a CoAl-LDH/Poly(vinyl alcohol) film, prepared for the first time by bottom-up layer by layer assembly (Figure 8b). A combination of high tensile strength and elastic modulus is obtained through an appropriate CoAl-LDH aspect ratio, which well compares to those of nacre and lamellar bone.

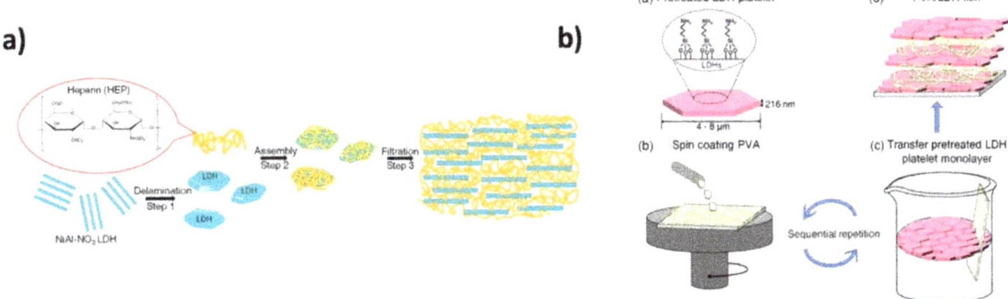

Figure 8. (**a**) Fabrication of the artificial nacre-like HEP/LDH film by three steps: (i) LDH delamination; (ii) assembly of exfoliated LDHs and HEP; and (iii) HEP-LDH hybrid building blocks are aligned by self-assembly induced by vacuum filtration. Reproduced with permission from Ref. [124]. Copyright year (2014) American Chemical Society. (**b**) Fabrication of artificial multilayered PVA/LDH hybrid films by: (i) modification of the surface of LDH platelets with slightly hydrophobic amine-terminated APTES; (ii) spin coating of a layer of PVA; (iii) gradual formation of Langmuir film at the air–water interface; and (iv) sequential repetition to fabricate multilayer films. Reproduced with permission from Ref. [125]. Copyright year (2014) American Chemical Society.

Meng et al. [126] reported another example of brick-and-mortar structure to enhance the mechanical performance of a composite. They developed a PTFE/MLDH (modified ZnAl-LDH)/GF (glass fiber) composite inspired by layer–layer structure and prepared via the multi-layered dipping method, which consists of GF acting as a reinforcement, PTFE acting as matrix and MLDH acting as the layer–layer structure. The results show that the tensile strength and the strain value of PTFE/MLDH/GF composites, with MLDH content around 1.6 wt%, increased to 163.57 MPa and 8.33%, respectively. These values resulted better than those of pure PTFE/GF composites, showing a tensile strength of 112.72 MPa and a fracture strain of 4.99%.

Interestingly, the use of LDH-based compounds is not only limited to mechanical properties, as mentioned above. Liu et al. [127] demonstrated that the combination of two or more diverse materials in a unique composite is a winning strategy, realizing a durable superhydrophobic sponge for oil and organic collector with magnetic-responsive and fire retarding properties. To achieve this result, by means of a two-step process, they combined polydopamine (PDA), ZnAl-LDH, Fe_3O_4 nanoparticles and n-octadecyl mercaptan (OM), acting, respectively, as a "bio-glue", a fire retardant agent, magnetic material and hydrophobic reagent. First, PDA-modified ZnAl-LDH and Fe_3O_4 were anchored on the skeleton of polyurethane (PU) sponge by self-polymerization. Next, OM containing thiol groups were covalently combined with PDA by means of Michael addition reaction. The resulting sponge (Figure 9a,b) exhibited remarkable hydrophobicity (water contact angle of 163°) and highly oleophilic (oil contact angle of 0°). As a result, it could absorb various kinds of oils and organics up to 53.6 times of its own weight, and the absorbed oils could be collected through a simple squeezing process (Figure 9c). In addition, the presence of magnetic material, Fe_3O_4, allows easy drive by an external magnetic bar, thus reducing any contaminations.

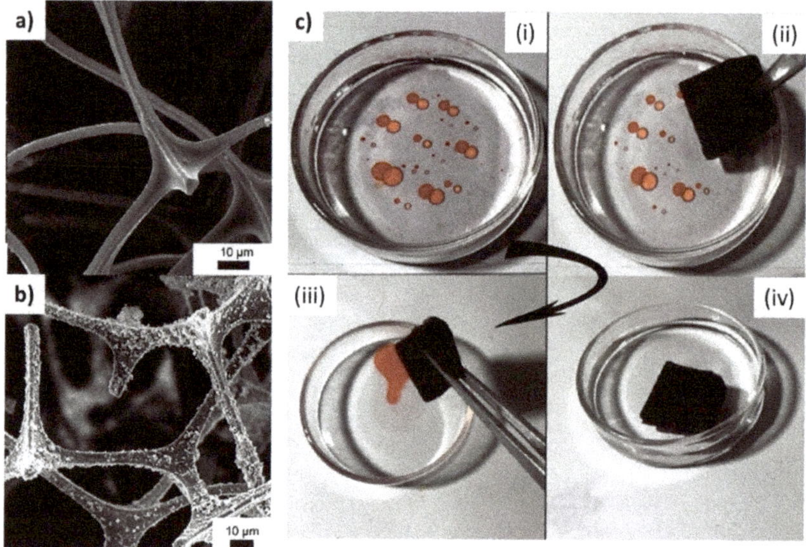

Figure 9. (**a**) SEM images of pristine PU sponge, (**b**) SEM images of PU-LDH-Fe_3O_4-PDA-OM sponge, (**c**) A series of photographs for the process of absorption and collection toluene (dyed with Sudan III) from the surface of water. Reprinted from Ref. [127] with permission from Elsevier.

The composite materials are also considerably advantageous in the realization of coating films in a large number of fields, leading to an increase in performance and multifunctionality. For example, Wang et al. [128] fabricated superhydrophobic MgAl-LDH coatings on medium density fiberboards with flame retardancy, obtained by the deposition of polydimethylsiloxane (PDMS) and 1H, 1H, 2H, 2H perfluorodecyltrichlorosilane (FDTS)-modified LDH particles. The PDMS@FDTS-Mg/Al LDH coating exhibited superhydrophobicity with a water contact angle of 155° and self-cleaning property. The flame retardant character was evaluated by limiting oxygen index (LOI) and cone calorimeter test. LOI value of superhydrophobic MDFs increased by 60.4% as compared to that of the pristine MDFs, from 24.0 to 38.5. The peak heat release rate (PHRR) and total heat release (THR) of MDFs coated with PDMS@FDTS-Mg/Al LDH reduced by 24.7% and 11.2% as compared to MDFs alone. This result demonstrates that the presence of inorganic coating improved the flame retardancy of the MDFs. Wu et al. [129], inspired by the micro- and nanostructure of lotus leaves, or the Cassie state surface, developed a self-layered coating on magnesium alloy through a hydrothermal treatment, followed by radiofrequency (RF) magnetron sputtering of polytetrafluoroethylene (PTFE). A super-hydrophobic (water contact angle of 170°) and acid resistant surface is created on the alloy. Recently, Yu et al. [130] reported on the possibility to use an environment-friendly LDH-base coating as high oxygen barrier coating in flexible food packaging. The coating film is realized by the reconstruction of MgAl-LDH from MgAl-layered double oxide (LDO) in concentrated aqueous glycine solutions, resulting in a coating solution of LDH and poly(vinyl alcohol) (PVA). It is worth observing the key role of the aspect ratio of reconstructed LDH in the coating permeability performance. Indeed, the relative permeability (P/P0) of the LDH/PVA coated layer decreases as the aspect ratio of LDH nanosheets are increased, from 0.0065 with aspect ratio 87 ± 17 to 0.0038 at 336 ± 170. This is explained by the highest aspect ratio of LDH nanosheets, which leads to better alignment in the coated layer. Thus, through the fine control of aspect ratio, the coating film reached an oxygen transmission rate <0.005 $mL{\cdot}m^{-2}{\cdot}day^{-1}$ at a film thickness of 1175 ± 101 nm.

5.2. Nanogenerators and Physical Sensors

In the era of smart devices and bioinspired sensor networks, LDHs and related compounds have been demonstrated to play multiple roles, starting, for example, from green nanogenerators for realizing self-powered sensors. In this field, it is worth to mention the work by Cui et al. [131], and those by Sun et al. [132] and Tian et al. [133], who realized a water-driven triboelectric nanogenerator (WD-TENG) and a natural water evaporation (NWE) driven generator, respectively, to harvest energy from water, the most abundant substance on our planet. Briefly, in the first approach, the realization followed a bottom-up strategy, by directly growing a well oriented MgAl-LDH nanosheet network on a metal substrate (Figure 10a). To reduce the surface energy, a further surface modification was performed via crosslinking of fluorine contained silanes, which in turn changed the morphology of the LDH network into a flower-like shape (Figure 10b). The fabricated WD-TENG relied on triboelectric effect, collecting energy from water droplets through the functionalized LDH film as triboelectric layer. The working mechanism is based on a contact-electrification and electrostatic induction at the liquid–solid interface (Figure 10c,d). In the second approach, a NWE-driven generator (NWEG) was realized via a spray-deposition of NiAl-LDHs on a polyethylene terephthalate (PET) substrate [132] (Figure 10e). It operates by means of an NWE-driven gradient of water which flows across the naturally formed surface-charged nanochannels between the LDH flakes, which were evenly spaced on the PET surface, with an ordered layer-by-layer stacking on the substrate (average channel width less than 50 nm) (Figure 10f,g). The performance of the NWEG was reasonably supposed to be regulated by the surface charge density (d_c), the hydrophilic character and the presence of nanochannels or pores in the NG active layer, which are easily tunable intrinsic properties in the case of LDHs. In a successive work [133] the same group reported on the relationship between d_c and the NWEG performance by precisely tuning d_c of NiAl-LDHs in the range of 2.52–4.59 e/nm^2, by adjusting the molar ratio of Al^{3+} to Ni^{2+}.

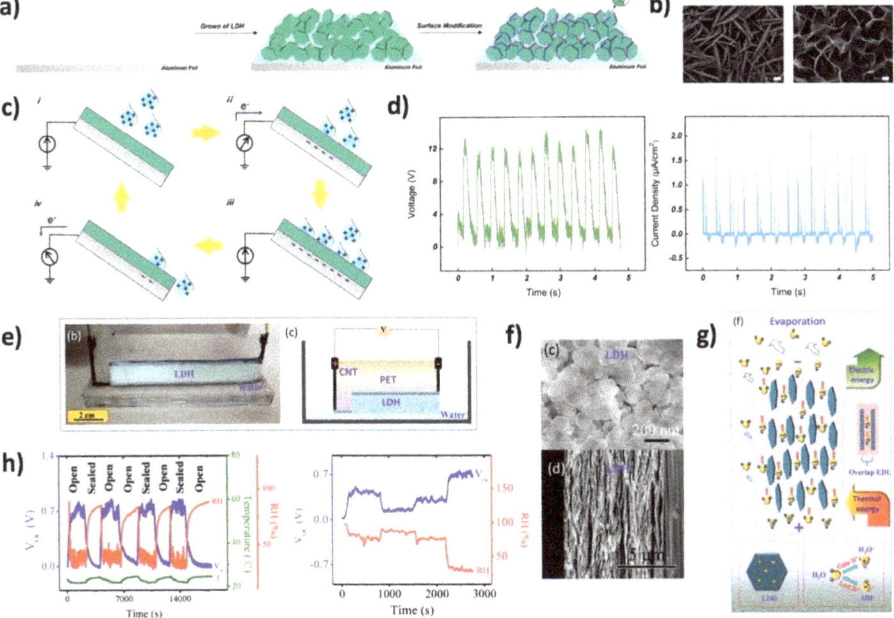

Figure 10. LDH-based water driven nanogenerators. (**a**) Schematic illustration of the fabrication process for the LDH-based triboelectric layer of the WD-TENG, showing (from left to right) the bare aluminum foil substrate and the MgAl-LDH nanosheet network before and after surface modification.

(**b**) Scanning electron microscope (SEM) images of the as grown (left) and surface modified (right) LDH film. (**c**) Schematic diagram of the modified LDH- (green) based WD-TENG collecting energy from water droplets. (**d**) Output voltage and current density of the WD-TENG under water droplets impacting. Reproduced from Ref. [131] with permission from Elsevier. (**e**) Photo (left) and schematic illustration (right) of the NiAl-LDH NWEG. (**f**) Top-view (top) and cross-sectional (bottom) SEM images of the NiAl-LDH film. (**g**) Schematic diagram of the working principle of the NWEG on a microscopic level. (**h**) (left) Evolution of the open circuit voltage (V_{OC}, blue line) in response to the variation of the relative humidity (RH, red line) due to periodic opening and sealing of the container. (right) Response of V_{OC} to periodic variations of RH between 96% and 28%. Reproduced from Ref. [132] with permission from Elsevier.

Another fundamental element in the fabrication of sensor systems is represented by the energy storage units. In a recent study, Xu et al. [134] reported on the realization of a coplanar asymmetric microscale hybrid device (MHD) supercapacitor based on MXene and CoAl-LDHs using a two-step screen-printing process. The LDH with its faradaic pseudo capacitance behavior was used as positive electrode material to enhance energy density and potential window. The resulting MHD exhibited very interesting electrical properties, such as outstanding cycling stability (92% retention of areal capacitance after 10,000 cycles), potential window extended to 1.45 V, and enhanced energy density (10.80 and 8.84 μWh cm^{-2} in 6 M KOH and with PVA-KOH, respectively). In a more recent work [135], a CoMn-LDH doped with polypyrrole (PPy) was used as positive electrode for asymmetric supercapacitor (ASC) with multilayer graphene as cathode, exhibiting strong electrochemical performance. This supercapacitor showed properties comparable to or better than those of similar devices recently reported, with good cycle performance (99.5% retention after 8000 cycles) and energy density as high as 29.6 Wh kg^{-1} at a power density of 0.5 kW kg^{-1}.

The application of LDH in the field of bioinspired sensors was successfully demonstrated by Ren et al. [136], who developed a skin-inspired sensor based on a multifunctional nanocomposite hydrogel consisting of sodium alginate/sodium polyacrylate/layered rare-earth hydroxide (LRH) (SA/PAAS/LRH), fabricated through a 3D printing system (Figure 11a,b). The device showed a promising multifunctional behavior, such as humidity-dependent electromechanical properties, a sensitivity to mechanical deformation, thanks to a strain-dependent conductivity (Figure 11c), and tunable fluorescence (Figure 11d), while maintaining the characteristics of transparency and stretchability indispensable for the realization of a device for human motion detection. Such versatility makes it a good candidate for future soft wearable equipment. The use of LDH was also demonstrated to drastically improve the performance of composite materials for sensors. For example, Beigi et al. [137] reported on a modified ionic polymer metal composite (IPCM) humidity-sensor based on a nafion polymeric matrix doped with CoAl-LDH nanoparticles, with a significant improvement of sensitivity and responsivity due to the intrinsic hydrophilic characteristic of LDH.

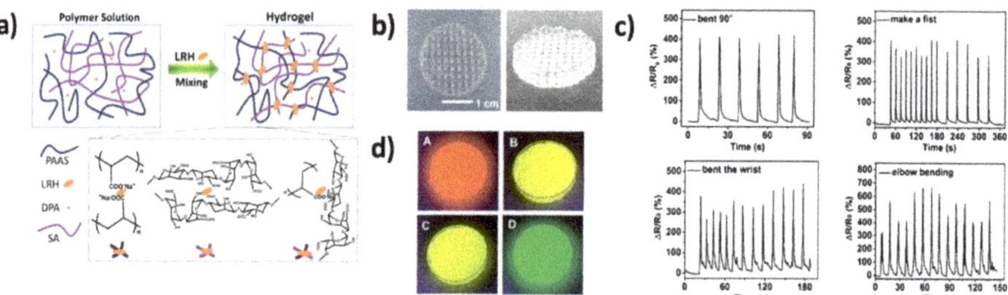

Figure 11. Skin-inspired multifunctional device. (**a**) Scheme of the preparation of SA/PAAS/LRH hydrogel. (**b**) Top and side photos of the printed hydrogel wafer with a grid structure. (**c**) Response to

sensor strain. First row: Sensor attached on the forefinger of a human hand responding to bending angles from 0° to 90° (left); and sensor attached on the back of the hand responding when making a fist (right). Second row: Relative changes of the hydrogel films in sensing wrist bending (left); and elbow bending (right). (**d**) Fluorescent colors of various hydrogels, with different rare-earth ratios, under 254 nm in wavelength ultraviolet irradiation. Reprinted with permission from Ref. [136] Copyright (2020) American Chemical Society.

5.3. Applications in Oxygen Evolution Reaction

The process of oxygen evolution reaction (OER), $4\,OH^- + \text{energy} \rightarrow O_2 + 2H_2O + 4e^-$, is one of the most critical steps of electrochemical water splitting, which is widely recognized as a promising and sustainable method to convert the intermittent electrical energy from the nature into stable and storable energy (hydrogen) [138,139]. However, OER, which is a multistep $4e^-$ process, is thermodynamically not favored and thus kinetically sluggish [140]. Indeed, an OER catalyst should be able to overcome both the activation energy barrier (E_a) and the standard free energy charge (ΔG^0 = 1.23 eV). In recent years, substantial research efforts have been devoted to develop novel non-noble metal-based active OER catalysts based on efficient, low-cost and earth-abundant substitutes for conventionally used precious metal compounds [141]. Different strategies have been conceived to improve the catalyst performance, which may be categorized on three scale levels [142]: (i) at the atomic-scale (e.g., alteration of oxidation state, doping, coordination and composition of metal composites); (ii) at the nano-scale, including different material combinations templated on nanostructures (e.g., nanowires, nanosheets and nanotubes), to increase OER activity by means of high surface area and number of active sites; and (iii) at the meso-scale, i.e., the creation of a porous supporting architecture to enhance mass transport to electrolytes and structural stability. As intriguingly noted in recent studies [142–144] such optimization problem presents many analogies with the demand of photosynthesis in the evolutionary development of plant leaves, like the need of maximized surface area to capture as much light as possible, and sufficient space between neighbors to promote good gas exchange and surface reactions. In addition the 1D hollow tubular structures under the leaves facilitate the transport of nutrients and water to each leaf [142].

In the last decade, LDHs have been proven to be highly active, cost-effective and durable OER catalysts, exhibiting electrocatalytic activity and stability for OER, comparable to or higher than commercial precious metal-based catalysts [145–147]. In their reports, the scholars leveraged all the above-mentioned optimization routes to improve OER activity of LDHs. In particular, in 2014, Lu et al. developed OER electrodes based on a three-dimensional (3D) porous film of vertically aligned NiFe-LDH nanoplates loaded on a nickel foam. Excellent OER performance was demonstrated, with a small onset overpotential (~230 mV), large anodic current density (30 mA cm^{-2}) and outstanding electrochemical durability, benefiting from the intrinsic high activity of the NiFe-LDH catalyst [148] and the unique 3D architecture, whose surface area was increased by the highly porous nickel foam. In the following, many research efforts were focused to improve the OER activity of LDHs (typically NiFe- and NiCo-LDHs), as increasing the number of active sites as well as increasing the activity of the individual active site. Song and Hu [149] and Liang et al. [150] demonstrated that liquid exfoliation of LDHs to single-layer nanosheets leads to greatly enhanced OER activity, due to an increase in the number of active edges sites and higher electronic conductivity, while preserving material composition. Next, Wang et al. [151] demonstrated that dry exfoliation of bulk CoFe LDHs into ultrathin LDH nanosheets through Ar plasma etching also resulted in the formation of multiple vacancies (including O, Co and Fe vacancies), thereby producing a dual effect to OER enhancement, due to the great number of exposed active sites in the 2D LDH nanosheets and their improved activity due to multiple vacancies. The dry-exfoliated CoFe-LDH performed very well in the OER with an overpotential of 266 mV to reach a current density of 10 mA cm^{-2} vs. 321 mV required for the untreated pristine CoFe-LDH. In a further report, the same group [152] dry-exfoliated bulk CoFe-LDHs in a N_2 plasma into edge-rich ultrathin nanosheets featuring, again, multiple vacancies, as well as nitrogen doping, which facilitated absorption of OER intermediates by altering the electronic density of the adjacent Co or Fe atoms.

The ultrathin N-doped CoFe LDH nanosheets loaded on Ni foam exhibited excellent OER performance, with an overpotential of 233 mV at a current density of 10 mA cm^{-2}. However, despite such atomic- to nano-scale optimization strategies, the OER performance of LDHs was limited by their intrinsically poor electrical conductivity. Some reports have proposed combining LDHs with conducive materials, e.g. carbon nanotubes [153], graphene [154], graphene oxide [155] and, more recently, silver [156] and CuO nanowires [157]. As for the latter, inspired by the monocot leaf structure in nature, Chen and coworkers conceived an engaging biomimetic nanoleaf based on ultrathin NiCo-LDH nanosheets in situ grown on Cu nanowires, forming a NiCo-LDH lamina featuring large electrochemical surface area (ECSA) and numerous active edge sites for OER reaction (see Figure 12). The CuO nanowires served as veins to support the LDH lamina, while providing mechanical support as well as improving the LDH conductivity, further increasing the OER activity. An enhanced OER performance was then achieved, significantly improved with respect to that of conventional NiCo-LDHs, with a quite small overpotential of 262 mV at 10 mA cm^{-2}, good stability and flexibility.

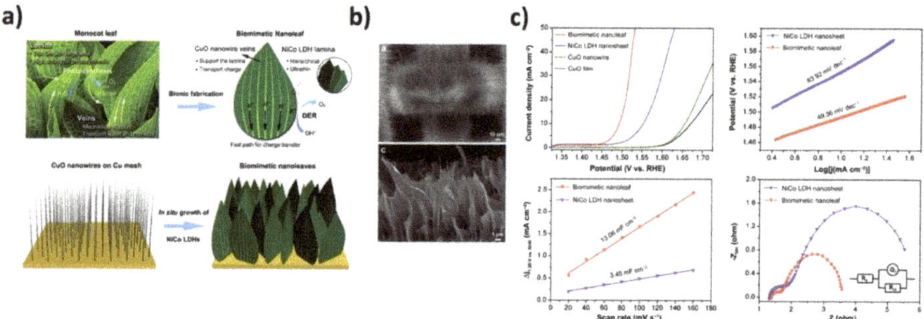

Figure 12. Biomimetic LDH nanoleaf for OER. (**a**) Schematic illustration of the biomimetic NiCo-LDH based nanoleaf, showing (top) a photograph of the monocot leaf and a sketch of the biomimetic nanoleaf, with indication of the OER process, and (bottom) the fabrication procedure, with in situ growth of NiCo-LDH on the CuO nanowires/Cu mesh substrate. (**b**) SEM images of the CuO nanowires on the Cu mesh (upper panel, scale bar 10 μm), and of the biomimetic nanoleaves (bottom) (scale bar 1 μm). (**c**) OER performance. First row: Linear sweep voltammetry (LSV) polarization curves (left); and corresponding Tafel plots (right). Second row: Capacitive current vs. scan rates (left); and Nyquist plots at an overpotential of 300 mV (0.1 Hz to 100 kHz) (right). Reprinted from Ref. [157] with permission from Elsevier.

5.4. Peroxidase-Like Activity

The peroxidase enzymes are widespread in natural systems from bacteria to plants and humans, given their fundamental role in decomposing hydrogen peroxide (H_2O_2), which is a toxin produced as a byproduct of oxygen during aerobic respiration processes. These proteins contain a heme group in their active site that uses hydrogen peroxide as the electron acceptor to catalyze oxidative reactions. Such electron transfer ability between reducing substrate and H_2O_2 can be conveniently mimicked by many metal oxide nanoparticles—e.g., iron oxides, cerium oxides, metal sulfides and carbon nanodots. In this regard, many reports have focused on the realization of hybrid structures in which LDHs increase the surface/volume ratio and favor the dispersion of the nanoparticles, thereby allowing the peroxidase activity to be enhanced, as demonstrated by Yang et al. with CoAl-LDH/MFe_2O_4 (M = Ni, Zn, Co) [158] or CoFe-LDH/CeO_2 hybrids [159], as well as core–shell Fe_3O_4/CoFe-LDH [160], finding applications in the field of analytical chemistry, especially for the determination of glucose, H_2O_2 and glucose and ascorbic acid, respectively. The mechanism of reaction is based on the electron transfer from the molecule 3,3′,5,5′-tetramethylbenzidine (TMB) to H_2O_2. In the presence of glucose oxidase enzyme, glucose is oxidized to gluconolactone and oxygen is reduced to H_2O_2. As a result, the higher is the glucose concentration, the higher is the concentration of H_2O_2 and, hence, the amount of oxidized TMB. Conversely, ascorbic acid is able to convert the oxidized form TMB to the reduced state. As a result, the obtained sensor is based on a colorimetric readout (see Figure 13a) and can be effectively employed in the field of low-cost glucose sensing.

Interestingly, LDH materials with intrinsic peroxidase activity were also engineered by simply introducing metal species featuring peroxidase-like activity mimicking the natural heme group. The first examples of such bioinspired systems were shown by Zhang et al. [161] who leveraged CoFe-LDH nanoplates to obtain a colorimetric sensor for H_2O_2 (based on the oxidation of TMB), reaching the optimal detection limit of 0.6 μM. Other similar examples are those from Su et al. [162], who leveraged NiCo-LDHs for acetylcholine detection (limit of detection equal to 1.62 μM), and from Zhan et al. [163], who investigated NiFe-LDHs as a sensor for H_2O_2, reaching the limit of detection of 4.4 ± 0.2 μM (see Figure 13b). Along with colorimetric detection, electrochemistry can also be applied to improve the detection limit of the sensor. In this regard, an example of an electrochemical sensor based on peroxidase mimicking LDHs worth mentioning was recently shown by Fazli et al. [164], who fabricated a PdAl-LDH/carboxymethyl cellulose (CMC) nanocomposite (CMC@Pd/Al-LDH) on a glassy carbon electrode to realize a sensor for H_2O_2 (limit of detection equal to 0.3 μM). Interestingly, this work shows how CMC is suitable for improving the sensitivity and the exposed active surface area, finally enabling a high number of available sites for electrochemical reactions. An intriguing application of peroxidase mimics for sensing acetylcholine (limit of detection equal to 1.7 μM) was reported by Wang et al. [165] who prepared NiAl-LDH/Carbon dot nanocomposites onto glassy carbon electrodes.

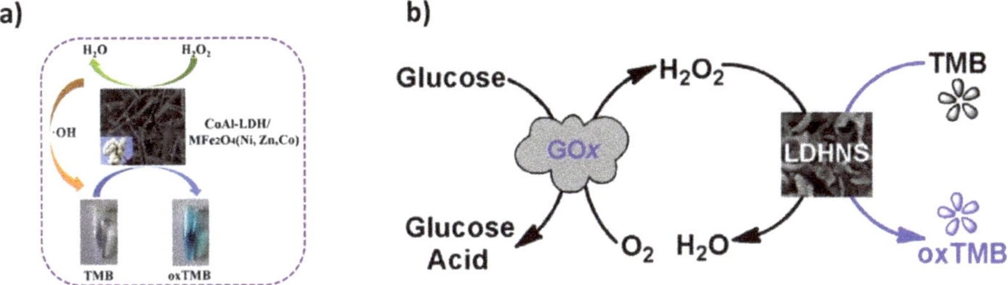

Figure 13. LDH-based peroxidase mimics. (**a**) CoAl-LDH/Fe_2O_4 hybrid materials catalyze the oxidation of TMB in presence of H_2O_2, mimicking the peroxidase activity. Reprinted from Ref. [158] with permission from Elsevier. (**b**) Intrinsic peroxidase-like activity in NiFe-LDH can be leveraged to build up colorimetric sensors for H_2O_2 and glucose. Reprinted from Ref. [163] with permission from Elsevier.

5.5. LDHs on Biotemplates for Bioremediation

The intrinsically high surface to volume ratio of LDHs, possibly enhanced by recently proposed ultrathin LDH synthesis routes [166–168], has been leveraged for many applications that require physical processes of absorption and interaction with molecules for bioremediation of contaminated sites, such heavy metals, toxic substances, pesticides or even favoring their photo-induced degradation [169]. An approach for further improving the efficiency of LDH-empowered devices for bioremediation is the realization of 3D biomimetic structures or the coupling with organic biotemplates (typically of plant origin) aiming at the realization of life-mimicking 3D high-surface hierarchical organization, miniaturization, eco-friendly characteristics and even specificity for the absorption of targeted molecular systems. For instance, typical examples of biotemplates are constituted by diatomaceous earth, leaves and cellulose fibers, which are prone to be easily functionalized by surface functionalization with LDHs.

A plethora of research efforts have been focused on the removal of oil, metal ions, pesticides and agrochemicals from contaminated water by using this combined approach. An interesting example comes from the research of Zhu et al. who showed a one-step synthesis of a biomimetic cactus-like hierarchical architecture for oil/water separation [170], based on a CoNi-LDH coated stainless steel mesh. The role of the 3D LDH morphology appears to be the key factor to induce an outstanding water locking capability into the cactus-like hierarchical structure. The trapped water is, in turn, able to form a stable water layer on the surface, ultimately forming a bioinspired barrier against oil penetration into the mesh, since the resulting surface would be superhydrophilic and underwater superoleophobic. The authors tested different types of oils (diesel oil, lubricating oil, silicon and n-hexane), finding excellent oil rejection ratio and high separation capability and outstanding recyclability over 20 cycles.

Abolghasemi and coworkers used the hierarchical structure of boehmite decorated with MgAl-LDH and porous carbon on a steel fiber for solid phase microextraction of fifteen different pesticides [171]. Interesting examples have also been provided for the removal of Congo red (CR), a dye which is widely employed in many textiles and biotechnological based industries. MgAl-LDH modified diatoms [172] and bioinspired magnetic $ZnFe_2O_4$ microspheres synthetized using pine pollen and covered with MgAl-LDH [173] have been used with this purpose. The presence of LDH significantly improved the absorption capacity of the composite material toward the dye, reaching values of about 300 mg/g in both cases. Whereas the 3D structure of the diatom triggered a multilayer CR dye adsorption, in the case of the magnetic $ZnFe_2O_4$ microspheres, the adsorption was better fitted following a Langmuir model. A very clever bioinspired approach for CR and Doxycycline (DC) removal and photodegradation under simulated sun light irradiation was proposed by Bing et al. [174], who combined photoactive bismute oxides, MgAl-LDH synthesized onto lotus pollen used as template, and calcination (C) treatments, for the fabrication of Bi_2O_3/Bi_2WO_6/MgAl-CLDH heterojunction hybrids with a 3D hierarchically porous structure. The adsorption of both CD and DC was very efficient, reaching 205.3 and 204.3 $mg{\cdot}g^{-1}$, respectively. Another relevant application of LDH triggered bioremediation comes from the examples in which toxic heavy metals ions can be absorbed on the LDH surfaces. To this aim, NiFe-LDHs [175] have been shown in combination with graphene oxide nanocomposite for the efficient removal of Pb(II) and Cd(II) ions from water, obtaining a maximum adsorption capacity of 986 and 971 $mg{\cdot}g^{-1}$, respectively, following a Langmuir model. A truly bioinspired system that efficiently removes Cu^{2+} ions is the one shown by Dou et al. [176]. The authors leveraged functionalization of MgAl-LDH surface by using the Kabachnik–Fields reaction. More specifically, the MgAl-LDHs were modified with polydopamine, which was further used as source of reactive amino groups used for functionalization with diethyl phosphite, terephthalaldehyde and thiourea, finally allowing the introduction of phosphate groups in the LDHs which are able to specifically adsorb Cu^{2+} ions, up to 105.44 $mg{\cdot}g^{-1}$. In another interesting report, Wang et al. [177] prepared a hybrid material that combined sulfide (derived from $(NH_4)_2MoS_4$) intercalated NiFe-LDHs with alginate for the extraction of Pb^{2+} in aqueous environments. The authors combined the high surface/volume ratio of the bioinspired alginate hydrogel with the LDH sulfide specific interaction with Pb^{2+}, leading to maximum adsorption capacity of about 18 $mg{\cdot}g^{-1}$.

An outstanding bioinspired example of phosphate removal from water was shown by Lai et al. [178] who prepared a highly porous composite combining graphene oxide/MgMn-LDH (GO/MgMn-LDH) onto *Garcinia subelliptica* leaves, that constituted a natural bio-template (see Figure 14). The authors grew MgMn-LDH in-situ on the leaf-templated GO (L-GO) to obtain L-GO/MgMn-LDH. Interestingly, after calcination at 300 °C (L-GO/MgMn-LDH-300), the flavonoids which derived from the leaves were able to intercalate into the LDHs, avoiding the collapse of its structure. In addition, these biomolecules provided an outstanding specificity towards the interaction with phosphate ions, which was quantified as 244.08 $mg{\cdot}g^{-1}$ at pH 3. The phosphate adsorption was very much dependent on the pH. Whereas optimal values were obtained at acidic pHs, the higher was the pH, the higher became the competition between OH^- groups and phosphate in the interaction with the LDHs, finally reducing the efficiency in phosphate adsorption. Another clever example of biotemplated approach for the removal of antibiotic, i.e., doxycycline, from water was shown by the intriguing approach of Bing et al. [179] who realized 3D hierarchical tubular micromotors from kapok fibers which were functionalized with Br-intercalated MgAl-LDH/Mn_3O_4 hybrid. The manganese oxide permitted to catalyze H_2O_2 decomposition generating oxygen bubbles for self-propulsion. The Br^- anions acted as initiator to form an imprinted polymer to specifically absorb doxycycline up to 224.23 $mg{\cdot}g^{-1}$.

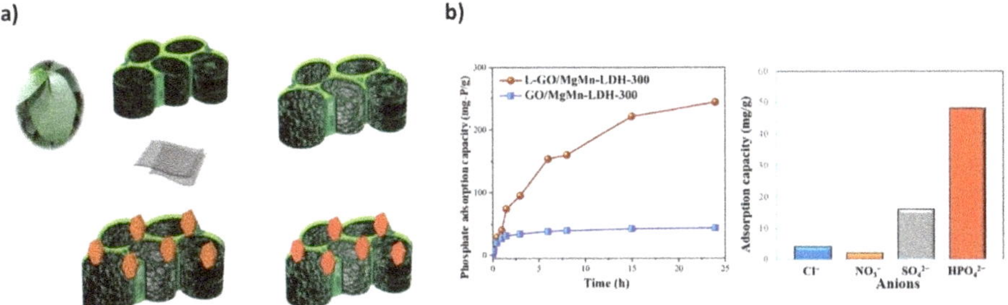

Figure 14. *Garcinia subelliptica* leaves templates for GO/MgMn-LDH based removal of phosphate anions. (**a**) Scheme of the templated assembly. (**b**) Phosphate anion adsorption capacity of L-GO/MgMn-LDH-300 compared to that of the calcined LDH composite without leaf-template (GO/MgMn-LDH-300) (left) and anion selectivity of L-GO/MgMn-LDH-300 (right). Figure reprinted from Ref. [178] with permission from Elsevier.

6. Conclusions and Perspectives

The field of LDHs is rapidly evolving from the initial investigations dealing with the materials discovery for new applications in chemistry, materials science and biomedicine. In fact, leveraging their outstanding simplicity in the synthesis, low cost, reusability and biocompatibility, LDHs have already proved to be a key player for in the field of nanotechnology.

Given their extraordinarily combination of properties and biocompatibility, as well as their capability to catalyze reactions under prebiotic conditions, LDH clays might be a material chosen by the molecular evolution of living systems. However, the approach correlating the existence of LDHs with the origin of life is very bold, and the question whether there could be a relationship between LDH clays (including green rust) and prebiotic synthesis remains unclear, thus continues being a hypothesis that would need to be addressed in depth in future studies. In general, it is believed that clays have been key systems for the development of prebiotic conditions, given their ability in concentrating and protecting life-essential molecules, creating autonomous semipermeable compartments, as well as catalyzing polymerization reactions. In this context, the specific role of LDHs is still underestimated, but it is growing steadily, especially considering the relevant role of these materials in the context of prebiotic peptides synthesis. Motivated by this, the review analyzes the currently state of the art concerning the interactions of life-relevant molecules (DNA, phospholipids, amino acids and carbohydrates) with LDHs, focusing on the structural point of view, in particular analyzing the possible formation of ordered and crystalline systems and, if possible, providing some insights into relevant biological role. Then, after briefly describing the fabrication of LDHs-based life-mimicking compartments, the review focuses on the technologically relevant applications of life-like and life-inspired devices. More specifically, we focus our attention on important applications where bioinspired LDHs play an important role, i.e., composites and coatings, physical and chemical sensors, catalysis and bioremediation.

The low cost and the eco-friendly synthetic approaches for LDHs make them suitable materials for relevant applications, as the formulation of paints, flame retardants, phytosanitary and pharmaceutical products, given their key role as dispersing or encapsulation agents for anionic molecular systems. To our knowledge, Kyowa Chemical Industry was the first to produce synthetic LDHs, since 1966, among their vast product portfolio, obtaining a good market success, especially in the field of resin stabilizers, residual catalyst removal and chlorine absorbers. The path to bring to the market the innovations and approaches reviewed in the present work is clearly challenging, and still many efforts are requested of researchers to facilitate a more extended development at a larger industrial scale. As a matter of fact, today, many hybrid organic–inorganic nanomaterials are entering a variety of markets [180]. New materials must aim toward higher levels of sophistication, be recyclable and environmentally-friendly and consume less energy or enable energy harvesting. We hope that the bioinspired approach showed in this review could trigger further technological transfer from academia to the industry, highlighting how bio-friendly LDH products can tackle the need of sustainable chemistry approaches for product manufacturing.

This review does not pretend to cover all the possible literature actually available in the field; however, it can be considered as an effort to review the most recent and exciting advancements in the field of bioinspired applications of LDHs. We hope that it can trigger future LDH-based studies, both on the fundamental science and on the applications, with the ultimate aim to favor a further development of LDHs-based nanotechnology under a novel eco-friendly, bioinspired perspective.

Author Contributions: Methodology, G.A. and R.P.; writing—original draft preparation, G.A. and G.P.; writing—review and editing, G.A. and G.P.; and supervision, G.A., G.P. and P.G.M. All authors have read and agreed to the published version of the manuscript.

Funding: The article processing charges were funded by SIMITECNO SRL, Sistemi e misure per tecnologie, Via Gallian, 62, 00133 Roma (www.simitecno.com).

Acknowledgments: We acknowledge the Università degli Studi di Palermo for hospitality and support.

Conflicts of Interest: The authors declare no conflict of interest.

References

1. Laipan, M.; Yu, J.; Zhu, R.; Zhu, J.; Smith, A.T.; He, H.; O'Hare, D.; Sun, L. Functionalized layered double hydroxides for innovative applications. *Mater. Horiz.* **2020**, *7*, 715–745. [CrossRef]
2. Cavani, F.; Trifirò, F.; Vaccari, A. Hydrotalcite-type anionic clays: Preparation, properties and applications. *Catal. Today* **1991**, *11*, 173–301. [CrossRef]
3. *Layered Double Hydroxides: Present and Future*; Rives, V., Ed.; Nova Science Publishers: Huntington, NY, USA, 2001; ISBN 978-1-59033-060-9.
4. *Layered Double Hydroxides*; Duan, X.; Evans, D.G., Eds.; Structure and Bonding; Springer-Verlag: Berlin/Heidelberg, Germany, 2006; Volume 119, ISBN 978-3-540-28279-2.
5. Arrabito, G.; Bonasera, A.; Prestopino, G.; Orsini, A.; Mattoccia, A.; Martinelli, E.; Pignataro, B.; Medaglia, P.G. Layered Double Hydroxides: A Toolbox for Chemistry and Biology. *Crystals* **2019**, *9*, 361. [CrossRef]
6. Choy, J.-H.; Park, M. Cationic and Anionic Clays for Biological Applications. In *Interface Science and Technology*; Wypych, F., Satyanarayana, K.G., Eds.; Elsevier: Amsterdam, The Netherlands, 2004; Volume 1, pp. 403–424, ISBN 1573-4285.
7. Taviot-Guého, C.; Prévot, V.; Forano, C.; Renaudin, G.; Mousty, C.; Leroux, F. Tailoring Hybrid Layered Double Hydroxides for the Development of Innovative Applications. *Adv. Funct. Mater.* **2018**, *28*, 1703868. [CrossRef]
8. Mohapatra, L.; Parida, K.M. A Review on Recent Progress, Challenges and Perspective of Layered Double Hydroxides as Promising Photocatalysts. *J. Mater. Chem. A* **2016**, *4*, 10744–10766. [CrossRef]
9. Costantino, U.; Leroux, F.; Nocchetti, M.; Mousty, C. LDH in Physical, Chemical, Biochemical, and Life Sciences. In *Developments in Clay Science*; Elsevier: Amsterdam, The Netherlands, 2013; Volume 5, pp. 765–791, ISBN 978-0-08-099364-5.
10. Cui, J.; Li, Z.; Liu, K.; Li, J.; Shao, M. A bifunctional nonenzymatic flexible glucose microsensor based on CoFe-Layered double hydroxide. *Nanoscale Adv.* **2019**, *1*, 948–952. [CrossRef]
11. Kura, A.U.; Hussein, M.Z.; Fakurazi, S.; Arulselvan, P. Layered double hydroxide nanocomposite for drug delivery systems; bio-distribution, toxicity and drug activity enhancement. *Chem. Cent. J.* **2014**, *8*, 47. [CrossRef]
12. Da Costa Fernandes, C., Jr.; Pinto, T.S.; Kang, H.R.; Magalhães Padilha, P.; Koh, I.H.J.; Constantino, V.R.L.; Zambuzzi, W.F. Layered Double Hydroxides Are Promising Nanomaterials for Tissue Bioengineering Application. *Adv. Biosyst.* **2019**, 1800238. [CrossRef]

13. Mishra, G.; Dash, B.; Pandey, S. Layered double hydroxides: A brief review from fundamentals to application as evolving biomaterials. *Appl. Clay Sci.* **2018**, *153*, 172–186. [CrossRef]
14. Lu, P.; Liu, Y.; Zhou, T.; Wang, Q.; Li, Y. Recent advances in layered double hydroxides (LDHs) as two-dimensional membrane materials for gas and liquid separations. *J. Membr. Sci.* **2018**, *567*, 89–103. [CrossRef]
15. Forano, C.; Bruna, F.; Mousty, C.; Prevot, V. Interactions between Biological Cells and Layered Double Hydroxides: Towards Functional Materials. *Chem. Rec.* **2018**, *18*, 1150–1166. [CrossRef]
16. Wang, J.; Zhang, T.; Li, M.; Yang, Y.; Lu, P.; Ning, P.; Wang, Q. Arsenic removal from water/wastewater using layered double hydroxide derived adsorbents, a critical review. *RSC Adv.* **2018**, *8*, 22694–22709. [CrossRef]
17. Errico, V.; Arrabito, G.; Plant, S.R.; Medaglia, P.G.; Palmer, R.E.; Falconi, C. Chromium inhibition and size-selected Au nanocluster catalysis for the solution growth of low-density ZnO nanowires. *Sci. Rep.* **2015**, *5*, 12336. [CrossRef] [PubMed]
18. Arrabito, G.; Errico, V.; Zhang, Z.; Han, W.; Falconi, C. Nanotransducers on printed circuit boards by rational design of high-density, long, thin and untapered ZnO nanowires. *Nano Energy* **2018**, *46*, 54–62. [CrossRef]
19. Bukhtiyarova, M.V. A review on effect of synthesis conditions on the formation of layered double hydroxides. *J. Solid State Chem.* **2019**, *269*, 494–506. [CrossRef]
20. Pushparaj, S.S.C.; Forano, C.; Prevot, V.; Lipton, A.S.; Rees, G.J.; Hanna, J.V.; Nielsen, U.G. How the Method of Synthesis Governs the Local and Global Structure of Zinc Aluminum Layered Double Hydroxides. *J. Phys. Chem. C* **2015**, *119*, 27695–27707. [CrossRef]
21. Jose, N.A.; Zeng, H.C.; Lapkin, A.A. Hydrodynamic assembly of two-dimensional layered double hydroxide nanostructures. *Nat. Commun.* **2018**, *9*, 4913. [CrossRef]
22. Bravo-Suárez, J.J.; Páez-Mozo, E.A.; Oyama, S.T. Review of the synthesis of layered double hydroxides: A thermodynamic approach. *Quím. Nova* **2004**, *27*. [CrossRef]
23. Tokudome, Y.; Morimoto, T.; Tarutani, N.; Vaz, P.D.; Nunes, C.D.; Prevot, V.; Stenning, G.B.G.; Takahashi, M. Layered Double Hydroxide Nanoclusters: Aqueous, Concentrated, Stable, and Catalytically Active Colloids toward Green Chemistry. *ACS Nano* **2016**, *10*, 5550–5559. [CrossRef]
24. Kayano, M.; Ogawa, M. Preparation of Large Platy Particles of Co-Al Layered Double Hydroxides. *Clays Clay Miner.* **2006**, *54*, 382–389. [CrossRef]
25. Ma, R.; Liu, Z.; Li, L.; Iyi, N.; Sasaki, T. Exfoliating layered double hydroxides in formamide: A method to obtain positively charged nanosheets. *J. Mater. Chem.* **2006**, *16*, 3809–3813. [CrossRef]
26. Wang, Q.; O'Hare, D. Recent Advances in the Synthesis and Application of Layered Double Hydroxide (LDH) Nanosheets. *Chem. Rev.* **2012**, *112*, 4124–4155. [CrossRef] [PubMed]
27. Liu, Z.; Ma, R.; Osada, M.; Iyi, N.; Ebina, Y.; Takada, K.; Sasaki, T. Synthesis, Anion Exchange, and Delamination of Co–Al Layered Double Hydroxide: Assembly of the Exfoliated Nanosheet/Polyanion Composite Films and Magneto-Optical Studies. *J. Am. Chem. Soc.* **2006**, *128*, 4872–4880. [CrossRef]
28. Liang, J.; Ma, R.; Iyi, N.; Ebina, Y.; Takada, K.; Sasaki, T. Topochemical Synthesis, Anion Exchange, and Exfoliation of Co–Ni Layered Double Hydroxides: A Route to Positively Charged Co–Ni Hydroxide Nanosheets with Tunable Composition. *Chem. Mater.* **2010**, *22*, 371–378. [CrossRef]
29. Carrasco, J.A.; Harvey, A.; Hanlon, D.; Lloret, V.; McAteer, D.; Sanchis-Gual, R.; Hirsch, A.; Hauke, F.; Abellán, G.; Coleman, J.N.; et al. Liquid phase exfoliation of carbonate-intercalated layered double hydroxides. *Chem. Commun.* **2019**, *55*, 3315–3318. [CrossRef] [PubMed]
30. Hibino, T.; Kobayashi, M. Delamination of layered double hydroxides in water. *J. Mater. Chem.* **2005**, *15*, 653–656. [CrossRef]
31. Wang, J.; Bao, W.; Umar, A.; Wang, Q.; O'Hare, D.; Wan, Y. Delaminated Layered Double Hydroxide Nanosheets as an Efficient Vector for DNA Delivery. *J. Biomed. Nanotechnol.* **2016**, *12*, 922–933. [CrossRef]
32. Zhang, Z.; Min, L.; Chen, P.; Zhang, W.; Wang, Y. Nature-inspired delamination of layered double hydroxides into monolayered nanosheets in water. *Mater. Lett.* **2017**, *195*, 198–200. [CrossRef]
33. Wang, Q.; O'Hare, D. Large-scale synthesis of highly dispersed layered double hydroxide powders containing delaminated single layer nanosheets. *Chem. Commun.* **2013**, *49*, 6301–6303. [CrossRef]
34. Ruengkajorn, K.; Erastova, V.; Buffet, J.-C.; Greenwell, H.C.; O'Hare, D. Aqueous immiscible layered double hydroxides: Synthesis, characterisation and molecular dynamics simulation. *Chem. Commun.* **2018**, *54*, 4394–4397. [CrossRef]

35. Guo, X.; Xu, S.; Zhao, L.; Lu, W.; Zhang, F.; Evans, D.G.; Duan, X. One-Step Hydrothermal Crystallization of a Layered Double Hydroxide/Alumina Bilayer Film on Aluminum and Its Corrosion Resistance Properties. *Langmuir* **2009**, *25*, 9894–9897. [CrossRef]
36. Scarpellini, D.; Falconi, C.; Gaudio, P.; Mattoccia, A.; Medaglia, P.G.; Orsini, A.; Pizzoferrato, R.; Richetta, M. Morphology of Zn/Al layered double hydroxide nanosheets grown onto aluminum thin films. *Microelectron. Eng.* **2014**, *126*, 129–133. [CrossRef]
37. Prestopino, G.; Arrabito, G.; Generosi, A.; Mattoccia, A.; Paci, B.; Perez, G.; Verona-Rinati, G.; Medaglia, P.G. Emerging switchable ultraviolet photoluminescence in dehydrated Zn/Al layered double hydroxide nanoplatelets. *Sci. Rep.* **2019**, *9*, 11498. [CrossRef]
38. Forticaux, A.; Dang, L.; Liang, H.; Jin, S. Controlled Synthesis of Layered Double Hydroxide Nanoplates Driven by Screw Dislocations. *Nano Lett.* **2015**, *15*, 3403–3409. [CrossRef]
39. Liu, J.; Li, Y.; Huang, X.; Li, G.; Li, Z. Layered Double Hydroxide Nano- and Microstructures Grown Directly on Metal Substrates and Their Calcined Products for Application as Li-Ion Battery Electrodes. *Adv. Funct. Mater.* **2008**, *18*, 1448–1458. [CrossRef]
40. Miyata, S. Anion-Exchange Properties of Hydrotalcite-Like Compounds. *Clays Clay Miner.* **1983**, *31*, 305–311. [CrossRef]
41. Meyn, M.; Beneke, K.; Lagaly, G. Anion-exchange reactions of layered double hydroxides. *Inorg. Chem.* **1990**, *29*, 5201–5207. [CrossRef]
42. Radha, A.V.; Vishnu Kamath, P.; Shivakumara, C. Mechanism of the anion exchange reactions of the layered double hydroxides (LDHs) of Ca and Mg with Al. *Solid State Sci.* **2005**, *7*, 1180–1187. [CrossRef]
43. Prasanna, S.V.; Kamath, P.V. Anion-Exchange Reactions of Layered Double Hydroxides: Interplay between Coulombic and H-Bonding Interactions. *Ind. Eng. Chem. Res.* **2009**, *48*, 6315–6320. [CrossRef]
44. Tavares, S.R.; Haddad, J.F.S.; Ivo, R.; Moraes, P.; Leitão, A.A. Computational exploration of the anion exchange on the basal surface of layered double hydroxides by molecular dynamics. *Appl. Surf. Sci.* **2020**, *513*, 145743. [CrossRef]
45. Debecker, D.P.; Gaigneaux, E.M.; Busca, G. Exploring, Tuning, and Exploiting the Basicity of Hydrotalcites for Applications in Heterogeneous Catalysis. *Chem. Eur. J.* **2009**, *15*, 3920–3935. [CrossRef] [PubMed]
46. Manohara, G.V.; Prasanna, S.V.; Kamath, P.V. Structure and Composition of the Layered Double Hydroxides of Mg and Fe: Implications for Anion-Exchange Reactions. *Eur. J. Inorg. Chem.* **2011**, *2011*, 2624–2630. [CrossRef]
47. Xu, M.; Wei, M. Layered Double Hydroxide-Based Catalysts: Recent Advances in Preparation, Structure, and Applications. *Adv. Funct. Mater.* **2018**, *28*, 1802943. [CrossRef]
48. Sideris, P.J.; Nielsen, U.G.; Gan, Z.; Grey, C.P. Mg/Al Ordering in Layered Double Hydroxides Revealed by Multinuclear NMR Spectroscopy. *Science* **2008**, *321*, 113–117. [CrossRef]
49. Vucelic, M. Cation Ordering in Synthetic Layered Double Hydroxides. *Clays Clay Miner.* **1997**, *45*, 803–813. [CrossRef]
50. Kim, D.; Huang, C.; Lee, H.; Han, I.; Kang, S.; Kwon, S.; Lee, J.; Han, Y.; Kim, H. Hydrotalcite-type catalysts for narrow-range oxyethylation of 1-dodecanol using ethyleneoxide. *Appl. Catal. A Gen.* **2003**, *249*, 229–240. [CrossRef]
51. Krivovichev, S.V.; Yakovenchuk, V.N.; Zhitova, E.S. Natural Double Layered Hydroxides: Structure, Chemistry, and Information Storage Capacity. In *Minerals as Advanced Materials II*; Krivovichev, S.V., Ed.; Springer Berlin Heidelberg: Berlin/Heidelberg, Germany, 2011; pp. 87–102, ISBN 978-3-642-20017-5.
52. Kuma, K.; Paplawsky, W.; Gedulin, B.; Arrhenius, G. Mixed-valence hydroxides as bioorganic host minerals. *Orig. Life Evol. Biosph.* **1989**, *19*, 573–601. [CrossRef]
53. Arrhenius, G.O. Crystals and Life. *HCA* **2003**, *86*, 1569–1586. [CrossRef]
54. Erastova, V.; Degiacomi, M.T.; Fraser, D.G.; Greenwell, H.C. Mineral surface chemistry control for origin of prebiotic peptides. *Nat. Commun.* **2017**, *8*, 2033. [CrossRef]
55. Greenwell, H.C.; Coveney, P.V. Layered Double Hydroxide Minerals as Possible Prebiotic Information Storage and Transfer Compounds. *Orig. Life Evol. Biosph.* **2006**, *36*, 13–37. [CrossRef]
56. Sanchez, C.; Arribart, H.; Giraud Guille, M.M. Biomimetism and bioinspiration as tools for the design of innovative materials and systems. *Nat. Mater.* **2005**, *4*, 277–288. [CrossRef] [PubMed]
57. Ruiz-Hitzky, E.; Darder, M.; Aranda, P.; Ariga, K. Advances in Biomimetic and Nanostructured Biohybrid Materials. *Adv. Mater.* **2010**, *22*, 323–336. [CrossRef] [PubMed]

58. Chen, P.-Y.; McKittrick, J.; Meyers, M.A. Biological materials: Functional adaptations and bioinspired designs. *Prog. Mater. Sci.* **2012**, *57*, 1492–1704. [CrossRef]
59. Zhang, C.; Mcadams, D.A.; Grunlan, J.C. Nano/Micro-Manufacturing of Bioinspired Materials: A Review of Methods to Mimic Natural Structures. *Adv. Mater.* **2016**, *28*, 6292–6321. [CrossRef]
60. Gong, C.; Sun, S.; Zhang, Y.; Sun, L.; Su, Z.; Wu, A.; Wei, G. Hierarchical nanomaterials *via* biomolecular self-assembly and bioinspiration for energy and environmental applications. *Nanoscale* **2019**, *11*, 4147–4182. [CrossRef]
61. Vijayan, P.P.; Puglia, D. Biomimetic multifunctional materials: A review. *Emergent Mater.* **2019**, *2*, 391–415. [CrossRef]
62. Ruiz-Hitzky, E.; Darder, M.; Wicklein, B.; Castro-Smirnov, F.A.; Aranda, P. Clay-based biohybrid materials for biomedical and pharmaceutical applications. *Clays Clay Miner.* **2019**, *67*, 44–58. [CrossRef]
63. Xu, S.; Zhao, J.; Yu, Q.; Qiu, X.; Sasaki, K. Effect of Natural Organic Matter Model Compounds on the Structure Memory Effect of Different Layered Double Hydroxides. *ACS Earth Space Chem.* **2019**, *3*, 2175–2189. [CrossRef]
64. Lambert, J.-F. Origins of life: From the mineral to the biochemical world. *BIO Web Conf.* **2015**, *4*, 00012. [CrossRef]
65. Orgel, L.E. The origin of life—A review of facts and speculations. *Trends Biochem. Sci.* **1998**, *23*, 491–495. [CrossRef]
66. Ferus, M.; Pietrucci, F.; Saitta, A.M.; Knížek, A.; Kubelík, P.; Ivanek, O.; Shestivska, V.; Civiš, S. Formation of nucleobases in a Miller–Urey reducing atmosphere. *Proc. Natl. Acad. Sci. USA* **2017**, *114*, 4306–4311. [CrossRef]
67. Clark, B.; Kolb, V. Comet Pond II: Synergistic Intersection of Concentrated Extraterrestrial Materials and Planetary Environments to Form Procreative Darwinian Ponds. *Life* **2018**, *8*, 12. [CrossRef]
68. Martin, W.; Russell, M.J. On the origins of cells: A hypothesis for the evolutionary transitions from abiotic geochemistry to chemoautotrophic prokaryotes, and from prokaryotes to nucleated cells. *Phil. Trans. R. Soc. Lond. B* **2003**, *358*, 59–85. [CrossRef]
69. Das, T.; Ghule, S.; Vanka, K. Insights Into the Origin of Life: Did It Begin from HCN and H_2O? *ACS Cent. Sci.* **2019**, *5*, 1532–1540. [CrossRef]
70. Meisner, J.; Zhu, X.; Martínez, T.J. Computational Discovery of the Origins of Life. *ACS Cent. Sci.* **2019**, *5*, 1493–1495. [CrossRef]
71. Rimola, A.; Sodupe, M.; Ugliengo, P. Role of Mineral Surfaces in Prebiotic Chemical Evolution. In Silico Quantum Mechanical Studies. *Life* **2019**, *9*, 10. [CrossRef]
72. Ponce, A. Radionuclide-induced defect sites in iron-bearing minerals may have accelerated the emergence of life. *Interface Focus* **2019**, *9*, 20190085. [CrossRef]
73. Sojo, V.; Herschy, B.; Whicher, A.; Camprubí, E.; Lane, N. The Origin of Life in Alkaline Hydrothermal Vents. *Astrobiology* **2016**, *16*, 181–197. [CrossRef]
74. Martin, W.; Baross, J.; Kelley, D.; Russell, M.J. Hydrothermal vents and the origin of life. *Nat. Rev. Microbiol.* **2008**, *6*, 805–814. [CrossRef]
75. Braterman, P.S.; Cairns-Smith, A.G. Photoprecipitation and the banded iron-formations—Some quantitative aspects. *Orig. Life Evol. Biosph.* **1987**, *17*, 221–228. [CrossRef]
76. Russell, M.J. Green rust: The simple organizing 'seed' of all life? *Life* **2018**, *8*. [CrossRef] [PubMed]
77. Bernal, J.D. The Physical Basis of Life. *Proc. Phys. Soc. Sect. A* **1949**, *62*, 537–558. [CrossRef]
78. Cairns-Smith, A.G. *Genetic Takeover and the Mineral Origins of Life*, 1st ed.; Cambridge Univ. Press: Cambridge, UK, 1987; ISBN 978-0-521-34682-5.
79. Ferris, J.P. Montmorillonite Catalysis of 30–50 Mer Oligonucleotides: Laboratory Demonstration of Potential Steps in the Origin of the RNA World. *Orig. Life Evol. Biosph.* **2002**, *32*, 311–332. [CrossRef] [PubMed]
80. Lahav, N.; White, D.; Chang, S. Peptide formation in the prebiotic era: Thermal condensation of glycine in fluctuating clay environments. *Science* **1978**, *201*, 67–69. [CrossRef]
81. Hanczyc, M.M.; Fujikawa, S.M.; Szostak, J.W. Experimental Models of Primitive Cellular Compartments: Encapsulation, Growth, and Division. *Science* **2003**, *302*, 618–622. [CrossRef]
82. Swadling, J.B.; Coveney, P.V.; Christopher Greenwell, H. Stability of free and mineral-protected nucleic acids: Implications for the RNA world. *Geochim. Cosmochim. Acta* **2012**, *83*, 360–378. [CrossRef]

83. Grégoire, B.; Greenwell, H.C.; Fraser, D.G. Peptide Formation on Layered Mineral Surfaces: The Key Role of Brucite-like Minerals on the Enhanced Formation of Alanine Dipeptides. *ACS Earth Space Chem.* **2018**, *2*, 852–862. [CrossRef]
84. Vasti, C.; Ambroggio, E.; Rojas, R.; Giacomelli, C.E. A closer look into the physical interactions between lipid membranes and layered double hydroxide nanoparticles. *Colloids Surf. B Biointerfaces* **2020**, *191*, 110998. [CrossRef]
85. Bernhardt, H. Making Molecules with Clay: Layered Double Hydroxides, Pentopyranose Nucleic Acids and the Origin of Life. *Life* **2019**, *9*, 19. [CrossRef]
86. Suen, N.-T.; Hung, S.-F.; Quan, Q.; Zhang, N.; Xu, Y.-J.; Chen, H.M. Electrocatalysis for the oxygen evolution reaction: Recent development and future perspectives. *Chem. Soc. Rev.* **2017**, *46*, 337–365. [CrossRef]
87. Lu, X.; Xue, H.; Gong, H.; Bai, M.; Tang, D.; Ma, R.; Sasaki, T. 2D Layered Double Hydroxide Nanosheets and Their Derivatives Toward Efficient Oxygen Evolution Reaction. *Nano-Micro Lett.* **2020**, *12*, 86. [CrossRef]
88. Wang, L.; Arrabito, G. Hybrid, multiplexed, functional DNA nanotechnology for bioanalysis. *Analyst* **2015**, *140*, 5821–5848. [CrossRef]
89. Yadav, M.; Kumar, R.; Krishnamurthy, R. Chemistry of Abiotic Nucleotide Synthesis. *Chem. Rev.* **2020**, *120*, 4766–4805. [CrossRef]
90. Baú, J.P.T.; Villafañe-Barajas, S.A.; da Costa, A.C.S.; Negrón-Mendoza, A.; Colín-Garcia, M.; Zaia, D.A.M. Adenine Adsorbed onto Montmorillonite Exposed to Ionizing Radiation: Essays on Prebiotic Chemistry. *Astrobiology* **2020**, *20*, 26–38. [CrossRef] [PubMed]
91. Li, Y.; Yan, Z.; Xiao, J.; Yue, T. DNA-conjugated layered double hydroxides penetrating into a plasma membrane: Layer size, thickness and DNA grafting density matter. *NanoImpact* **2020**, *18*, 100222. [CrossRef]
92. Murase, N.; Gonda, K. Adsorption of Liposomes by Clay. *J. Biochem.* **1982**, *92*, 271–273. [CrossRef]
93. Konnova, S.A.; Sharipova, I.R.; Demina, T.A.; Osin, Y.N.; Yarullina, D.R.; Ilinskaya, O.N.; Lvov, Y.M.; Fakhrullin, R.F. Biomimetic cell-mediated three-dimensional assembly of halloysite nanotubes. *Chem. Commun.* **2013**, *49*, 4208–4210. [CrossRef]
94. Itoh, T.; Ohta, N.; Shichi, T.; Yui, T.; Takagi, K. The Self-Assembling Properties of Stearate Ions in Hydrotalcite Clay Composites. *Langmuir* **2003**, *19*, 9120–9126. [CrossRef]
95. Aisawa, S.; Kudo, H.; Hoshi, T.; Takahashi, S.; Hirahara, H.; Umetsu, Y.; Narita, E. Intercalation behavior of amino acids into Zn–Al-layered double hydroxide by calcination–rehydration reaction. *J. Solid State Chem.* **2004**, *177*, 3987–3994. [CrossRef]
96. Wang, J.; Zhang, W.; Hao, L.; Sun, J.; Zhang, W.; Guo, C.; Mu, Y.; Ji, W.; Yu, C.; Yuan, F. Amino acid–intercalated layered double hydroxide core @ ordered porous silica shell as drug carriers: Design and applications. *J. Mater. Res.* **2019**, *34*, 3747–3756. [CrossRef]
97. Paecht-Horowitz, M. The possible role of clays in prebiotic peptide synthesis. *Orig. Life* **1974**, *5*, 173–187. [CrossRef]
98. Yadollahi, M.; Namazi, H.; Barkhordari, S. Preparation and properties of carboxymethyl cellulose/layered double hydroxide bionanocomposite films. *Carbohydr. Polym.* **2014**, *108*, 83–90. [CrossRef] [PubMed]
99. Zhang, L.; Chen, K.; He, L. Super-reinforced photothermal stability of cellulose nanofibrils films by armour-type ordered doping Mg-Al layered double hydroxides. *Carbohydr. Polym.* **2019**, *212*, 229–234. [CrossRef]
100. Liang, B.; Wang, J.; Shu, Y.; Yin, P.; Guo, L. A biomimetic ion-crosslinked layered double hydroxide/alginate hybrid film. *RSC Adv.* **2017**, *7*, 32601–32606. [CrossRef]
101. Ai, Y.; Xie, R.; Xiong, J.; Liang, Q. Microfluidics for Biosynthesizing: From Droplets and Vesicles to Artificial Cells. *Small* **2020**, *16*, 1903940. [CrossRef]
102. Ma, Q.; Song, Y.; Sun, W.; Cao, J.; Yuan, H.; Wang, X.; Sun, Y.; Shum, H.C. Cell-Inspired All-Aqueous Microfluidics: From Intracellular Liquid–Liquid Phase Separation toward Advanced Biomaterials. *Adv. Sci.* **2020**, *7*, 1903359. [CrossRef] [PubMed]
103. Arrabito, G.; Reisewitz, S.; Dehmelt, L.; Bastiaens, P.I.; Pignataro, B.; Schroeder, H.; Niemeyer, C.M. Biochips for Cell Biology by Combined Dip-Pen Nanolithography and DNA-Directed Protein Immobilization. *Small* **2013**, *9*, 4243–4249. [CrossRef] [PubMed]
104. Arrabito, G.; Schroeder, H.; Schröder, K.; Filips, C.; Marggraf, U.; Dopp, C.; Venkatachalapathy, M.; Dehmelt, L.; Bastiaens, P.I.H.; Neyer, A.; et al. Configurable Low-Cost Plotter Device for Fabrication of Multi-Color Sub-Cellular Scale Microarrays. *Small* **2014**, *10*, 2870–2876. [CrossRef] [PubMed]

105. Bracha, D.; Karzbrun, E.; Daube, S.S.; Bar-Ziv, R.H. Emergent Properties of Dense DNA Phases toward Artificial Biosystems on a Surface. *Acc. Chem. Res.* **2014**, *47*, 1912–1921. [CrossRef]
106. Arrabito, G.; Cavaleri, F.; Montalbano, V.; Vetri, V.; Leone, M.; Pignataro, B. Monitoring few molecular binding events in scalable confined aqueous compartments by raster image correlation spectroscopy (CADRICS). *Lab Chip* **2016**, *16*, 4666–4676. [CrossRef]
107. Arrabito, G.; Cavaleri, F.; Porchetta, A.; Ricci, F.; Vetri, V.; Leone, M.; Pignataro, B. Printing Life-Inspired Subcellular Scale Compartments with Autonomous Molecularly Crowded Confinement. *Adv. Biosyst.* **2019**, *3*, 1900023. [CrossRef] [PubMed]
108. Cui, Y.; van Duijneveldt, J.S. Microcapsules Composed of Cross-Linked Organoclay. *Langmuir* **2012**, *28*, 1753–1757. [CrossRef]
109. Subramaniam, A.B.; Wan, J.; Gopinath, A.; Stone, H.A. Semi-permeable vesicles composed of natural clay. *Soft Matter* **2011**, *7*, 2600. [CrossRef]
110. Li, M.; Harbron, R.L.; Weaver, J.V.M.; Binks, B.P.; Mann, S. Electrostatically gated membrane permeability in inorganic protocells. *Nat. Chem.* **2013**, *5*, 529–536. [CrossRef] [PubMed]
111. Kumar, B.V.V.S.P.; Patil, A.J.; Mann, S. Enzyme-powered motility in buoyant organoclay/DNA protocells. *Nat. Chem.* **2018**, *10*, 1154–1163. [CrossRef] [PubMed]
112. Arrabito, G.; Pignataro, B. Solution Processed Micro- and Nano-Bioarrays for Multiplexed Biosensing. *Anal. Chem.* **2012**, *84*, 5450–5462. [CrossRef] [PubMed]
113. Choi, H.W.; Zhou, T.; Singh, M.; Jabbour, G.E. Recent developments and directions in printed nanomaterials. *Nanoscale* **2015**, *7*, 3338–3355. [CrossRef]
114. Micciché, C.; Arrabito, G.; Amato, F.; Buscarino, G.; Agnello, S.; Pignataro, B. Inkjet printing Ag nanoparticles for SERS hot spots. *Anal. Methods* **2018**, *10*, 3215–3223. [CrossRef]
115. Arrabito, G.; Errico, V.; De Ninno, A.; Cavaleri, F.; Ferrara, V.; Pignataro, B.; Caselli, F. Oil-in-Water fL Droplets by Interfacial Spontaneous Fragmentation and Their Electrical Characterization. *Langmuir* **2019**, *35*, 4936–4945. [CrossRef]
116. Zhang, Y.; Evans, J.R.G. Morphologies developed by the drying of droplets containing dispersed and aggregated layered double hydroxide platelets. *J. Colloid Interface Sci.* **2013**, *395*, 11–17. [CrossRef]
117. Zhang, Y.; Evans, J.R.G. Alignment of layered double hydroxide platelets. *Colloids Surfaces A Physicochem. Eng. Asp.* **2012**, *408*, 71–78. [CrossRef]
118. Zheng, Y.-M.; Li, N.; Zhang, W.-D. Preparation of nanostructured microspheres of Zn–Mg–Al layered double hydroxides with high adsorption property. *Colloids Surfaces A Physicochem. Eng. Asp.* **2012**, *415*, 195–201. [CrossRef]
119. Chen, Y.; Jing, C.; Zhang, X.; Jiang, D.; Liu, X.; Dong, B.; Feng, L.; Li, S.; Zhang, Y. Acid-salt treated CoAl layered double hydroxide nanosheets with enhanced adsorption capacity of methyl orange dye. *J. Colloid Interface Sci.* **2019**, *548*, 100–109. [CrossRef] [PubMed]
120. Sun, J.; Bao, B.; He, M.; Zhou, H.; Song, Y. Recent Advances in Controlling the Depositing Morphologies of Inkjet Droplets. *ACS Appl. Mater. Interfaces* **2015**, *7*, 28086–28099. [CrossRef]
121. Zhu, H.; Huang, S.; Yang, Z.; Liu, T. Oriented printable layered double hydroxide thin films via facile filtration. *J. Mater. Chem.* **2011**, *21*, 2950. [CrossRef]
122. Vijayamma, R.; Kalarikkal, N.; Thomas, S. Layered double hydroxide based nanocomposites for biomedical applications. In *Layered Double Hydroxide Polymer Nanocomposites*; Elsevier: Amsterdam, The Netherlands, 2020; pp. 677–714, ISBN 978-0-08-102261-0.
123. Chatterjee, A.; Bharadiya, P.; Hansora, D. Layered double hydroxide based bionanocomposites. *Appl. Clay Sci.* **2019**, *177*, 19–36. [CrossRef]
124. Shu, Y.; Yin, P.; Wang, J.; Liang, B.; Wang, H.; Guo, L. Bioinspired Nacre-like Heparin/Layered Double Hydroxide Film with Superior Mechanical, Fire-Shielding, and UV-Blocking Properties. *Ind. Eng. Chem. Res.* **2014**, *53*, 3820–3826. [CrossRef]
125. Shu, Y.; Yin, P.; Liang, B.; Wang, H.; Guo, L. Bioinspired Design and Assembly of Layered Double Hydroxide/Poly(vinyl alcohol) Film with High Mechanical Performance. *ACS Appl. Mater. Interfaces* **2014**, *6*, 15154–15161. [CrossRef]
126. Meng, Y.; Zhang, B.; Su, J.; Han, J. Bioinspired Design of LDH-Based Mobile Building Materials with Enhanced Mechanical and Ultraviolet-Shielding Performance. *Macromol. Mater. Eng.* **2019**, *304*, 1900276. [CrossRef]

127. Liu, P.; Zhang, Y.; Liu, S.; Zhang, Y.; Du, Z.; Qu, L. Bio-inspired fabrication of fire-retarding, magnetic-responsive, superhydrophobic sponges for oil and organics collection. *Appl. Clay Sci.* **2019**, *172*, 19–27. [CrossRef]
128. Wang, Z.; Shen, X.; Qian, T.; Xu, K.; Sun, Q.; Jin, C. Fabrication of Superhydrophobic Mg/Al Layered Double Hydroxide (LDH) Coatings on Medium Density Fiberboards (MDFs) with Flame Retardancy. *Materials* **2018**, *11*, 1113. [CrossRef] [PubMed]
129. Wu, H.; Shi, Z.; Zhang, X.; Qasim, A.M.; Xiao, S.; Zhang, F.; Wu, Z.; Wu, G.; Ding, K.; Chu, P.K. Achieving an acid resistant surface on magnesium alloy via bio-inspired design. *Appl. Surf. Sci.* **2019**, *478*, 150–161. [CrossRef]
130. Yu, J.; Buffet, J.-C.; O'Hare, D. Aspect Ratio Control of Layered Double Hydroxide Nanosheets and Their Application for High Oxygen Barrier Coating in Flexible Food Packaging. *ACS Appl. Mater. Interfaces* **2020**, *12*, 10973–10982. [CrossRef]
131. Cui, P.; Wang, J.; Xiong, J.; Li, S.; Zhang, W.; Liu, X.; Gu, G.; Guo, J.; Zhang, B.; Cheng, G.; et al. Meter-scale fabrication of water-driven triboelectric nanogenerator based on in-situ grown layered double hydroxides through a bottom-up approach. *Nano Energy* **2020**, *71*, 104646. [CrossRef]
132. Sun, J.; Li, P.; Qu, J.; Lu, X.; Xie, Y.; Gao, F.; Li, Y.; Gang, M.; Feng, Q.; Liang, H.; et al. Electricity generation from a Ni-Al layered double hydroxide-based flexible generator driven by natural water evaporation. *Nano Energy* **2019**, *57*, 269–278. [CrossRef]
133. Tian, J.; Zang, Y.; Sun, J.; Qu, J.; Gao, F.; Liang, G. Surface charge density-dependent performance of Ni–Al layered double hydroxide-based flexible self-powered generators driven by natural water evaporation. *Nano Energy* **2020**, *70*, 104502. [CrossRef]
134. Xu, S.; Dall'Agnese, Y.; Wei, G.; Zhang, C.; Gogotsi, Y.; Han, W. Screen-printable microscale hybrid device based on MXene and layered double hydroxide electrodes for powering force sensors. *Nano Energy* **2018**, *50*, 479–488. [CrossRef]
135. Guo, Y.; Zhang, S.; Wang, J.; Liu, Z.; Liu, Y. Facile preparation of high-performance cobalt–manganese layered double hydroxide/polypyrrole composite for battery-type asymmetric supercapacitors. *J. Alloys Compd.* **2020**, *832*, 154899. [CrossRef]
136. Ren, Y.; Feng, J. Skin-Inspired Multifunctional Luminescent Hydrogel Containing Layered Rare-Earth Hydroxide with 3D Printability for Human Motion Sensing. *ACS Appl. Mater. Interfaces* **2020**, *12*, 6797–6805. [CrossRef]
137. Beigi, F.; Mousavi, M.S.S.; Manteghi, F.; Kolahdouz, M. Doped nafion-layered double hydroxide nanoparticles as a modified ionic polymer metal composite sheet for a high-responsive humidity sensor. *Appl. Clay Sci.* **2018**, *166*, 131–136. [CrossRef]
138. Grätzel, M. Photoelectrochemical cells. *Nature* **2001**, *414*, 338–344. [CrossRef] [PubMed]
139. Gray, H.B. Powering the planet with solar fuel. *Nat. Chem.* **2009**, *1*, 7. [CrossRef] [PubMed]
140. Zhang, M.; de Respinis, M.; Frei, H. Time-resolved observations of water oxidation intermediates on a cobalt oxide nanoparticle catalyst. *Nat. Chem.* **2014**, *6*, 362–367. [CrossRef] [PubMed]
141. Liu, Y.; Li, Q.; Si, R.; Li, G.-D.; Li, W.; Liu, D.-P.; Wang, D.; Sun, L.; Zhang, Y.; Zou, X. Coupling Sub-Nanometric Copper Clusters with Quasi-Amorphous Cobalt Sulfide Yields Efficient and Robust Electrocatalysts for Water Splitting Reaction. *Adv. Mater.* **2017**, *29*, 1606200. [CrossRef] [PubMed]
142. Wang, Y.; Jiang, K.; Zhang, H.; Zhou, T.; Wang, J.; Wei, W.; Yang, Z.; Sun, X.; Cai, W.-B.; Zheng, G. Bio-Inspired Leaf-Mimicking Nanosheet/Nanotube Heterostructure as a Highly Efficient Oxygen Evolution Catalyst. *Adv. Sci.* **2015**, *2*, 1500003. [CrossRef] [PubMed]
143. Wei, W.; He, W.; Shi, B.; Dong, G.; Lu, X.; Zeng, M.; Gao, X.; Wang, Q.; Zhou, G.; Liu, J.-M.; et al. A bio-inspired 3D quasi-fractal nanostructure for an improved oxygen evolution reaction. *Chem. Commun.* **2019**, *55*, 357–360. [CrossRef] [PubMed]
144. Trogadas, P.; Coppens, M.-O. Nature-inspired electrocatalysts and devices for energy conversion. *Chem. Soc. Rev.* **2020**. [CrossRef]
145. Xu, S.-M.; Pan, T.; Dou, Y.-B.; Yan, H.; Zhang, S.-T.; Ning, F.-Y.; Shi, W.-Y.; Wei, M. Theoretical and Experimental Study on $M^{II}M^{III}$-Layered Double Hydroxides as Efficient Photocatalysts toward Oxygen Evolution from Water. *J. Phys. Chem. C* **2015**, *119*, 18823–18834. [CrossRef]
146. Dionigi, F.; Strasser, P. NiFe-Based (Oxy)hydroxide Catalysts for Oxygen Evolution Reaction in Non-Acidic Electrolytes. *Adv. Energy Mater.* **2016**, *6*, 1600621. [CrossRef]

147. Sun, H.; Yan, Z.; Liu, F.; Xu, W.; Cheng, F.; Chen, J. Self-Supported Transition-Metal-Based Electrocatalysts for Hydrogen and Oxygen Evolution. *Adv. Mater.* **2020**, *32*, 1806326. [CrossRef]
148. Lu, Z.; Xu, W.; Zhu, W.; Yang, Q.; Lei, X.; Liu, J.; Li, Y.; Sun, X.; Duan, X. Three-dimensional NiFe layered double hydroxide film for high-efficiency oxygen evolution reaction. *Chem. Commun.* **2014**, *50*, 6479–6482. [CrossRef] [PubMed]
149. Song, F.; Hu, X. Exfoliation of layered double hydroxides for enhanced oxygen evolution catalysis. *Nat. Commun.* **2014**, *5*, 4477. [CrossRef] [PubMed]
150. Liang, H.; Meng, F.; Cabán-Acevedo, M.; Li, L.; Forticaux, A.; Xiu, L.; Wang, Z.; Jin, S. Hydrothermal Continuous Flow Synthesis and Exfoliation of NiCo Layered Double Hydroxide Nanosheets for Enhanced Oxygen Evolution Catalysis. *Nano Lett.* **2015**, *15*, 1421–1427. [CrossRef] [PubMed]
151. Wang, Y.; Zhang, Y.; Liu, Z.; Xie, C.; Feng, S.; Liu, D.; Shao, M.; Wang, S. Layered Double Hydroxide Nanosheets with Multiple Vacancies Obtained by Dry Exfoliation as Highly Efficient Oxygen Evolution Electrocatalysts. *Angew. Chem. Int. Ed.* **2017**, *56*, 5867–5871. [CrossRef] [PubMed]
152. Wang, Y.; Xie, C.; Zhang, Z.; Liu, D.; Chen, R.; Wang, S. In Situ Exfoliated, N-Doped, and Edge-Rich Ultrathin Layered Double Hydroxides Nanosheets for Oxygen Evolution Reaction. *Adv. Funct. Mater.* **2018**, *28*, 1703363. [CrossRef]
153. Gong, M.; Li, Y.; Wang, H.; Liang, Y.; Wu, J.Z.; Zhou, J.; Wang, J.; Regier, T.; Wei, F.; Dai, H. An Advanced Ni–Fe Layered Double Hydroxide Electrocatalyst for Water Oxidation. *J. Am. Chem. Soc.* **2013**, *135*, 8452–8455. [CrossRef]
154. Long, X.; Li, J.; Xiao, S.; Yan, K.; Wang, Z.; Chen, H.; Yang, S. A Strongly Coupled Graphene and FeNi Double Hydroxide Hybrid as an Excellent Electrocatalyst for the Oxygen Evolution Reaction. *Angew. Chem. Int. Ed.* **2014**, *53*, 7584–7588. [CrossRef]
155. Ma, W.; Ma, R.; Wang, C.; Liang, J.; Liu, X.; Zhou, K.; Sasaki, T. A Superlattice of Alternately Stacked Ni–Fe Hydroxide Nanosheets and Graphene for Efficient Splitting of Water. *ACS Nano* **2015**, *9*, 1977–1984. [CrossRef]
156. Zhang, X.; Marianov, A.N.; Jiang, Y.; Cazorla, C.; Chu, D. Hierarchically Constructed Silver Nanowire@Nickel–Iron Layered Double Hydroxide Nanostructures for Electrocatalytic Water Splitting. *ACS Appl. Nano Mater.* **2020**, *3*, 887–895. [CrossRef]
157. Chen, B.; Zhang, Z.; Kim, S.; Baek, M.; Kim, D.; Yong, K. A biomimetic nanoleaf electrocatalyst for robust oxygen evolution reaction. *Appl. Catal. B Environ.* **2019**, *259*, 118017. [CrossRef]
158. Yang, W.; Li, J.; Liu, M.; Ng, D.H.L.; Liu, Y.; Sun, X.; Yang, J. Bioinspired hierarchical CoAl-LDH/MFe2O4(Ni, Zn, Co) as peroxidase mimics for colorimetric detection of glucose. *Appl. Clay Sci.* **2019**, *181*, 105238. [CrossRef]
159. Yang, W.; Li, J.; Yang, J.; Liu, Y.; Xu, Z.; Sun, X.; Wang, F.; Ng, D.H.L. Biomass-derived hierarchically porous CoFe-LDH/CeO2hybrid with peroxidase-like activity for colorimetric sensing of H2O2 and glucose. *J. Alloys Compd.* **2020**, *815*, 152276. [CrossRef]
160. Yang, W.; Li, J.; Wang, M.; Sun, X.; Liu, Y.; Yang, J.; Ng, D.H.L. A colorimetric strategy for ascorbic acid sensing based on the peroxidase-like activity of core-shell Fe3O4/CoFe-LDH hybrid. *Colloids Surf. B Biointerfaces* **2020**, *188*, 110742. [CrossRef] [PubMed]
161. Zhang, Y.; Tian, J.; Liu, S.; Wang, L.; Qin, X.; Lu, W.; Chang, G.; Luo, Y.; Asiri, A.M.; Al-Youbi, A.O.; et al. Novel application of CoFe layered double hydroxide nanoplates for colorimetric detection of H2O2 and glucose. *Analyst* **2012**, *137*, 1325. [CrossRef]
162. Su, L.; Yu, X.; Qin, W.; Dong, W.; Wu, C.; Zhang, Y.; Mao, G.; Feng, S. One-step analysis of glucose and acetylcholine in water based on the intrinsic peroxidase-like activity of Ni/Co LDHs microspheres. *J. Mater. Chem. B* **2017**, *5*, 116–122. [CrossRef]
163. Zhan, T.; Kang, J.; Li, X.; Pan, L.; Li, G.; Hou, W. NiFe layered double hydroxide nanosheets as an efficiently mimic enzyme for colorimetric determination of glucose and H2O2. *Sensors Actuators B Chem.* **2018**, *255*, 2635–2642. [CrossRef]
164. Fazli, G.; Esmaeilzadeh Bahabadi, S.; Adlnasab, L.; Ahmar, H. A glassy carbon electrode modified with a nanocomposite prepared from Pd/Al layered double hydroxide and carboxymethyl cellulose for voltammetric sensing of hydrogen peroxide. *Microchim. Acta* **2019**, *186*, 821. [CrossRef] [PubMed]

165. Wang, L.; Chen, X.; Liu, C.; Yang, W. Non-enzymatic acetylcholine electrochemical biosensor based on flower-like NiAl layered double hydroxides decorated with carbon dots. *Sensors Actuators B Chem.* **2016**, *233*, 199–205. [CrossRef]
166. Zhao, Y.; Chen, G.; Bian, T.; Zhou, C.; Waterhouse, G.I.N.; Wu, L.-Z.; Tung, C.-H.; Smith, L.J.; O'Hare, D.; Zhang, T. Defect-Rich Ultrathin ZnAl-Layered Double Hydroxide Nanosheets for Efficient Photoreduction of CO_2 to CO with Water. *Adv. Mater.* **2015**, *27*, 7824–7831. [CrossRef]
167. Liu, P.F.; Yang, S.; Zhang, B.; Yang, H.G. Defect-Rich Ultrathin Cobalt–Iron Layered Double Hydroxide for Electrochemical Overall Water Splitting. *ACS Appl. Mater. Interfaces* **2016**, *8*, 34474–34481. [CrossRef]
168. Nie, Q.; Ma, J.; Xie, Y. A new synthesis method of ultrathin Zn-Al layered double hydroxide with super adsorption capacity. *IOP Conf. Ser. Earth Environ. Sci.* **2019**, *300*, 052003. [CrossRef]
169. Zhao, G.; Zou, J.; Chen, X.; Yu, J.; Jiao, F. Layered double hydroxides materials for photo(electro-) catalytic applications. *Chem. Eng. J.* **2020**, *397*, 125407. [CrossRef]
170. Zhu, L.; Li, H.; Yin, Y.; Cui, Z.; Ma, C.; Li, X.; Xue, Q. One-step synthesis of a robust and anti-oil-fouling biomimetic cactus-like hierarchical architecture for highly efficient oil/water separation. *Environ. Sci. Nano* **2020**, *7*, 903–911. [CrossRef]
171. Abolghasemi, M.M.; Amirifard, H.; Piryaei, M. Bio template route for fabrication of a hybrid material composed of hierarchical boehmite, layered double hydroxides (Mg-Al) and porous carbon on a steel fiber for solid phase microextraction of agrochemicals. *Microchim. Acta* **2019**, *186*, 678. [CrossRef] [PubMed]
172. Sriram, G.; Uthappa, U.T.; Losic, D.; Kigga, M.; Jung, H.-Y.; Kurkuri, M.D. Mg–Al-Layered Double Hydroxide (LDH) Modified Diatoms for Highly Efficient Removal of Congo Red from Aqueous Solution. *Appl. Sci.* **2020**, *10*, 2285. [CrossRef]
173. Sun, Q.; Tang, M.; Hendriksen, P.V.; Chen, B. Biotemplated fabrication of a 3D hierarchical structure of magnetic ZnFe2O4/MgAl-LDH for efficient elimination of dye from water. *J. Alloys Compd.* **2020**, *829*, 154552. [CrossRef]
174. Bing, X.; Li, J.; Liu, J.; Cui, X.; Ji, F. Biomimetic synthesis of Bi2O3/Bi2WO6/MgAl-CLDH hybrids from lotus pollen and their enhanced adsorption and photocatalysis performance. *J. Photochem. Photobiol. A Chem.* **2018**, *364*, 449–460. [CrossRef]
175. Baruah, A.; Mondal, S.; Sahoo, L.; Gautam, U.K. Ni-Fe-layered double hydroxide/N-doped graphene oxide nanocomposite for the highly efficient removal of Pb(II) and Cd(II) ions from water. *J. Solid State Chem.* **2019**, *280*, 120963. [CrossRef]
176. Dou, J.; Chen, J.; Huang, Q.; Huang, H.; Mao, L.; Deng, F.; Wen, Y.; Zhu, X.; Zhang, X.; Wei, Y. Preparation of polymer functionalized layered double hydroxide through mussel-inspired chemistry and Kabachnik–Fields reaction for highly efficient adsorption. *J. Environ. Chem. Eng.* **2020**, *8*, 103634. [CrossRef]
177. Wang, J.; Yang, Q.; Zhang, L.; Liu, M.; Hu, N.; Zhang, W.; Zhu, W.; Wang, R.; Suo, Y.; Wang, J. A hybrid monolithic column based on layered double hydroxide-alginate hydrogel for selective solid phase extraction of lead ions in food and water samples. *Food Chem.* **2018**, *257*, 155–162. [CrossRef]
178. Lai, Y.-T.; Huang, Y.-S.; Chen, C.-H.; Lin, Y.-C.; Jeng, H.-T.; Chang, M.-C.; Chen, L.-J.; Lee, C.-Y.; Hsu, P.-C.; Tai, N.-H. Green Treatment of Phosphate from Wastewater Using a Porous Bio-Templated Graphene Oxide/MgMn-Layered Double Hydroxide Composite. *iScience* **2020**, *23*, 101065. [CrossRef] [PubMed]
179. Bing, X.; Zhang, X.; Li, J.; Ng, D.H.L.; Yang, W.; Yang, J. 3D hierarchical tubular micromotors with highly selective recognition and capture for antibiotics. *J. Mater. Chem. A* **2020**, *8*, 2809–2819. [CrossRef]
180. Sanchez, C.; Belleville, P.; Popall, M.; Nicole, L. Applications of advanced hybrid organic–inorganic nanomaterials: From laboratory to market. *Chem. Soc. Rev.* **2011**, *40*, 696–753. [CrossRef] [PubMed]

Review

Layered Double Hydroxides: A Toolbox for Chemistry and Biology

Giuseppe Arrabito [1,*], Aurelio Bonasera [1], Giuseppe Prestopino [2,*], Andrea Orsini [3], Alessio Mattoccia [2], Eugenio Martinelli [4], Bruno Pignataro [1] and Pier Gianni Medaglia [2]

1 Department of Physics and Chemistry Emilio Segrè, University of Palermo, 90128 Palermo, Italy
2 Department of Industrial Engineering, University of Rome "Tor Vergata", 00133 Rome, Italy
3 Università degli Studi Niccolò Cusano, Via Don Carlo Gnocchi, 3, 00166 Rome, Italy
4 Department of Electronic Engineering, University of Rome Tor Vergata, 00133 Rome, Italy
* Correspondence: giuseppedomenico.arrabito@unipa.it (G.A.); giuseppe.prestopino@uniroma2.it (G.P.)

Received: 20 June 2019; Accepted: 13 July 2019; Published: 15 July 2019

Abstract: Layered double hydroxides (LDHs) are an emergent class of biocompatible inorganic lamellar nanomaterials that have attracted significant research interest owing to their high surface-to-volume ratio, the capability to accumulate specific molecules, and the timely release to targets. Their unique properties have been employed for applications in organic catalysis, photocatalysis, sensors, drug delivery, and cell biology. Given the widespread contemporary interest in these topics, time-to-time it urges to review the recent progresses. This review aims to summarize the most recent cutting-edge reports appearing in the last years. It firstly focuses on the application of LDHs as catalysts in relevant chemical reactions and as photocatalysts for organic molecule degradation, water splitting reaction, CO_2 conversion, and reduction. Subsequently, the emerging role of these materials in biological applications is discussed, specifically focusing on their use as biosensors, DNA, RNA, and drug delivery, finally elucidating their suitability as contrast agents and for cellular differentiation. Concluding remarks and future prospects deal with future applications of LDHs, encouraging researches in better understanding the fundamental mechanisms involved in catalytic and photocatalytic processes, and the molecular pathways that are activated by the interaction of LDHs with cells in terms of both uptake mechanisms and nanotoxicology effects.

Keywords: layered double hydroxides; cellular biology; catalysis; DNA; drug delivery; hydrotalcite; osteogenesis; photocatalysis; RNA.

---

1. Introduction

Nanostructured materials or nanomaterials (NMs) represent an important area of research and a technological sector with full expansion for many different applications. They have long been considered of paramount importance due to their tunable physicochemical properties such as melting point, wettability, electrical and thermal conductivity, catalytic activity, light absorption, and scattering, which ultimately result in better performance compared to their bulk counterparts. In general, NMs are described as materials with length of 1–100 nm in at least one dimension, and can be roughly classified according to their:

i) *Composition* (carbon-based, inorganic-based, organic-based, composite),
ii) *Dimension (0D, 1D, 2D, 3D),*
iii) *Origin* (natural, synthetic),
iv) *Synthesis method* (top-down, bottom-up/self-assembly).

Of particular importance is the classification according to their size [1]. *0D nanomaterials* are systems that are at nanoscale in all their *x*, *y*, *z* sizes. Remarkable examples include carbon nanodots [2] and metallic nanoparticles (NPs) [3]. *1D nanomaterials* have one dimension greater than nanoscale, e.g., length on the micron scale in one direction only (for instance carbon nanotubes [4], silicon nanowires [5], or ZnO nanowires [6,7]). They have shown important applications as it both interconnects and has key units with a nanoscale dimension in electronics and optoelectronics. On the other hand, *2D nanomaterials* have at least two dimensions on the micron scale: Examples include nanoplatelets and nanoribbons with length or diameter on the micron scale and a nanoscale thickness. In particular, the synthesis and the application of 2D nanomaterials have become a fundamental research area in materials science, due to their many intriguing low-dimensional features that are different from the bulk properties, making them more attractive for subsequent utilization as key building blocks of nanodevices [8]. Besides the basic understanding of new physico-chemical phenomena, 2D NMs are also particularly interesting for investigating and developing novel applications in sensors, photocatalysts, nanocontainers, nanoreactors, and templates for 2D structures of other materials [9]. Three-dimensional nanomaterials (3D) can be defined as materials whose characteristic x, y, z dimensions are well beyond the nanoscale (i.e., >100 nm). These systems are typically not included in the category of nanostructured materials, unless their internal structure is nanostructured. Typical examples of this class of materials are the following: dispersions of nanoparticles, bundles of nanowires, as well as multi-nanolayers, nanocrystalline, or nanoporous materials [10].

Among NMs, layered double hydroxides (LDHs) represent an emerging class of 2D layered materials belonging to the group of hydrotalcite-like (HT) compounds, or anionic clays [11]. LDH structure can be described based on the stacking of charged brucite-like layers consisting of a divalent metal ion M^{2+} (e.g., Ca^{2+}, Zn^{2+}, Mg^{2+}, Ni^{2+}), octahedrally coordinated to six OH^- hydroxyl groups, in which part of the divalent cations (M^{2+}) are substituted by the trivalent ions M^{3+} (e.g., Al^{3+} Fe^{3+}, Cr^{3+}, In^{3+}). Such replacement leads to the formation of positively charged layers, whose net charge is compensated, to maintain the global electroneutrality, by the presence of exchangeable anions (A^{n-}, such as hydroxyl groups, nitrates, carbonates, and sulfates) in between the layers conjointly with water molecules. The general formula of LDHs is the following: $[M_{1-x}{}^{2+}\ M_x{}^{3+}(OH)_2]^{x+}\cdot[A_{x/n}]^{n-}\cdot mH_2O$, where M^{2+} and M^{3+} are the divalent and trivalent metal ions, respectively, A^{n-} are inorganic or organic anions, m is the number of interlayer water, and $x = M^{3+}/(M^{2+}+M^{3+})$ is the layer charge density, or molar ratio. LDHs are characterized by a low dimensional opened structure that is suitable for physico-chemical intercalation and adsorption processes with a large variety of molecules ranging from organic molecules to biomacromolecules.

The conventional approach to the synthesis of LDHs is based on the coprecipitation method [12–14], which can be briefly described as follows. The addition of a base to a water solution containing the salts of two different metals, namely M^{2+} and M^{3+}, causes the precipitation of the metal hydroxides and the formation of LDHs. The precipitates are collected, washed and then dried to be deposited on a solid substrate, or dispersed in solution phase. In some applications it is necessary to produce homogenous films onto solid substrates [15,16]. For this, two different research groups [17,18] demonstrated the possibility of growing stable films of well-formed interconnected LDH nanoplatelets directly onto aluminum surfaces, by immersing aluminum thick foils in a water solution of zinc nitrate or acetate. Differently from the conventional growth method, a single salt is employed in the growth solution providing the divalent metal (Zn^{2+}), while the trivalent ion (Al^{3+}) is provided by the sacrificial aluminum foil, which behaves as both reactant and substrate, finally improving the adhesion as well. In this regard, the growth of regular ZnAl LDH nanoplates [19] on Al-coated silicon substrates was demonstrated, along with the influence of the reacting aluminum layer thickness on the final morphology and composition of the LDH nanostructures. These kind of LDHs found application as gas sensors [20], Li-ion battery electrodes [16], for enhanced oxygen evolution catalysis [21], and surface enhanced fluorescence [22]. LDH systems are subjected to the so-called "memory effect", i.e., for suitably low calcination temperatures, the resulting phase can be restored to that of LDH by rehydration in water. Remarkably, during the rehydration new anionic species can be intercalated and functionalized, obtaining desired physicochemical properties [23]. In the following, the M^{2+}:M^{3+}

stoichiometric ratio in LDH will be precisely mentioned and discussed, only if definitely stated by the authors to be meaningful for the application of the LDHs. Otherwise, the LDH formula will be simply indicated as $M^{2+}M^{3+}$ LDH (see Figure 1).

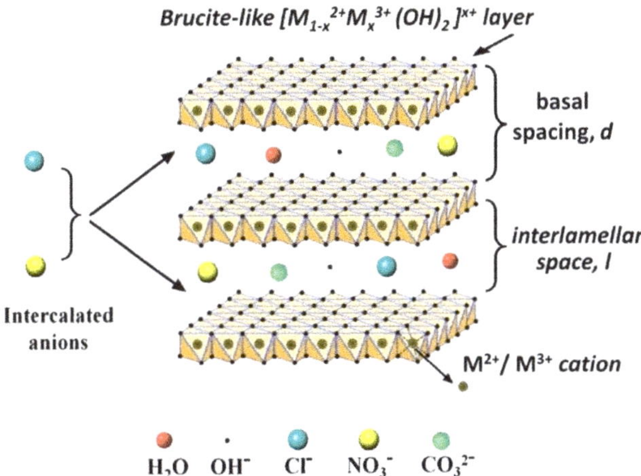

Figure 1. Schematic view of the $M^{2+}M^{3+}$ LDH general structure, with Cl^-, NO_3^- and CO_3^{2-} anions intercalated in the brucite-like structure. Other possible chemical species eventually present in the interlamellar space are reported, i.e., hydroxyl groups and interlayer water molecules. The figure is reprinted from Ref. [24] under a Creative Commons Attribution Non-Commercial No Derivatives License (CC BY-NC-ND).

The aim of this review is to introduce the emerging role of LDH materials in chemistry and biology, providing the reader with a broad up-to-date view of the role of these materials in these fields. Indeed, given the widespread contemporary interest in these issues, the topic is certainly hot, and from time-to-time, it is necessary to review the recent progresses. Differently to the previously published reviews, the aim of the present work is to provide some fundamental references concerning earlier investigations and also providing recent researches that are establishing new trends. As to the chemistry-related topics, the interest for LDHs is reviewed especially focusing on relevant organic chemistry reactions and recent advances in photocatalysis, with the ultimate goal to elucidate the key role of LDHs that is rarely investigated in literature. As to the biology-related field, LDHs are reviewed focusing within the well-explored applications in electrochemical and optical biosensors, along with DNA delivery. In addition to these, the review summarizes novel and barely explored topics, namely optical sensors, RNA delivery, cellular differentiation, and fundamental biological pathways relevant to the LDHs uptake into cells. The critical and unprecedented overview provided by the present review is hopefully able to motivate new researchers interest in these emerging fields.

Initially, we will discuss the immense potentiality of these 2D materials in organic chemistry, in particular focusing on their role as catalysts in highly relevant chemical reactions, namely the Baeyer–Villiger reaction, Knoevenagel reaction, Michael addition, and transesterification (Section 2). We will then focus on the outstanding role as photocatalytic centers, discussing their applications to organic molecule degradation, water splitting reaction, CO_2 conversion/reduction, and other examples (Section 3). Subsequently, we will analyze the intriguing role of these materials in biology, where their biocompatibility and the suitability as powerful drug vectors have triggered an enormous research interest in the last few years (more than 1000 papers published in 2018 regarding drug and DNA delivery, according to data extracted from the Scopus database). In this regard, we will provide a focus on the research results of the last two years, regarding applications of LDHs as biosensors (Section 4), especially focusing on the novel exciting applications as optical sensors. We will then analyze their important role as nucleic acid delivery agents (Section 5), providing an unprecedented view on the novel applications in siRNA therapy. Lastly, a specific view on LDHs as key materials in cellular differentiation, drug delivery, and contrast agents will also be provided (Section 6).

2. LDH Applications in Organic Catalysis

This section will focus on the role of LDH-based materials in organic chemistry, mainly as catalyst for the preparation of fine chemicals, intermediates, and valuable molecules. The interest for LDHs in chemical synthesis is due to some common features they share with classic heterogeneous catalysts; once a reaction has been catalyzed by LDHs, these layered materials can be easily removed from the crude reaction mixture (e.g., by filtration) and recovered for next reaction cycles, with significant simplification of the work-up procedure for the chemist. Among the plethora of available heterogeneous catalysts, LHDs have some additional appealing features, such as the simplicity of their preparation from cheap precursors, the involvement of non-harmful precursors, low toxicity of their possibly produced decomposition products, and no need of expensive rare elements. For details on synthesis routes, the reader is advised to go back to the previous Section, however some other aspects relevant to LDH fabrication strategies will be briefly discussed here as well, in order to clarify why scientists devote increasing attention towards specific LDH compounds.

Chemists' interest over clays does not go so back in time: early in the 80s the works of Kruissink and Reichle on methanation [25] and acetone oligomerization [26], respectively, pointed out that LDHs could be employed as precursors for catalytic species, after thermal degradation. In those works, the properties of the pristine LDHs were not addressed, but shortly after Martin and Pinnavaia revealed how LDHs could be useful also in their pristine form [27]. From there on, the number of publications on catalytic LDHs quickly grew. The aforementioned paper not only had a historical relevance, but it allows us to discuss the different strategies that researchers have when dealing with LDHs as catalysts (see Figure 2):

i) *LDH-materials as supporting surface.* The insertion of new species adds a reactivity that pristine LDHs do not possess at all. LDHs role lies in the chemical confinement, enhancement of catalyst selectivity, but no active participation to the reaction steps,
ii) *LDH-materials as source of catalytically active species.* LDHs are induced to transform in a new material with the required catalytic properties,
iii) *LDH-materials as absorbers.* The catalytic species are adsorbed in between the interlayer galleries, and can be lately exchanged with the external environment. LDHs must be previously exposed to an excess of the species to be adsorbed, which are lately released. It is evident that in this case, LDHs act as a reservoir, or create a catalytic environment/condition, but the active role is due to another atomic/molecular object,
iv) *LDH-materials as they are.* The catalytic properties belong to the lattice as it is, and are not subject to the inclusion of other species.

We are going to restrict our discussion on the latter case. Indeed, in the first case, LDH materials derive their reactivity from the hosted species, and the LDH crystal lattice has a relatively marginal role, boosting up the catalytic activity of the absorbed species but not participating directly. Several reviews have already been published focusing over specific examples where carbon nanoforms [28,29], metal nanoparticles [29,30], organic guests [31,32], and oxometalates [33,34] are used for the preparation of interesting LDH-based hybrids. In the second group, LDHs typically undergo thermal stress, with a consequent loss of structural features and transformation into mixed metal oxides (MMOs) [35,36]. Such a procedure is useful when the lattice is limiting the interaction between the metal centers and the reactants, thus the catalytic behavior arises when the layered clay framework, that is the key feature of a LDH-material, is lost. MMOs are used for a wide range of reactions, mostly in the gas phase [37–40], and various alkylation protocols, condensations and transesterifications [41]. The easiest way to reduce LDHs into MMOs is calcination, but this procedure should not be considered as a protocol since it points only to the loss of LDH structural features. Firstly, it could be a simple way to remove excess of water, which is detrimental for several catalytic routes [42]. Secondly, operating a strict control over calcination temperature, the so-prepared MMOs can regenerate LDHs after rehydration [43–46]. This is the so-called "memory effect", and published data reported that calcination temperature (T_c) should be below 500 °C, as well as compatible with thermal stability of eventual other species introduced/grafted on LDHs [47].

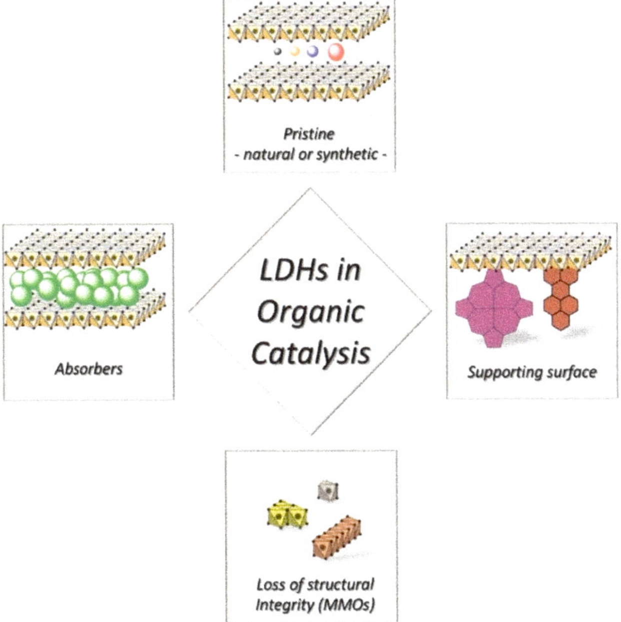

Figure 2. Representation of different possible approaches for the development of LDH-based catalysts for organic synthesis.

The reactivity intrinsically ascribed to LDHs is related to their layered structure formed by cations, anions and structural water molecules, but there are other ruling parameters; the most crucial ones are summarized here:

i) Considerable anion-exchange capability, due to the freedom that anions have to travel through the interlayer aqueous medium, the absence of strong localized interactions and specific steric constrains [48–50],
ii) Pronounced basicity, due to either structural hydroxyl anions (Brønsted theory, hydrated material) and O^{2-}–M^{2+} pairs (Lewis theory, water-free calcined material) [51],
iii) Nature of metal cations, presence of more than two metal species within the lattice (see below).

Intriguingly, the metal cation composition strongly contributes to the specific reactivity of LDHs, but it tunes as well the strength of basic moieties. The incorporation of specific metals in the octahedral layer can produce LDH compounds with unexpected and novel properties, as well as modulate the existing chemical reactivity [51,52]. It is worthy to mention literature where the arrangement of cations within the LDH lattice is discussed. As a matter of fact, the ordering of cations is believed to have crucial effects on many of the physicochemical properties of LDHs, affecting the charge density of the metal hydroxide sheets, and the overall bonding, mobility, orientation, and reactivity of the chemical species in the interlayer spaces and on the surface [53–58]. It is therefore desirable to have homogeneous distribution of cations without segregation of phases with specific composition, which would result in varied reactivity, and in catalytic processes, generation of nonspecific products [59]. Similarly, the composition of the anionic species can be varied [60]. Some other clear advantages bring researchers to consider LDHs as catalysts of choice instead of more traditional ones [57,58]. Besides their activity and/or selectivity, in heterogeneous catalysis LDHs offers the opportunity of minor production of waste and an easier removal of the catalytic species from the crude reaction mixture. One example is given by the work of Lü et al., who fabricated LDH-based catalytic films over porous alumina membrane [61]. Interestingly, the authors revealed superior activity of the catalytic film in the aldol condensation of acetone, compared to the LDH-powders prepared under identical conditions. The authors also pointed out that those LDH-crystallites grown over the alumina support retained hexagonal plate-shape perpendicularly on the substrate, which favored the diffusion of the reactants throughout the substrate, despite the more disordered organization of the LDH-powders.

In the next paragraphs, we will consider some organic reactions of interest, and we will try to give an overview of the most relevant and recent contributions of LDH compounds in organic synthesis. Our aim is to give to the general audience, not only to organic chemists, the feeling why so many researchers focused on specific reaction pathways. We will report, as well, some examples regarding industrially relevant molecules and/or precursors whose synthesis can be improved applying LDHs chemistry. Our attention regards the following classes:

i) Baeyer–Villiger reaction,
ii) Knoevenagel reaction,
iii) Michael addition,
iv) Transesterification (biodiesel production).

We would like to recommend a review by Sels et al. [62], whose analysis embraced several classes of reactions, although the reader should notice that no distinction was made among LDHs, MMOs derived from LDHs, intercalated materials and else. With an insight into life sciences, it is worth to point out that the synthesis of therapeutic molecules (i.e., drugs, active compounds) typically requires a high number of transformation and purification steps. A straightforward strategy to improve the sustainability of complex chemical synthesis is to consider how nature synthesizes biologically active compounds using enzymatic catalysis. Unfortunately, enzymes that catalyze C-C and C -N bond forming reactions employed in the synthesis of active pharmaceutical ingredients (such as Knoevenagel condensation, Henry reaction, Michael addition, and Friedel-Crafts alkylation) are quite rare [63]. In this regard, chemists have looked into more robust and cost-effective catalytic systems that can be adapted to a broad range of reaction conditions, in order to decrease the production costs of synthesis.

2.1. *Baeyer–Villiger Reaction*

The Baeyer–Villiger (BV) reaction (see Figure 3) is an oxidative pathway bringing to the formation of esters and lactones by oxidation of carbonyl compounds with a peroxide derivative [64]. The reaction takes the name of its two developers, Adolf von Baeyer (previously known for the synthesis of indigo) and his student Victor Villiger, reporting the use of Caro's acid ($KHSO_5$) as a new oxidant for the conversion of cyclic ketones to the corresponding lactones [65]. This versatile protocol was revised in the last century of applications [66]; the interest towards this reaction increased when the work by Fried et al. provided the first definitive evidence of lactone formation in living organisms according to the Baeyer–Villiger pathway [67]. Moreover, the enzymatic B-V oxidation was also explored to facilitate biotransformation of sterically demanding ketones [68]. These evidences opened the way to the search of new oxidants, with major urge in the last 20 years due to the necessity of greener protocols [69,70].

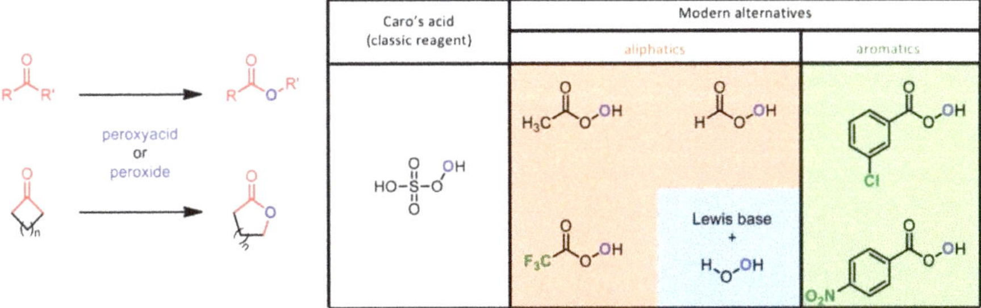

Figure 3. Baeyer-Villiger reaction protocols, and the evolution of commercially available oxidative reagents.

Some interesting papers came out in the middle of the nineties from Kaneda et al., who reported heterogeneous Baeyer-Villiger oxidation of ketones catalyzed by MgAl-CO_3 LDHs using an oxidant combination of molecular oxygen and aldehydes [71]. They also reported higher activity for oxidation of five-membered ring ketones than for that of six-membered ones, with improved yields using mild reaction conditions (40 °C). Successive work reported multi-metallic MgAl LDHs containing Fe ions efficiently oxidizing various cyclic ketones in the same mild conditions, while the exchange of Fe with Cu gave superlative yields with bicyclic ketones [72]. The intuition of the beneficial effects due to the presence of iron in the LDH lattice was also confirmed by Kawabata and coworkers, in the frame of a wider investigation including LDHs with different metal species, such as Fe, Co, Ni, or Cu [73]. An attempt to adapt the procedure to other commercially available oxidants was reported as well [74].

A step towards greener chemistry with no or less harmful chemicals was detailed by Pillai et al., who reported Sn-doped LDHs as efficient and relatively cheap catalysts for BV oxidation reactions, in combination with classic H_2O_2 as oxidizing agent [75]. The mechanism of oxidation was tentatively explained as due to the carbonyl group activation of ketone by the Sn present in the interstitial spaces of LDHs, followed by a subsequent nucleophilic attack by the active peroxide species (peroxycarboximidic acid by reaction of acetonitrile and H_2O_2); the so formed Criegee adduct rearranges to give the corresponding lactone as a final product. Lately, Jiménez-Sanchidrián et al. [76] confirmed the choice of Sn as a primary element to be considered in the design of BV-active LDHs, with reactions running under very mild conditions (atmospheric pressure and a temperature of 70 °C), conversion yields sometimes higher than 80%, and 100% selectivity after 6 h. Among their findings, it is worth noting the evidence that solids containing Zr promoted the decomposition of the hydrogen peroxide and hence adversely influenced the oxidation reaction.

Recent investigations by Olszówka et al. showed that the catalytic performance of MgAl LDHs in cyclohexanone oxidation with H_2O_2 /nitrile system could be significantly improved by the use of Mg-rich catalysts (Mg/Al = 3.69) [77]. Yields up to 50% ε-caprolactone could be achieved in a single-phase reaction medium based on the acetonitrile solvent. The use of acetonitrile rather than the more expensive, more toxic and more difficult to handle benzonitrile, represented a greener and a more economically viable option. Successive structural studies revealed that a lowering of the microcrystalline domains was beneficial for the BV reaction as it improved catalyst's selectivity [78]. This effect was attributed to the observed higher hydrophilicity of less crystalline materials. While it was argued that the enhanced hydrophilicity of poorly crystalline HT samples facilitates the approach and activation of H_2O_2 on surface basic centers, it makes the catalysts more prone to random interactions with the organic substrate, and to its subsequent non-selective transformation [79].

2.2. Knoevenagel Reaction

The Knoevenagel condensation (see Figure 4) is one of the most diffused tools in the hands of organic chemists. Such versatile reaction can be briefly defined as the product of interaction between a carbonyl compound (aldehyde or ketone) and any compound having an active methylene group. When we say active, we mean that the presence of ancillary electron withdrawing moieties (nitro, cyano, or acyl group) weakens C-H bonds in close proximity. The reaction terminates with a dehydration reaction in which a molecule of water leaves the molecular skeleton. The primary product of a Knoevenagel reaction is usually an α,β-unsaturated ketone [80]. The Knoevenagel condensation is frequently used when building biologically active scaffolds, and it finds application in the synthesis of several biologically-active molecules [81], such as steroids [82].

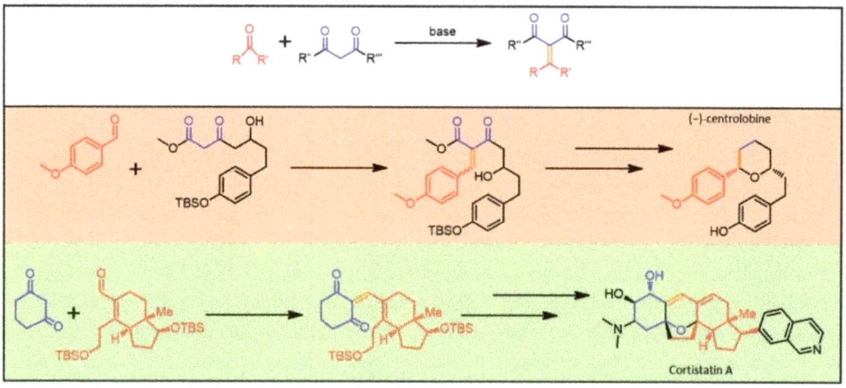

Figure 4. Knoevenagel reaction general scheme. Middle and bottom: examples of some pharmaceutically relevant molecules, whose synthesis involves Knoevenagel condensation. In evidence, the newly formed chemical bond (orange) and the residues belonging to the starting carbonyl (red) and 1,3-diketo (blue) moieties.

An interesting application of LDHs in Knoevenagel reaction goes back to 1999, when Rousselot et al. reported the synthesis of mixed Ga/Al-containing LDHs using different counter-ions ($CO_3{}^{2-}$, F^-, and NO^{3-}) and doped with different divalent and trivalent cations (Cu^{2+}, Mg^{2+}, Zn^{2+}, Al^{3+}, Ga^{3+}, Mn^{3+}, and Sc^{3+}) [83]. In their systematic study, they reported on the relative amounts of metallic ions necessary for achieving the best performing material. Besides the structural characterization, the most interesting results regarded the reactivity performance for the Knoevenagel condensation of ethyl cyanoacetate with benzaldehyde. Both calcined and uncalcined materials showed high reactivity; a rehydration process of the calcined samples during the catalytic reaction may explain the similarity. Water, formed together with ethyl cyanocinnamate, would react with the catalysts to give back the original LDH structure.

Fluorinated LDHs (LDH-F) were later developed by Choudary and collaborators in the effort to obtain a more basic (and efficiently) Knoevenagel catalyst [84]. The catalytic clay was prepared starting from a MgAl–NO_3 LDH, which underwent calcination (450 °C) followed by rehydration in presence of KF aqueous solution. The highly polarized basic fluoride ions displayed high catalytic activity both in Knoevenagel condensation (tested the reaction between 2-methoxybenzaldehyde and malononitrile) and 1,4-Michael addition (tested the reaction between acetylacetone and methyl vinyl ketone), under mild liquid phase conditions. The other advantages of LDH-F include easy separation of the catalyst by simple filtration, high atom economy to enable waste minimization, reduced corrosion, and reusability. Such features are certainly appealing for the industrial world converting to greener protocols.

In the general attempt of finding the best combination of experimental parameters, a step forward came from Constantino et al. [85]. In their work, they focused on NiAl-CO_3 LDHs applied to the Knoevenagel reaction involving malononitrile and ethylcyanoacetate, in neat conditions, since the reagents could work as a mixing liquid phase. Mild reaction conditions (60 °C) were sufficient to bring the reaction to completion. Interestingly, during preparation of the heterogeneous catalyst, they performed a thermal pre-treatment (150 °C per 1 h) claiming that the excellent reaction yields could be due to the anhydrous catalyst that should be able to co-intercalate water molecules generated in the Knoevenagel condensation, and withdraw the reaction equilibrium to the formation of the adduct. Dimethylmalonate, having methylene groups of very low acid strength ($pK_a > 13$) did not give the benzaldehyde adduct, opening the room to chemoselective reactions. Shortly later, the results from Li et al. confirmed the intuition that calcined materials could work in a more profitable way in the context of Knoevenagel protocols [86]. In their study, carbonate-containing LDHs with different combinations of Al^{3+}, In^{3+}, and Mg^{2+} cations were investigated in the condensation of ethyl cyanoacetate with benzaldehyde. The structural characterization pointed out that other relevant parameters should be taken in account, i.e., porosity, surface area, and basicity of the materials, which always exhibited higher values in the case of the more performing calcined materials. Lei et al. moved little further, pointing at the crystallinity of the material [87]. For their MgAl-OH LDHs obtained by calcination/rehydration of MgAl-CO_3 LDH-precursors with high crystallinity, a base-acid catalytic mechanism was proposed to interpret the catalytic behavior. Activated MgAl LDHs synthesized by urea hydrolysis showed a much higher activity in aldol and Knoevenagel reactions than the corresponding material synthesized by the co-precipitation method, suggesting that the presence of acid-base hydroxyl pairs was required, and that catalytic activity was maximized for highly crystalline structures with an ordered array of surface hydroxyl groups. Other groups pursued different approaches towards the improvement of catalytic performance, not related to structural features. Among them, one option was to maximize the basicity of the material, by the introduction of new chemical motifs. In a recent example [88], MgAl LDH-nanosheets were functionalized by grafting aminopropyltriethoxysilane (APTS) onto the clay surface, introducing free amino groups. The prepared MgAl-NH_2 LDH-nanosheets exhibited excellent performance in the Knoevenagel condensation compared to the homogeneous catalyst APTS and the heterogeneous catalysts with Al_2O_3 and SiO_2. These results find reasonable justification in the high surface area of the hybrid nanosheets, and synergistic effects among MgAl LDHs, the grafted -NH_2 groups, and reaction substrates.

Some recent reports tried to conjugate the advantages of heterogeneous catalysis with greener approaches. An example is grafting of LDHs with ionic liquids (ILs). Khan et al. opened the way testing widely described MgAl LDHs (Mg:Al = 3:1), synthesized following published procedures and avoiding pre-treatments, immersed in ILs [89]. Besides the advantageous efficiency of LDH-catalysts and the presence of ILs as a safe and reusable media, they reported also significant alteration in diastereoselectivity in the case of the nitroaldol reaction with nitroethane. In a recent report, Li et al. grafted ILs with different length of their alkyl chains (IL-Cn with n = 4, 8, or 12) onto MgAl-NO_3 LDHs, revealing that the grafting approach helped to adjust the distribution of basic sites, and also induced flexibility to the catalyst, allowing easy accessibility of the active centers by the substrates [90]. Application of LDH-ILs-Cn for Knoevenagel condensation of various aldehydes with ethyl cyanoacetate/malononitrile results in excellent yields, high selectivity, and efficacy in aqueous solution at room temperature, and recyclability as well.

It is here worth mentioning a less conventional approach proposed by Zhou and coworkers, who reported on LDHs developed for the catalysis of two different reactions, which happened in sequence in their case, but that could be easily controlled by the presence/absence of specific chemicals [91]. In more detail, they investigated NiGa LDHs, which could catalyze oxidation reactions and Knoevenagel condensations; in the specific case, alcohols were oxidized to the corresponding carbonyl compound (in the study, benzyl alcohol to benzaldehyde), which reacted in the next step with a partner molecule (benzoylacetonitrile). The introduction of the second partner necessarily ruled the occurrence of the second step, otherwise the simple oxidation would result at the end of the treatment. A good tolerance for the catalyst to various substrates, and excellent recyclability were also reported.

2.3. Michael Addition

The Michael addition reaction (see Figure 5) is generally described as a base-catalyzed addition of a nucleophile, such as an enolate anion, to an activated α,β-unsaturated carbonyl-containing compound, resulting in the formation of a new C-C bond [92]. Since the pioneering work by Arthur Michael [93], this versatile protocol was widely developed for several different fields, ranging from basic research [94] to industrial applications [95].

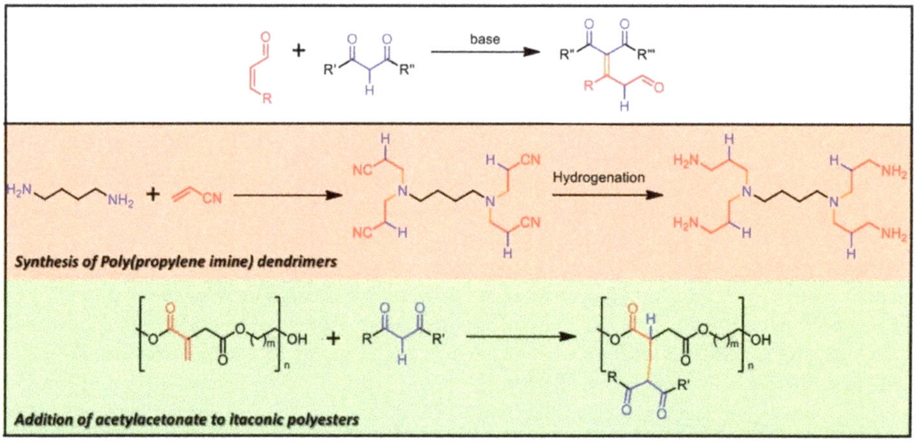

Figure 5. Michael Addition Reaction General Scheme. Middle and bottom: examples of some industrially relevant materials, whose synthesis takes advantage of Michael Addition. In evidence, the newly formed chemical bond (orange) and the residues belonging to the starting carbonyl (red) and 1,3-diketo (blue) moieties.

LDHs play again an interesting role in the recent development of new synthetic routes. A first combination of LDH chemistry and Michael additions came with the work by Choudary et al. [96]. They reported selective 1,4-additions on methyl vinyl ketone, methyl acrylate, simple and substituted chalcones by donors such as nitroalkane, malononitrile, diethylmalonate, cyanoacetamide, and thiols, catalyzed by MgAl LDHs. In their work LDHs, as synthesized or just calcined, showed no relevant activity for the reactions described here, but only the material obtained by decarbonation and subsequent rehydration was an efficient and very selective catalyst. Products of undesirable side reactions resulting from 1,2-addition, polymerization and bis-addition were not observed. Later [97], they discussed the activity of MgAl-OtBu LDHs prepared by incorporating tert-butoxide anion into the interlayer of the clay structure to enhance the basicity of the material. The results were encouraging; reaction yields were reported always to be >85% with mild reaction conditions (room temperature in methanol). Compared with already reported procedures based on non-LDH catalysts, Michael reaction does not entail anymore very long reaction times (in the worst presented case, 2 h vs. 75 h for traditional protocols), high catalyst loading and low yields of the adducts. Waste minimization without any side reactions, use of non-toxic and inexpensive catalysts and recyclability of catalyst are other advantageous features of this procedure. Survey over several reaction parameters (solvent, temperature, and time) was reported by Naciuk et al. [98] who investigated Michael reaction for the synthesis of γ-aminobutyric acid (GABA). Another study on the Michael additions of 2-methylcyclohexane-1,3-dione, 2-acetylcyclopentanone, and 2-acetylcyclohexanone to methyl vinyl ketone interestingly revealed that Mg/Al molar ratio was strongly related to the catalytic performance, pointing out a nonlinear correlation between the Mg content and catalytic activity [99]. This finding would suggest that pure MgO would be more efficient than MgAl LDHs, as they tested and reported, but it causes consecutive reactions of the Michael addition products, which detrimentally reduce product selectivity and yield. Kaneda et al. reported further insights into the reaction mechanism and the quantification of LDH basic sites [100]. LDH compounds were also used as catalysts in green protocols for Aza-Michael addition. In this regard, in the effort to determine the LDH-catalyst with the highest activity and selectivity, Kantam et al. discussed the reaction between dibutylamine and methyl acrylate (Aza-Michael reaction, due to the amino-group acting as nucleophile) comparing several LDH compounds obtained from calcination and rehydration of different precursors, as well as pure metal oxides [101]. CuAl LDHs proved to be the best candidates, with HT catalyst at room temperature in very good yields. The Cu-Al hydrotalcite showed enhanced activity over the other tested solid catalysts, and it was reused for several cycles with consistent activity and selectivity. One last interesting report concerns LDH-catalysts with anchored L-proline [102]; the catalytic activity was comparable to what reported for other LDHs, but the novelty lay in the asymmetric induction on the Michael addition products. Reaction between β-nitrostyrene and acetone with an inversion in the asymmetric induction was observed when compared to the reaction using pure L-proline catalysis, opening the way to the production of enantiomers using catalysts where the chiral (and expensive) enantiomer is recycled for both the two reactions.

2.4. Transesterification (Biodiesel Production)

The transesterification reaction is one of the universal and well-known tool in the hands of organic chemists. It consists in the exchange of the alkoxy-group of an ester with another one generated by an alcohol; the reaction is reversible, and can be acid-base catalyzed [103]. Besides basic research, it has a never ending number of industrial applications, such as in plastics and polymer technologies [104,105], and biodiesel conversion of biomasses [106].

The importance of basic sites for the catalysis of transesterification induced the scientific community to verify the suitability and performance of LDH-materials. Cantrell et al. reported a deeply detailed investigation over a series of MgAl-CO_3 LDHs [107]. All materials revealed to be effective catalysts for the liquid phase transesterification of glyceryl tributyrate with methanol for biodiesel production. The rate increased steadily with Mg content, with the Mg rich $Mg_{2.93}Al$ catalyst an order of magnitude more active than MgO (comparable results were lately reported by Xie et al. and Zeng et al., in two separate studies using as starting biomass soybean and rape oils, respectively [108,109]). Pure Al_2O_3 (completely inert) was investigated as well for a complete comparison. Their structural investigation resulted in a correlation between reaction rates and intralayer electron density, which could be associated to increased basicity. Some other interesting aspects were revealed by Liu et al. [110]. In their investigation on poultry fat transesterification with methanol, they determined that rehydration of the calcined catalyst before reaction using wet nitrogen decreased catalytic activity, and also, methanol had to be contacted with the catalyst before the reaction took place, otherwise catalyst activity was seriously impaired by strong adsorption of triglycerides on the active sites. Both increased temperature and methanol-to-lipid molar ratio favorably affected the reaction rate. Navajas et al. confirmed these observations about rehydration, and achieved outstanding sunflower oil conversions up to 96% c with 2% w/w of catalyst [111].

Kondawar et al. worked on glycerol transesterification preparing MgAl LDHs, pure or doped with Ca and La [112] (see Figure 6). While the study also interested the simple metal oxides, and CaO showed the highest activity, the formation of soluble calcium glycerate prevented its recovery and recyclability. For this reason, the study justified Ca-doped LDHs as a good compromise between performance and regenerability of the catalyst. Higher activity for Ca-based materials could be traced in the higher basicity obtained for these entries.

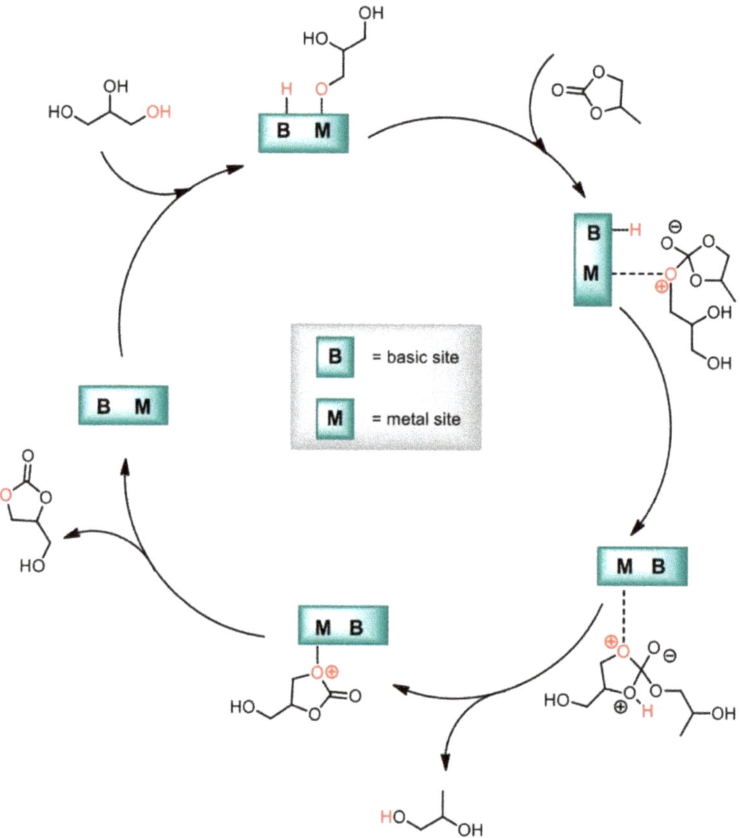

Figure 6. Mechanism of solid-base-catalyzed transesterification of glycerol. Adapted from Ref. [112]. Copyright 2017 American Chemical Society.

Looking for alternative approaches, Yagiz et al. developed MgAl-NO_3 LDHs functionalized with lipase enzymes [113]. The influence of temperature, pH, time and particle sizes was comparatively evaluated on enzyme activity, and the optimum was reached with the immobilization of 13 mg/g of enzyme, working at pH 8.5 and 4 °C. It is reported that the immobilized lipase on HT yielded a lipolytic activity equivalent to 36% of initial activity of lipase, with the relevant advantage of an easy removal of the catalytic hybrid clay at the end of the process and possible recyclability.

3. LDHs Applications in Photocatalysis

In the previous Section, we have discussed in detail some classic organic reactions, and the contribution that LDH-based compounds offer as a catalyst in the development of alternative, greener, and/or more efficient protocols.

A separate section is here dedicated to LDHs in photocatalysis motivated by the following reasons. First, LDHs have been always tested and studied for several applications, e.g., catalysis, with a regular production of literature; in the case of photocatalytic applications, the trend in the publications appears quite anomalous. The first evidences came late in the 80s, with two pioneering reports from the research unit lead by Pinnavaia [114,115]; after that, a negligible number of publications (4 papers in 15 years) was recorded, until the new century coming, with a boom of reports starting from 2008 when two well recognized papers from the team of Vansant received attention from the scientific community [116,117]. Thereafter, the urge to find answers to the global climate and environmental crisis pushed the accelerator for a new reinterpretation of LDH chemistry. One of the most active researchers in the field of sunlight-driven water splitting, Michael Grätzel, advertised NiFe LDHs as efficient photoactive catalyst electrodes [118]. The volume of publications released in the time window 2016-2019 on photocatalytically active LDHs grew to >150 items. A second reason led us to the creation of a separate section in the present review: although several detailed reviews are available in the recent literature, for example the report from Mohapatra and Parida [119], the number of new available papers calls for a new up-to-date overview.

This section will address three fields of applications, which are hot topics in the recent literature:

i) organic molecules degradation,
ii) water splitting reaction,
iii) CO_2 conversion/reduction,
iv) other examples, which do not belong to the previously mentioned ones.

3.1. Organic Molecules Degradation

Dye molecules are intimately connected with history of humankind, from the prehistoric age to the industrial civilization. Nowadays, although countless applications of dyes, their ubiquitous presence calls for deeper investigations over their implications in environment pollution [120,121]. Many reports closely connect dyes to insurgency of allergic phenomena [122]. It sounds like a paradox, the search for colorants with magnificent properties, for example persistence, ended up in molecules that are difficult to be degraded. The reason why is that chemists pursued different approaches for solving this urgent issue [123–125].

In the last decade LDHs found increasing application in this variegated field, starting from the first report, again from the team of Vansant [126], with increasing attention gained after the publication by Parida and Mohapatra [127], dealing with efficient and cheap ZnFe-CO_3 LDHs. Herein, attention is given to specific LDH compositions focusing on the most recent literature. Zn and Al are recurrent elements in the photocatalytic HT clays formulation [128–135]. Many works report that pristine LDHs can be very efficient without any further treatment, such as calcination, providing decoloration percentage of 90% for solutions of methylene blue (MB) after 1 h irradiation time (better than that of commercial ZnO nanoparticles) [128]. In order to improve the photodegradation performance, additives can be included in the LDH-lattice; cerium doping appeared an interesting option to be considered, as it provided an efficiency 1.5 times higher than pure ZnAl LDH [131] (see Figure 7).

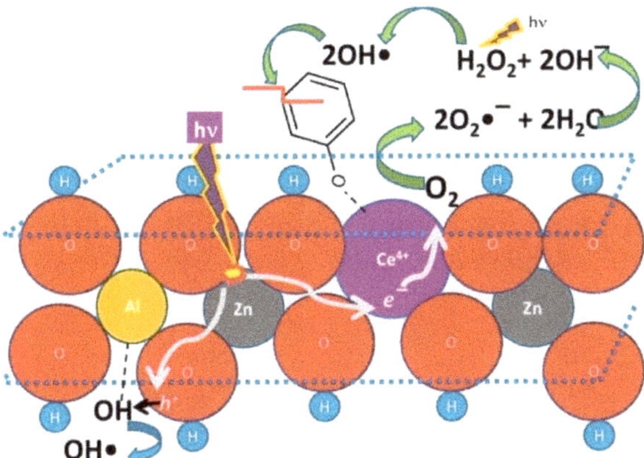

Figure 7. Proposed photocatalytic mechanism for the cerium presence in ZnAl LDH materials and its role as charge separator. Adapted from Reference [131]. Copyright 2016 Elsevier.

A proposed operation mechanism considers Ce^{4+} cations able to: i) Collect e^- photogenerated by the LDH lattice; ii) convert to $^*Ce^{3+}$; and iii) transfer the excitation energy (and electron) to an O_2 adsorbed molecule. The so-generated O_2^- radicals can react with water molecules and generate H_2O_2; UV light will break the peroxide and generate highly reactive OH radicals, which are deputed to decompose the organic pollutant (e.g., phenol). In a similar way, lanthanum provided beneficial effect in MB photodegradation operated by doped ZnCr LDHs [130]. However, the integration of TiO_2 in LDHs [134] looks more appealing, since it has the relevant advantage to be a well-known photosensitizer with an extremely low price on the market, making its choice more affordable and convenient for large scale production/application. Another interesting manipulation is the anionic exchange and decoration with Ag_2CO_3, which provided higher selectivity over anionic dyes, as revealed in comparative tests between MB and Red X-3B [129].

In order to find alternative materials where synergic effects could be helpful, Mohamed et al. reported a study over polypyrrole nanofibers coated with ZnFe LDHs [135]. The novel hybrid material was tested for the photocatalytic removal of Safranin dye, or Basic Red 2, a biological stain. The adsorption capability was reported to be about 22% higher than the naked nanofibers, and 31% higher than the pristine ZnFe LDHs. The photocatalytic removal was improved by 42% and 54% higher than the nanofibers and the starting LDHs, respectively.

As a general remark, to be kept in mind throughout the following sections, almost all the specific photocatalytic reactions that we decided to mention in the present work deal with Zn(II)-based LDHs. While the presence of Zn(II) is constant in the reported examples, where Zn seems to be a component of fundamental importance, some works focus their attention over different elemental species. One example is reported by Timár et al. [136]; in their work, the photocatalytic degradation of MB was reported using Mn_2Cr LDHs as photocatalysts. The performance of such LDHs was comparable with commercially available Degussa P25 TiO_2, and remained unaltered over five consecutive runs. Further analyses over materials derived from calcination of the pristine LDHs showed inferior activity; changes in the performance could be traced in the gradual collapse of the layered structure, until no activity could be observed for the photocatalytically inactive double oxide.

3.2. Water Splitting Reaction

LDHs have emerged as highly active photocatalysts for water splitting reaction due to some appealing features. The large surface area and semiconductor characteristics captured the attention of several groups. Some limitations stimulated researchers to exploit modifications of LDH-based photocatalysts. Among them, Fu et al. reported on the doping procedure of ZnCr LDHs with terbium cations, demonstrating an improved photocatalytic performance [137]. Photoluminescence and photoelectrochemistry measurements on the ZnCrTb LDH samples revealed a more efficient charge carrier separation and higher injection efficiency, compared to the pristine non-doped ZnCr LDHs. Optimal performances were observed for a specific doping level (0.5%), with a 2-fold increase in photocatalytic O_2 production efficiency. Oxygen generation through photocatalytic water splitting under visible light irradiation was studied by Gomes Silva et al., who investigated a series of ZnTi, ZnCe, and ZnCr LDHs at different Zn-to-metal atomic ratio (from 4:2 to 4:0.25) [138]. ZnCr LDHs proved once again to be the best matching case, with an atomic ratio of 4:2. The authors focused on the apparent quantum yields for oxygen generation, with values well comparable with previous results by Fu et al. It is worthwhile to point out that these quantum yields (60.9% and 12.2% at 410 nm and 570 nm, respectively) were among the highest values ever determined with visible light for solid materials in the absence of a light harvesting dye. Interestingly, in a field of rapid growth, it is appreciable that different research groups arrived to the confirmation of similar findings, further supporting and consolidating the trend in photocatalytic exploitation of LDHs.

Lee and coworkers followed a different approach to the improvement of photocatalytic performances, with specific focus over the other half-reaction, bringing the evolution of H_2 [139]. In their case, the material consisted of a combination of graphitic carbon nitride (g-C_3N_4), which acted as a support for ZnCr LDH nanocrystals. Structural characterization and investigation of surface area confirmed the localization of the nanocrystals in the mesopores of the graphene-like lattice; an efficient electronic coupling between both the two components of the hybrid material gave rise to the observed enhanced visible light absorptivity and suppression of electron–hole recombination.

In two different reports [140,141], photoanodes were successfully prepared by the integration of bismuth vanadate $BiVO_4$ with Co-based LDHs (see Figure 8). The design presented by Vo and collaborators was more complex, bringing to the preparation of a CoMnZn trimetallic anionic clay, but several details are herein discussed. The evidences suggested synergistic effect of the three-metal composition towards the enhanced photoelectrochemical performance. Surface modification of photoanodes with trimetallic hydroxides greatly improved the migration of holes from bismuth vanadate to LDH, facilitating fast separation and transport of holes, thus retarding the recombination of photogenerated charges.

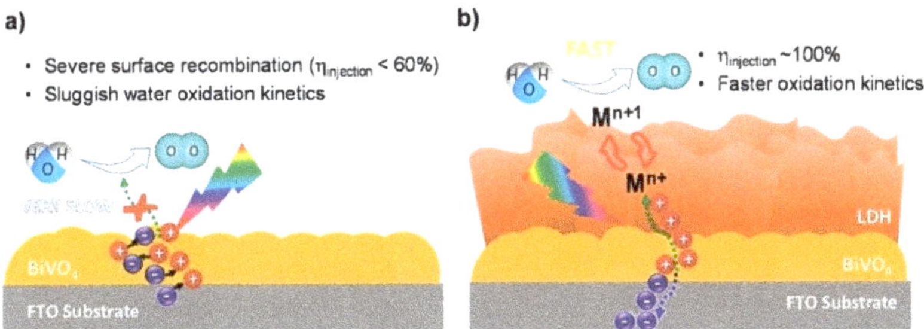

Figure 8. Schematic representation of charge carrier dynamics in water oxidation on BVO (**a**) and BVO/CMZ-LDH (**b**) photoelectrodes. Adapted from Reference [141]. Copyright 2019 Elsevier.

3.3. CO_2 Conversion/ Reduction

Photocatalytic conversion of CO_2 into alcohols is a chemical pathway widely pursued in order to answer two relevant issues. From one side, to find application to an inert molecule which is the last step of carbon oxidative chain, and a dangerous greenhouse gas as well. From the other side, it is a way to produce liquid fuels in an environmentally compatible manner and in a less energy demanding way. Several different approaches and materials were tested [142–144], and LDHs resulted to be an intriguing option also in this case. Some interesting achievements concerned MgAl LDHs, which are the simplest, most common, and deeply studied HT compounds. Their high CO_2 adsorption performance, even at room temperature, was already reported [145]. Among the latest reports, Flores-Flores et al. focused on the development of a microwave-assisted protocol for catalytically active LDHs [146]. As the authors claimed, MW-irradiation was necessary to produce samples with high crystallinity, which had a strong impact over methanol production rate. During the photocatalytic experiments, under halogen lamp irradiation, MgAl LDHs showed higher selectivity for CH_3OH production in liquid phase than in gas phase, due to a more negative flat band potential to carry out the CO_2 reduction. Iguchi et al. played with the composition of MgAl LDHs, introducing fluorine species in the lattice (e.g., AlF_6^{3-} anions) [147]. In their reports, they observed that the reduction cascade led to the selective production of CO under UV-light irradiation. A highly selective compound for CO_2 photoreduction into methanol employed ZnCuGa-CO_3 LDHs [148]. In the paper, the photocatalytic material was required to be heated at 150 °C in vacuum in order to reduce the content of interlayer water by 31%; after this stage, if LDHs never got in contact with air prior to the photoreduction tests, methanol production selectivity was verified to be >97% in all the studied cases.

A change in the divalent cation (Ni instead of Mg) was demonstrated to modify the outcome of the catalysis; e.g., the resulting NiAl LDHs showed 80% selectivity towards CO among the reduction products after 20 h of UV light irradiation. The substitution of metal cation came from another broad investigation over 16 different kinds of transition metal containing $M^{2+}M^{3+}$ LDHs (M^{2+} = Co, Ni, Cu,

Zn; M^{3+} = V, Cr, Mn, Fe) applied to the photocatalytic conversion of CO_2 to CO in an aqueous solution of NaCl [149]. A *summa* of these results can be finally seen in the investigation reported by Tokudome et al., where NiAl LDHs were tested as nanocrystals [150]; the remarkable rate of photocatalytic CO_2 reduction by the nano LDH catalyst was reported to be almost one order of magnitude (7 times) higher than that measured by means of LDH catalyst prepared through conventional methods. It looks like the research unit made a step backward, ignoring the best performances offered by NiV LDHs, which were supposed to benefit from the implementation of nanoforms. It can be argued, however, that vanadium is more expensive and exotic than well established (and cheap) aluminum.

Crystallinity is a recurrent keyword in the work of Zhao et al. [151]. In their work, they compared the synthetic protocols for MgAlTi LDHs (co-precipitation, co-precipitation + hydrothermal, and co-precipitation + calcination + reconstruction). Crystalline TiO_2 domains were present in the LDHs obtained by hydrothermal or reconstruction processes. The material hydrothermally treated at 150-200°C demonstrated the highest CO production due to a well-balanced TiO_2 crystallinity and specific surface area. Compared with commercial TiO_2-P25 nanoparticles, MgAlTi LDHs demonstrated 2 to 4 times higher catalytic activity in CO_2 photoreduction to CO.

Another example calls back what was already described for LDH application in water splitting reaction. The work from Tonda et al. exploited the possibility to take advantage of a two-component hybrid combining g-C_3N_4 with LDHs [152] (see Figure 9). In the present case, a detailed characterization and description of the photocatalytic mechanism in the g-C_3N_4/NiAl-LDH heterojunction culminated in the observation of a CO production rate 5 times higher than that of pure g-C_3N_4, and 9 times higher than what was revealed for pure NiAl LDHs.

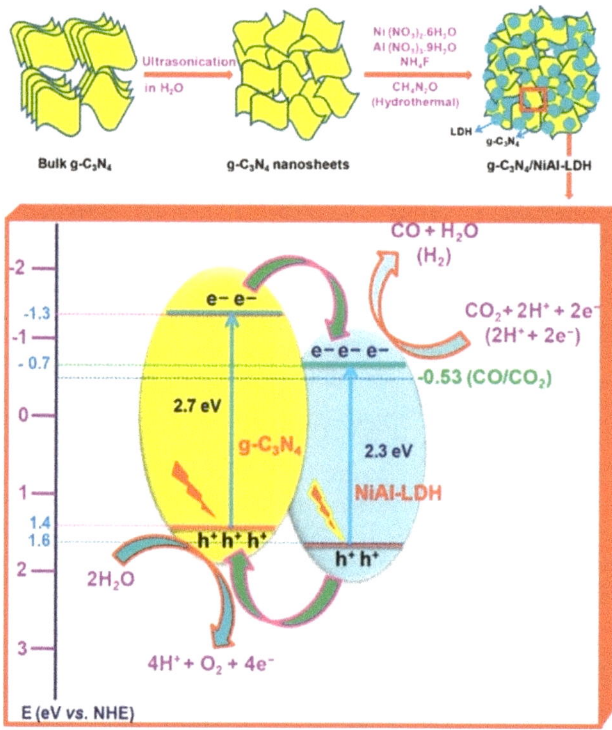

Figure 9. Top. Schematic illustration of the synthesis process of g-C_3N_4/NiAl-LDH hybrid heterojunctions. Bottom. Proposed mechanism for CO_2 photoreduction in the g-C_3N_4/NiAl-LDH heterojunctions. Adapted from Reference [152]. Copyright 2018 American Chemical Society.

3.4. Others

In this section, a couple of interesting papers focusing on other relevant topics are discussed. In the previous Section, several words have been spent over transesterification as an important route to produce biodiesel. Again, LDH materials are used, as well, in photocatalytic applications close to the petroleum chemistry, as Gao et al. reported [153] (see Figure 10). LDHs were employed in the mitigation of polluting emissions of diesel oil, mostly related to the high concentration of sulphur-containing hydrocarbons [154]. Their LaZnAl-MoO_4 LDHs were reported to promote desulfurization of diesel oil under UV irradiation, favoring the oxidation of dibenzothiophene (DBT) to the corresponding sulphone. The conversion rate, based on the quantity of sulfone that remained adsorbed to the LDH surface, was determined as being up to 84% in 1 h.

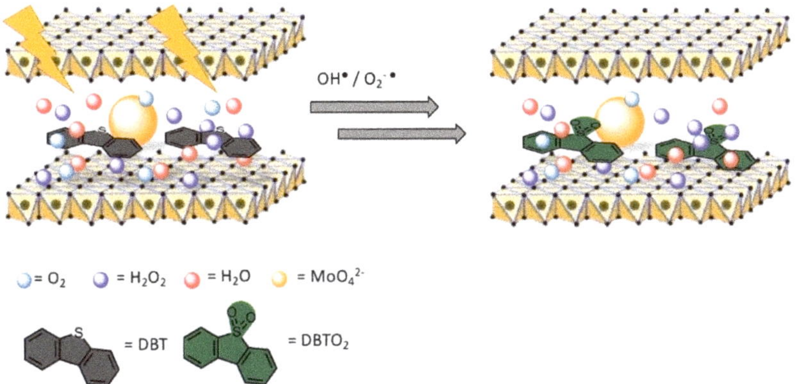

Figure 10. Catalytic oxidation of dibenzothiophene (DBT) on $LaZnAl_3$-MoO_4 LDH-catalyst. Adapted from Reference [153]. Copyright 2018 Taylor and Francis.

To justify their evidences, they claimed that MoO_4^{2-} anions increased the interlayer space promoting the adsorption of dibenzothiophene (DBT), which is one of the most relevant sources of sulphur in the oil. Synergistically, MoO_4^{2-} acted as the active sites for the oxidation of DBT, resulting in the high desulfurization efficiency. These statements were justified by comparing the molibdate LDH-photocatalyst with an equivalent one with carbonate anions, which showed inferior desulfurization performance. The presence of La^{3+} cations was mandatory, since they brought more positive charge to the brucite-like sheets, leading to an improved adsorption of molibdates on the surface of the layers.

Photocatalytic LDHs found application in the improvement of another fundamental process: N_2 fixation. The reaction consists in the reduction of gaseous N_2 to ammonia, NH_3; it is an essential mechanism for the production of nitrogen containing biological molecules [155], as well as industry relevant products, such as fertilizers [156]. The Haber-Bosch process [157], one of the most important processes of human history, produces ammonia under extreme conditions cycle (400–500 °C, 200–250 bar), leaving room for improvements and discovery of more convenient alternatives. Zhao et al. reported the photocatalytic activity of ultrathin LDH nanosheets (NSs) made with different combinations of di- and trivalent metal cations (M^{2+} = Mg, Zn, Ni, Cu; M^{3+} = Al, Cr) [158]. The most promising samples were the CuCr-LDH NSs; photocatalytic reduction of N_2 to NH_3 was observed in water at 25 °C under visible-light irradiation, with quantum yield of about 0.44% at 380 nm and 0.10% at 500 nm. Monochromatic light of wavelength 500 nm afforded a NH_3 evolution rate of about 7.1 µmol L^{-1}, which was a great achievement considering that the authors were far away from the UV and significantly closer to the maximum of solar emission spectrum. The photocatalytic activity was attributed to the distorted structure in the LDH NSs, which is supposed to enhance N_2 chemisorption and to promote NH_3 formation.

4. LDH Applications in Biosensors

LDHs have been largely employed as efficient biosensors, owing to their excellent biocatalytic properties and to the possibility of producing hybrid materials with enzymes. Starting from the first example of urea biosensors based on the immobilization of urease into oppositely charged clays [159], most of the research in the field of LDH-based biosensors was devoted to the fabrication of oxidoreductase enzyme/LDH amperometric biosensors, where typically employed enzymes were transketolase, acetylcholinesterase, horseradish peroxidase, and glucose oxidase [160]. The researchers also showed the possibility to prepare hybrid LDHs containing redox active molecules as enzyme immobilization matrices, such as anthraquinone sulfonate, ferrocene, and 2,2′-azinobis 3-ethylbenzothiazoline-6-sulphonate.

In general, the class of enzyme-based biosensors is featured with high costs and low stability, being their response potentially affected by factors such as temperature, pH, and ionic strength [161,162]. For these reasons, recent research efforts were focused on the fabrication of enzyme-free biosensors mostly based on the functionalization of electrodes with functional nanomaterials, benefiting from low costs, rapid response, high sensitivity [163], and from the possibility to enhancing electrode activity providing much more accessible exposed active sites as well as to provide convenient ion/electron transport channels for electrochemical detection of analyte molecules. In this scenario, LDHs have been explored as convenient materials for the fabrication of enzyme-less glucose biosensors. As a remarkable example, Cui et al. [164] reported on the fabrication of a bifunctional non-enzymatic flexible glucose microsensor based on CoFe LDHs by directly growing a CoFe layered double hydroxide nanosheet array (LDH-NSA) on a Ni wire. The LDH system showed high sensitivity and high selectivity in electrochemical and colorimetric detection of glucose, with linear ranges from 10 to 1000 mM and from 1 to 20 mM, and detection limits of 0.27 μM and 0.51 μM, respectively. The electrocatalytic glucose oxidation mechanism of CoFe LDH was tentatively explained by the authors, as follows:

$$\text{LDH-Co(II)} + \text{OH}^- \rightarrow \text{LDH(OH}^-\text{)Co(III)} + \text{e}^-$$
$$\text{LDH(OH}^-\text{)Co(III)} + \text{glucose} \rightarrow \text{LDH-Co(II)} + \text{gluconolactone}$$

As further examples, Ai et al. modified glassy carbon electrodes with a NiAl LDH composite with chitosan, obtaining a good linear range (0.01–10 mM) for glucose detection [165]. Li et al. employed NiAl LDHs onto titanium electrodes, obtaining a detection limit of 5 μM and a linear range up to 10.0 mM [166]. More recently, the preparation of a glassy carbon electrode modified with a composite material based on gold nanoparticles decorated with NiAl LDHs and single-walled carbon nanotubes permitted to obtain a wide linear range from 10 μM to 6.1 mM [167]. The authors ascribed the good detection performances to the combined effects of enhanced electrical conductivity deriving from 3D network formed by carbon nanotubes, good accessibility to active reaction sites from NiAl LDH and more electron transfer passages provided by Au nanoparticles. Besides glucose, non-enzymatic LDHs-based sensors have also found applications in the detection of drugs, such as Terazosin hydrochloride, an Alpha-adrenergic Blocking Agent, by means of MgAl LDHs [168], or for the detection of nitrite ions from solution by NiFe LDHs fabricated onto carbon cloth substrates [169].

As another pivotal example, Asif et al. demonstrated the easy fabrication of hybrid LDH nanosheets with layers of reduced graphene oxide for electrochemical simultaneous determination of dopamine, uric acid and ascorbic acid [170]. They produced such 2D composite material by following a coprecipitation route in which the ZnNiAl LDH and GO precursors were added dropwise to a formamide solution with continuous stirring. In another synthesis procedure, they directly mixed ZnNiAl LDH and GO supernatant solutions and added hydrazine in order to obtain reduced GO. The latter synthesis approach allowed to self-assemble a periodic superlattice compound by integrating positively charged semiconductive sheets of a ZnNiAl LDH and negatively charged layers of reduced graphene oxide. By operating at typical working potentials of −0.10 V, +0.13 V, and +0.27 V vs. saturated calomel electrode, the obtained lower detection limits for ascorbic acid, dopamine, and uric acid were 13.5 nM, 0.1 nM, and 0.9 nM, respectively. The authors also showed the possibility to track the dopamine released from human neuronal functioning neuroblastoma cell line SH-SY-5Y after being stimulated by highly K^+ buffer.

Along with electrochemical sensors, many efforts focused on the preparation of LDH-based sensors with fluorescence readout. Recently, Liu et al. demonstrated a fluorometric displacement assay for measuring the concentration of adenosine triphosphate (ATP) using layered cobalt(II) double hydroxide nanosheets [171]. In particular, they used a dye-labeled oligonucleotide adsorbed on the LDHs. The adsorption led to the complete and fast quenching of the green fluorescence of the label. In presence of ATP in the solution phase, the DNA oligonucleotide was rapidly detached from the LDH because of the stronger affinity of ATP for LDH, finally leading to the restoration of the green fluorescence signal. This remarkable effect was used by the authors to produce an assay showing a linear response in the 0.5–100 μM ATP concentration range and a 0.23 μM lower detection limit, allowing for the determination of ATP in spiked serum samples. In a similar approach, Abdolmohammad-Zadeh et al. showed a fluorescent sensor based on nanostructured MgAl LDH intercalated with salicylic acid (SA) for sensing the ferric ions in solution [172]. The calibration graph was linear in the concentration range of 0.07–100 μmol/L, along with a detection limit of 26 nmol/L. The authors demonstrated the intercalation of salicylic acid into the layers of the host Mg-Al LDH matrix by showing an increased interlayer spacing as measured by XRD analysis. The fluorescence intensity of salicylic acid was increased by its intercalation into LDHs, given the effect of confinement, which reduced the interaction between salicylic acid and the solvent. In presence of Fe^{3+} ions, the fluorescence signal of MgAl LDHs intercalated with SA decreased with increasing of Fe^{3+} ions concentration because of the formation of a stable complex with salicylic acid.

Among optical based techniques, chemiluminescence is also a powerful approach for biomolecular detection given its high sensitivity and low cost in comparison to fluorescence [173]. Many reports have shown the possibility to employ LDHs for the realization of chemiluminescent glucose sensors. For instance, Wang et al. demonstrated that MgAl LDHs can be supporting materials for immobilizing luminol reagent, by triggering luminol chemiluminescence in a moderately acid pH (5.8). In the presence of horseradish peroxidase, the luminol-LDH hybrid was able to produce chemiluminescence signal to glucose in the range of 0.005–1.0 mM, along with a detection limit for glucose equal to 0.1 μM [174]. In another paper, Pourfaraj et al. demonstrated that CoNi LDHs exhibit catalytic activities towards the luminol-H_2O_2 reaction. Under optimum conditions, the chemiluminescence intensity was linear in the range 0.1–12μM of H_2O_2 concentrations along with a detection limit (S/N=3) of 0.05 μM [175]. More recently, Pan et al. demonstrated the possibility to use the LDH–luminol–H_2O_2 system-based chemiluminescence platform for sensing carminic acid, a colorant used in food additives [176]. The detection principle consists of two steps: First, LDH adsorbs carminic acid onto the surface, then the carminic acid quenches the chemiluminescence of the LDH–luminol–H_2O_2 system by resonance energy transfer, reduction of reactive oxygen species, and occupation of the brucite-like layers. The authors demonstrate the possibility to obtain a linear response to the analyte in the concentration range from 0.5 to 10 μM, along with a limit of detection equal to 0.03 μM.

Telomeric DNA could also be a target analyte as recently shown by Haarone et al. [177], who produced a MgAl LDHs intercalated with DAPI (4′,6-diamidino-2-phenylindole), which is a fluorescent molecule that strongly binds to adenine–thymine rich regions in DNA sequences. DAPI could assemble into the LDH nanosheets by blending with single strand DNA via the co-assembly method (see Figure 11). The DAPI containing LDH was able to detect long single strand DNA/telomere sequences simply by changing the single strand DNA sequence. The authors demonstrated a dynamic range within the 3–20 μg/mL range along with a detection limit of 20 μg/mL.

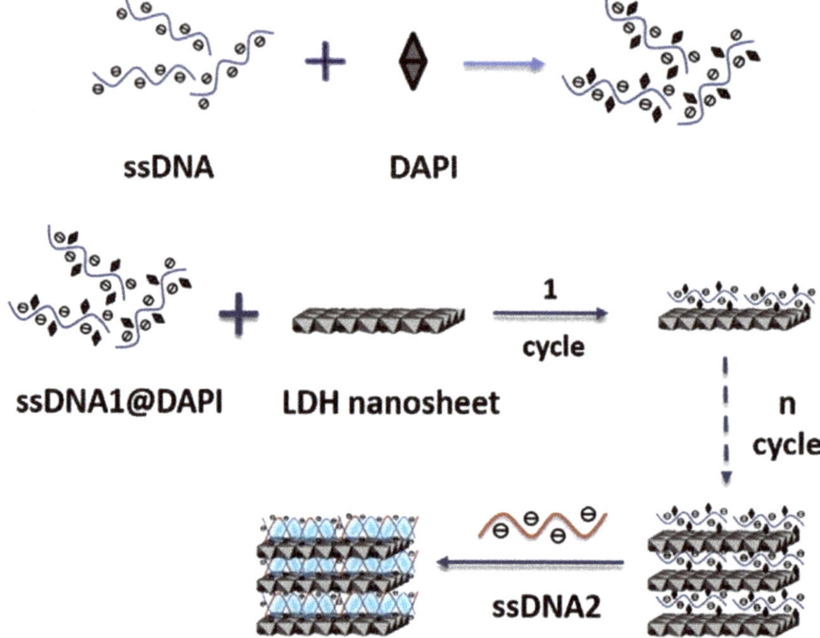

Figure 11. Scheme of the assembly of ssDNA blending into DAPI/LDH ultrathin nanosheets. Figure readapted from Ref. [177], with permission from Elsevier.

Finally, LDHs could be applied for guanine detection by means of silver nanoclusters (AgNCs) stabilized by nuclear fast red (NFR) sodium salt on Mg_2Al LDH nanosheets [178]. Due to the confinement effects in 2D layered LDH nanosheets, the fluorescence intensity and photostability of the AgNCs-NFR/LDH compound were significantly improved if compared with those of the AgNCs-NFR based solution. By introducing Cu^{2+} ions as a modulator, AgNCs-NFR/LDHs were successfully applied for determining guanine in the concentration range of 10 μM–100 μM, along with a detection limit equal to 1.85 μM. Recently, dopamine biosensors for real time detection of dopamine (detection limit down to 2 nM) from live cells (human neuronal functioning neuroblastoma cell line SH-SY 5Y) were realized in the form of composite materials produced by exfoliated charged nanosheets of LDHs and graphene [179]. Similarly, carbon nanotubes loaded onto a CuMn LDH nanohybrid allowed for a high-sensitivity electrochemical detection of H_2S from live A375 cells (detection limit equal to 0.3 nM) [180].

5. LDHs Applications in Nucleic Acids (DNA, RNA) Delivery

Besides sensing molecules and nucleic acids in biological samples, LDHs can find important applications in life sciences given their ability to deliver biomolecules to biological systems. LDHs are efficient drug delivery systems, possess good biocompatibility, high drug-loading density, high drug-transportation efficiency, low toxicity to target cells, or organs, offering excellent protection to loaded molecules from undesired enzymatic degradation [160,181]. DNA is the carrier of the genetic information in living cells. The structure of DNA consists in two single strands of DNA, which can bind together by hydrogen bonding between complementary pairing bases (adenine to thymine and guanine to cytosine), thus forming the well-known DNA double helix structure. DNA delivery to cells represents one of the most investigated applications for the biological sciences. In this regard, the successful DNA delivery to cells strictly requires that the foreign DNA material must remain stable within the host cell by fusing into its genome or retaining the ability to integrate intracellularly replicate. This requires foreign DNA to be delivered by a suitable vector, which has the capability to enter the host cell and accurately deliver the DNA molecule to the cell's genome, being the vectors employed for gene delivery sorted into recombinant viruses and synthetic vectors. Similar to DNA, RNA is self-assembled as a chain of nucleotides, with the only difference that uracil is used instead of thymine. RNA has many important roles in biological organisms, such as conveying the genetic information and directing protein synthesis. In the last ten years, RNA molecules were being processed via the RNA interference (RNAi) pathway, which consists in silencing the expression of genes with complementary nucleotide sequences by degrading the mRNA after transcription from DNA, ultimately preventing translation. In this regard, the small interfering RNA (or silencing RNA, siRNA) therapy is hampered by the barriers for siRNA bioavailability to enter into cells cytoplasm and exert their gene silencing activity. In this scenario, LDHs have been demonstrated as an ideal synthetic vector for both DNA and RNA molecules, due to a well-elucidated adsorption mechanism, taking into account that the phosphate backbone of the DNA polymer coordinates with the metal cations of the LDH lattice via the ligand-exchange process [182,183].

The formation of LDH-DNA hybrids for DNA delivery can be obtained either through the incorporation of small DNA molecules and antisense oligonucleotides into the LDH matrix by a simple ion-exchange reaction [184], or by a more general coprecipitation route involving the in situ formation of LDH layers of various ionic compositions around intercalated DNA [185]. In particular, the feasibility of the latter approach was demonstrated by Desigaux et al. [185] that demonstrated the intercalation of DNA into the LDH matrix by the net increase of the interlayer distance, from ~0.77 nm for all nitrate parent LDHs to ~2.11 nm, ~1.80 nm, and ~1.96 nm for DNA molecules complexed with Mg_2Al, Mg_2Fe, and Mg_2Ga, respectively (see Figure 12). Choy et al. were the first that showed the possibility to intercalate c-myc antisense oligonucleotide (As-myc) into MgAl LDH nanoparticles by anion exchange [186].

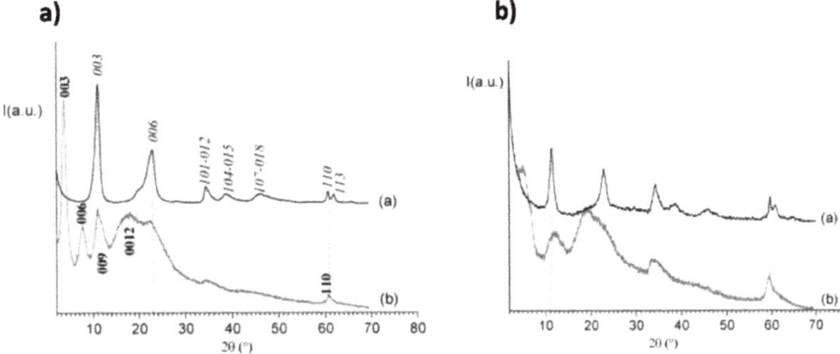

Figure 12. Powder X-ray diffraction patterns in the 2θ range 2°–70° of LDH/DNA Hybrids. (**a**) Mg_2Al/NO_3 (spectrum a) and Mg_2Al/DNA (spectrum b). (**b**) Mg_2Fe/NO_3 (spectrum a) and Mg_2Fe/DNA (spectrum b). Reprinted with permission from Ref. [185]. Copyright (2006) American Chemical Society.

The results from cellular internalization experiments demonstrated a strong inhibition of the proliferation of HL-60 cancer cells exposed to As-myc–LDH hybrids, reaching 65% growth inhibition compared with untreated cells. In a more recent report, ZnAl LDH nanoparticles were loaded with the plasmid pCEP4 to permit the expression of the Cdk9 gene in C2C12 myoblasts cells, as confirmed by PCR and Western blotting results [187]. They were also loaded with valproate and methyldopa drugs, allowing for sustained pH-triggered drug delivery. With the aim to combine dissipative molecular-dynamics simulations and experimental work, Li et al. [188] demonstrated that delaminated LDH-DNA bioconjugates can penetrate the membrane walls of plant cells (BY-2 cells). The authors delaminated the LDH nanoparticles by intercalating lactate into the layers of LDH nanoparticles. They also demonstrated that only the DNA-LDH-DNA sandwich complex was efficiently taken by the cells, whereas LDH-DNA-LDH sandwich complex and the DNA-LDH complex were not compatible for intracellular delivery, mainly due to the fact that the hydrophobic sequences of DNA provided a driving force for penetration. For example, the DNA-LDH-DNA sandwich complex was gradually internalized into the membrane to minimize exposure of the DNA hydrophobic sequences in the hydrophilic solvent.

Along with the above reported methods, DNA oligonucleotides can be also immobilized by silanization of LDHs in aqueous suspension as demonstrated by Ádok-Sipiczki et al. [189]. In particular, the 3-aminopropyltriethoxysilane (APTES) was employed as linker. APTES covalently attached to the MgAl LDHs. In turn, the amino group of the APTES linker reacted with the activated 5′ carboxylic functional group of the nucleic acid strand, for the activation of which 1-ethyl-3-(3-dimethylaminopropyl) carbodiimide hydrochloride salt (EDC) and N-hydroxysuccinimide (NHS) were used. The EDC reacts with the carboxylic group of the nucleic acid to form the O-acylisourea mixed anhydride. The addition of a selected nucleophile such as NHS, which reacts faster than the competing acyl transfer, generates an intermediate compound being active enough to couple with the amino group, finally also preventing any possible side reactions. The authors conclude that the covalent linkage of the nucleic acids confers to this model nanoparticulate system promising properties and potential for applications as therapeutic agents, since the DNA could be taken into the cells allowing for intracellular delivery.

In a similar way to DNA, also RNA complexed with LDHs nanoparticles can be delivered to cells, being the application of siRNA the main driving force for developing this type of molecular delivery. For instance, Wong et al. [190] reported the delivery of siRNA to primary neuron cells resulting in the silencing of gene expression. This possibility has important applications for potential diseases such as the treatment of neurodegenerative conditions (e.g., Huntington's disease). Interestingly, authors found that it was possible to produce Mg_2Al LDHs for simultaneously deliver anticancer drug 5-fluorouracil (5-FU) and Cell Death siRNA for effective cancer treatment, by loading the siRNA on the surface of LDH nanoparticles and the 5-FU into interlayer spacing [191]. Compared to treatment with only CD-siRNA or 5-FU, the combination of the two different molecular systems led to an enhanced cytotoxicity to three cancer cell lines, i.e. MCF-7, U2OS, and HCT-116. This interesting synergic effect was ascribed to a coordinated mitochondrial damage process. In another interesting approach, Park et al. [192] showed the possibility to obtain an efficient in vivo and in vitro delivery system for Survivin siRNA assembled with either passive MgAl LDHs, with a particle size of 100 nm, or active LDHs conjugated with a cancer overexpressing receptor targeting ligand, folic acid (LDHFA). These routes allowed targeting the tumor by either EPR-based clathrin-mediated or folate receptor-mediated endocytosis. The authors also showed the ability to induce potent gene silencing at mRNA and protein levels in vitro, and achieved a 3.0-fold higher suppression of tumor volume than LDH/Survivin in an in vivo tumor mouse model. Another interesting example of RNA intercalation in LDHs was provided by Acharya et al. for potential application in neurodegenerative diseases [193]. They employed short hairpin RNA (shRNA) intercalated in MgAl LDH nanoparticles at the mass ratio of (1:75), leading to the formation of shRNA-plasmid-LDH nanoconjugates, with an average size of 40–60 nm, which were efficiently transfected into mammalian neuroblastoma cells (SH-SY5Y), observing a maximum internalization of ~26% at 24 h. This, in turn, led to a significant downregulation of the protein alpha TNF, demonstrating the efficient functionality of the delivered shRNA.

In the case of CaAl LDH nanoparticles, the pH at which these nanomaterials are prepared can affect the efficiency of RNA uptake [194] (see Figure 13). Rahaman et al. showed that the involvement of carbonate ion as impurity was critical during the preparation of CaAl LDHs (see Figure 13). The authors prepared CaAl LDHs at two different pHs, i.e., pH 8.5 vs. pH 12.5, finding that at the higher pH, more carbonate ions were intercalated into the CaAl LDH structure, leading to lower intercalated RNA, as demonstrated by FTIR and XRD characterizations. In particular, after intercalation of shRNA the *d* spacing of the basal plane (0 0 2) increased from 8.61 Å in the phase pure CaAl LDH to 20.30 Å in shRNA sample. Interestingly, these data were nicely correlated with cellular uptake using colon cancer cell line (HCT 116), since LDHs prepared at pH 8.5 led to a significantly higher uptake (9.34%) in comparison to LDHs prepared at pH 12.5 (3.54%).

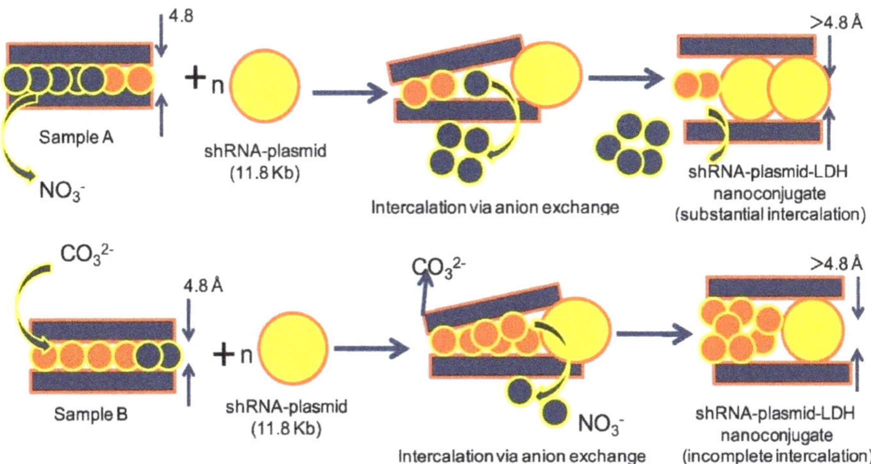

Figure 13. Schematic representation of the mechanism of intercalation of anionic shRNA-plasmid, in presence of NO_3^-/CO_3^{2-}, in cationic layers of samples A (precipitated at pH 8.5) and B (precipitated at pH 12.5). Reprinted from Ref. [194], Copyright (2019), with permission from Elsevier.

In more recent reports, the researchers strived to the aim of improving the efficacy of LDH delivery of RNA in comparison with other nanomaterials. For example, Li et al. [195] demonstrated that siRNA could be more efficiently delivered to osteosarcoma (U2OS) cells by mannose-conjugated SiO_2 coated LDH nanocomposites (Man-SiO2@LDH) compared to unmodified LDH NPs. The enhanced uptake was attributed to the active mannose-receptor interaction-mediated endocytosis, considering also the specific strong binding affinity towards lectin that is expressed on the cancer cell membrane.

Wu et al. [196] compared LDHs and lipid-coated calcium phosphate nanoparticles (LCPs) as effective vectors for siRNA delivery into suspended T lymphocytes (EL4) for silencing the target PD-1 gene. They found that LCPs showed a higher cellular uptake and higher PD-1 gene silence efficiency in mouse T cell line EL4 (about 70%) in comparison to LDHs (40%). They ascribed this difference to the smaller size of the LCPs with respect to LDHs (50 nm vs. 100–200 nm) and to the higher efficiency to get endosome escape when LCPs were dissolved in the endosomes, leading to more sustained RNA release in the cytosol. From this report, it appears that it is still necessary to obtain a better understanding of the optimal LDH:DNA ratio parameters for the delivery of functional siRNA using LDH nanoparticles. In this regard, Wu et al. [197] demonstrated that an optimal LDH/gene mass ratio was around 20:1 in terms of cellular uptake amount of gene segments, whereas the ratio was around 5:1 in terms of target gene silencing efficacy in MCF-7 cells. The authors ascribed these interesting results to a reasonable trade-off between DNA loading on the LDHs and dissolution rate. Cellular internalization of LDH NPs is basically driven by clathrin-mediated endocytosis. Once entered into the cells, LDH nanoparticles tend to be dissolved (at higher rate within the endosomes) and release the siRNA. The efficacy of RNA-induced silencing is therefore dependent upon the loading amount of dsDNA/ siRNA per LDH particle, the number of LDH particles, and the suitable release rate of dsDNA/siRNA. The 5:1 ratio seems to be a reasonable compromise between these different factors.

6. LDHs Applications in Cell Biology: From Cellular Differentiation to Cancer Therapy

In general, 2D clay materials have been demonstrated as biocompatible [198,199], suitable systems for synthetic biology [200]. By employing micron sized colloidal objects these compounds have been effectively used for in vitro reconstituting cellular motility [201]. In the specific case of LDHs, these clays are starting to be more and more investigated as to their direct interaction with living cells [202,203].

In particular, the toxicity of LDHs to cells has been thoroughly investigated [204]. As expected from other types of clays, their toxicity potential is dose and time dependent with particle sizes, their shapes and surface charge being determinant features for cellular uptakes. The reticular endothelial system is able to sequestrate LDHs systems, especially those with sizes greater than 50 nm. Notably, LDHs with sizes between 50 and 200 nm show concentration dependent uptake, whereas sizes 350 nm and above are not concentration dependent [205]. LDHs enter cells through endocytotic pathway; however the hexagonal shaped LDH crystallite was found to be distributed within the nucleus of the cells [205]. These results suggested the promising drug delivery potential of LDH at the cellular level without damaging cell structure. In the biofluids, LDHs produce some tissue and cell friendlily by-products (H_2O, Mg^{2+}, Al^{3+}, Zn^{2+}) under physiological conditions [206]. In turn, this effect can mitigate the acidification tendencies in endosomes and lysosome of cells after the nanocomposite uptake, ultimately leading to a sustained and pH dependent drug release that is also beneficial for reducing drug toxicity [204].

6.1. Cellular Differentiation

In a very recent work, LDHs have been reported as a promising material for bioengineering application, showing that LDH nanomaterials modulate cell adhesion, proliferation, and migration and demonstrating their suitability in the biomaterial field and in the specific context of tissue bioengineering [207]. In the following, LDH interaction with cells will be investigated focusing on the biocompatibility, the latest applications in cellular differentiation, and the applications in cancer therapy. A specific investigation regarding the LDH biocompatibility was recently carried out by Cunha et al. [208]. In their research, adult female Wistar rats were subjected to the intramuscular implantation of two types of LDHs of magnesium/ aluminum (Mg_2Al-Cl) and zinc/ aluminum (Zn_2Al-Cl). Interestingly, the LDHs did not lead to any sign of inflammatory reactions; on the contrary, they promoted collagen adsorption. More specifically, Mg_2Al-Cl promoted multiple collagen invaginations (mostly collagen type-I), whereas Zn_2Al-Cl induced collagen type–III. Li et al. reported on the fabrication of layered double hydroxide/poly-dopamine composite coating with surface heparinization onto Mg alloys for application in endothelialization and hemocompatibility [209]. The LDHs were directly grown onto the Mg sample by reported hydrothermal methods [210] in which Mg alloy plates are placed in a Teflon liner which contain an aluminum nitrate solution at 120 °C for 12 h. The MgAl LDH was then coated with poly-dopamine and heparin. The resulting composite material was tested for its corrosion resistance and the ability to enhance, with respect to the bare Mg surface, the adherence, proliferation and migration rate of the human umbilical vein endothelial cells and inhibited the platelets adhesion.

The interaction between cells and nanomaterials can be employed for tuning differentiation switches of stem cells, according to precise biomechanical or biochemical triggers [211,212]. LDHs were demonstrated as biocompatible materials by Ramanathan et al. [213]. In their experiments, they prepared Poly(3-Hydroxybutyric acid)-Poly (N-vinylpyrrolidone) fibers loaded with MgAl LDHs and coated over the hydroxyapatite pellets for the realization of bone grafts; the resulting systems showed good compatibility towards MG63 human osteosarcoma 3AB-OS cancer stem cell lines. LDH nanocomposites with arginylglycylaspartic acid and kartogenin were incorporated into a gel together with tonsil-derived mesenchymal stem cells [214]. Interestingly, the cells were observed to aggregate in the nanocomposite system. By analysis of mRNA content, chondrogenic biomarkers of type II collagen and transcription factor SOX 9 significantly increased, ultimately facilitating a chondrogenic differentiation.

In a more in-depth investigation, Kang et al. [215] studied the effect of MgAl LDHs and ZnAl LDHs in promoting the dynamic expression of genes in osteogenic differentiation pathways. LDHs determined a cytoskeleton rearrangement on the pre-osteoblasts, dependent on the modulating of Cofilin and PP2A phosphorylations, and a significant upmodulation of the osteoblastic classical marker genes (Runt-related transcription factor 2, Osterix, and Osteocalcin), while requiring the activation of c-Jun N-terminal Kinase JNK and extracellular signal–regulated kinases genes. Very recently, a thorough investigation of the LDH-induced inflammatory landscape onto osteoblasts was carried out by da Silva Feltran et al. [216]. In their report, they demonstrated that MgAl LDHs and ZnAl LDHs challenged on the MC3T3-E1 pre-osteoblasts inducing a down-modulation of major pro-inflammatory genes related to cytokines, precisely the tnfα, nfkb, il18, il6, and il1ß genes. On the other hand, an up-modulation of anti-inflammatory cytokines il6, il10, and tgfß genes was observed (see Figure 14). The authors concluded that the effects observed by LDHs could be ascribed to the activation of the Sonic hedgehog signalling in the osteoblasts, permitting the obtaining of a better comprehension of the molecular mechanisms driving the later LDH-induced osteoblast differentiation.

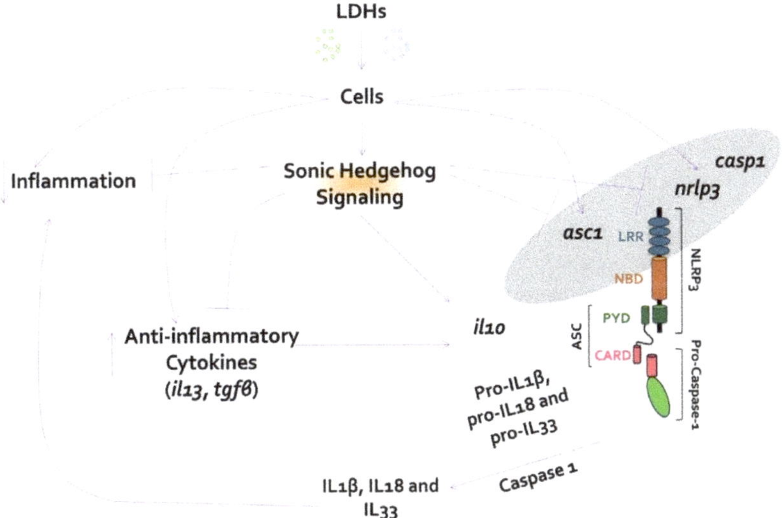

Figure 14. Role of LDHs in inducing the anti-inflammatory Sonic hedgehog acute inflammatory molecular pathway. Reprinted from Ref. [216], Copyright (2019), with permission from Elsevier.

6.2. Contrast Agents

Manganese-based LDHs have been demonstrated as powerful magnetic contrast agents with outstanding pH response and high relaxivity. Indeed, Li et al. [217] reported that, in comparison with free Mn^{2+} ions, the Mn-based LDH systems (prepared from isomorphic substitution from Mg_3Al LDHs) showed a higher longitudinal relaxivity (9.48 mm^{-1} s^{-1} at pH 5.0, and 6.82 mm^{-1} s^{-1} at pH 7.0 vs. 1.16 mm^{-1} s^{-1} at pH 7.4 of free Mn^{2+} ions). Intriguingly, they ascribed these results to the unique microstructural coordination environment surrounding Mn atoms in the Mn LDH framework, as demonstrated by EXAFS measurements. They also observed that the variation of Mn···(OH_2)x coordination in Mn-LDHs in acidic buffer (pH 5.0) led to the excellent T1-weighted magnetic resonance imaging performance. Afterwards, they proved the intracellular uptake of the Mn-LDHs on the B16F10 cancer cells and the outstanding potentiality as in vivo tumor imaging systems (see Figure 15).

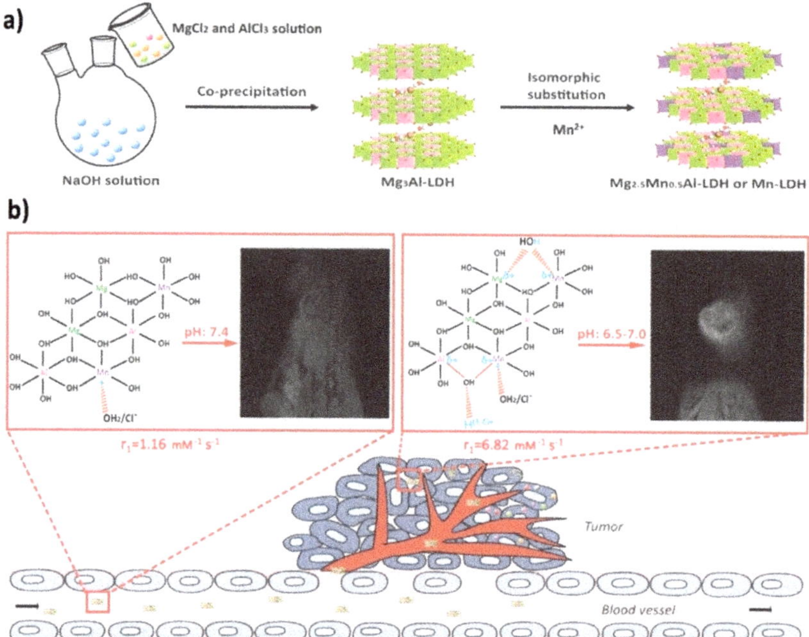

Figure 15. (**a**) Scheme of the $Mg_{2.5}Mn_{0.5}Al$ LDH (Mn-LDH) from co-precipitation and isomorphic substitution. (**b**) Application of the Mn LDH as magnetic contrast agents showing an ultrasensitive pH response and high relaxivity. Reproduced from Ref. [217]. Copyright © 2014 by John Wiley and Sons, Inc. Reprinted with permission from John Wiley and Sons, Inc.

6.3. Drug Delivery

LDHs are reported as very efficient drug nanovehicles [218,219] since, in comparison to other inorganic nanovehicles, including silica and gold nanoparticles, quantum dots, and carbon nanotubes, they are featured with excellent biocompatibility [205], high drug loading capacity [220], and pH-responsive property [221], with biodegradability in the cellular cytoplasm [222]. Such outstanding properties make LDHs an efficient non-viral drug delivery vehicle, and also a reservoir for bioactive or bio-fragile molecules. Note that the intercalated drugs can be released either by deintercalation through anionic exchange with the surrounding anions (such as Cl^- and phosphate), or through the acidic dissolution of LDH hydroxide layers. The predominant mechanism depends on the pH value and the nature of the drug. LDH internalization into cells is mainly carried out by clathrin-mediated endocytosis [223], in which the material to be internalized is surrounded by an area of cell membrane, which buds off inside the cell to form a vesicle containing the ingested material. However, following the reports from Bao et al. [224] for LDH nanosheets-mediated molecular delivery to plant cells, the typical inhibitors of endocytosis and low temperature incubation did not prevent LDH internalization, meaning that the penetration of the plasma membrane can also occur via non-endocytic pathways (see Figure 16).

Many reports have demonstrated the immense potentiality of these materials in drug loading and sustained release. For instance, model antioxidant drugs such as carnosine and gallic acid can be intercalated into MgAl LDHs by ion exchange and coprecipitation. The drugs are released in a pH = 7.4 phosphate buffered saline medium, following a gradual and biphasic release [225]. Drug release from LDHs can be triggered by three different mechanisms: ion exchange, desorption and weathering. Whereas ion exchange only depends on the anion nature, the desorption from LDHs is dependent on the pH of the medium. Taking ibuprofen as a model drug, anion exchange triggers ibuprofen release in the intestinal medium, whereas surface reactions mediated by solid weathering determine the release in acid media [226]. Many recent examples report on the fabrication of drug-LDHs nanohybrids. Yasaei et al. [227] demonstrated the direct coprecipitation in the presence of simvastatin (a drug used for bone regeneration) and the ion exchange of nitrate and carbonate containing LDHs in a simvastatin solution. The authors found a higher drug loading and more sustained drug releases in the case of nitrate-based LDHs. Bouaziz et al. found similar efficient loading and sustained release of nisin (among the most widely used bacteriocin peptides used in food safety) onto zinc-aluminum LDHs [228]. Similarly, pH-sensitive bead systems composed of folic acid intercalated into LDHs and chitosan represent an ideal drug delivery system for folic acid release at simulated conditions similar to the gastrointestinal tract [229]. Among the possible investigated drug systems, alendronate represents a relevant example, being an anti-resorptive and bone-remodeling drug with poor intracellular permeability, which in turn results in low drug efficacy. In this regard, Piao et al. [230] demonstrated that MgAl LDHs were able to obtain a 10-fold enhancement of the cellular uptake efficacy of alendronate in MG63 cells with respect to the bare alendronate. Notably, LDH itself did not show any effect on proliferation and osteogenic differentiation of MG63 cells (see Figure 17).

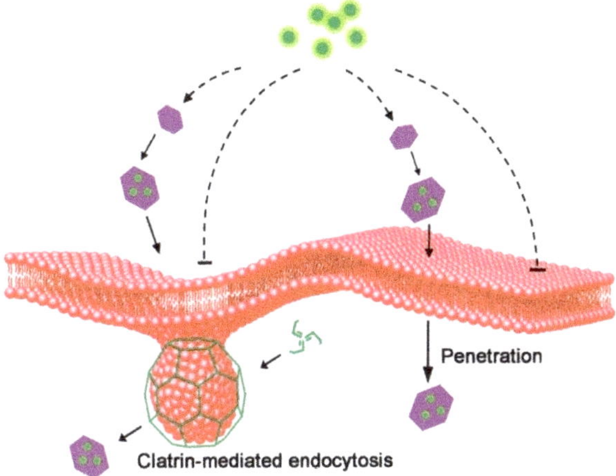

Figure 16. Mechanisms of LDH-lactate internalization (clathrin-mediated endocytosis) for molecule delivery into plant cells. Figure reproduced from Ref. [224] under the terms of the Creative Commons Attribution Non-Commercial License.

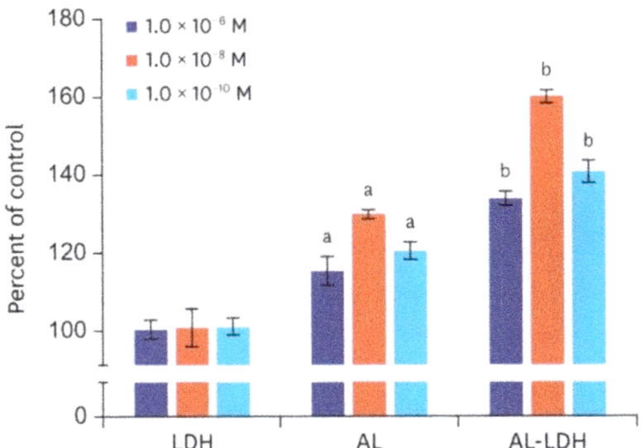

Figure 17. Experimental results showing the biocompatibility of pristine MgAl LDHs (LDH), Alendronate (AL) and Alendronate-LDH nanohybrids (AL-LDH). Figure reproduced from Ref. [230] under the terms of the Creative Commons Attribution Non-Commercial License.

LDHs have also shown important applications in chemotherapy, since they can improve treatment of tumors by facilitating pH-triggered drug delivery. For instance, Khorsandi et al. [231] investigated upon the cellular responses to curcumin-LDH hybrid nanoparticles following a photodynamic treatment of MDA-MB-231 human breast cancer cells. For this purpose, the human breast cancer cells were treated with curcumin-LDH NPs and were then irradiated, demonstrating that the curcumin-LDH hybrid had a cytotoxic and antiprolifrative effect due to the generation of reactive oxygen species, which led to autophagy and apoptosis. Recently, Liu et al. [232] demonstrated a very intriguing pH-triggered hydrazone-carboxylate complex of doxorubicin, a model chemotherapy drug, which was encapsulated in MgAl LDH NPs via a ion-exchange processes. These nanohybrids were internalized by clathrin-dependent endocytosis and then shifted to lysosomes, where hydrazone-Dox complexes were released (due to the low pH of this compartment), ultimately resulting in free cytosolic doxorubicin which in turn facilitated the cell death (MCF-7 and HeLa cells) via cathepsin-mediated cell apoptosis. A more complex drug-nanohybrid was realized by Wen et al. [233] with the aim to produce a pH-triggered drug delivery system consisting of folic-acid (FA) functionalized tellurium nanodots (Te NDs) which were in-situ synthesized in paclitaxel (PTX)-loaded MgAl LDHs. The latter were, in turn, gated onto mesoporous silica nanoparticles (MSNs). The structural properties of the resulting nanosystem were characterized by SEM, XRD, XPS and FT-IR methods. Interestingly, these nanosystems combined the ability of Te nanodots as phototherapeutic agent and the pH-triggered release of paclitaxel to human cervical cancer cells HeLa and human embryonic kidney cells 293T (see Figure 18).

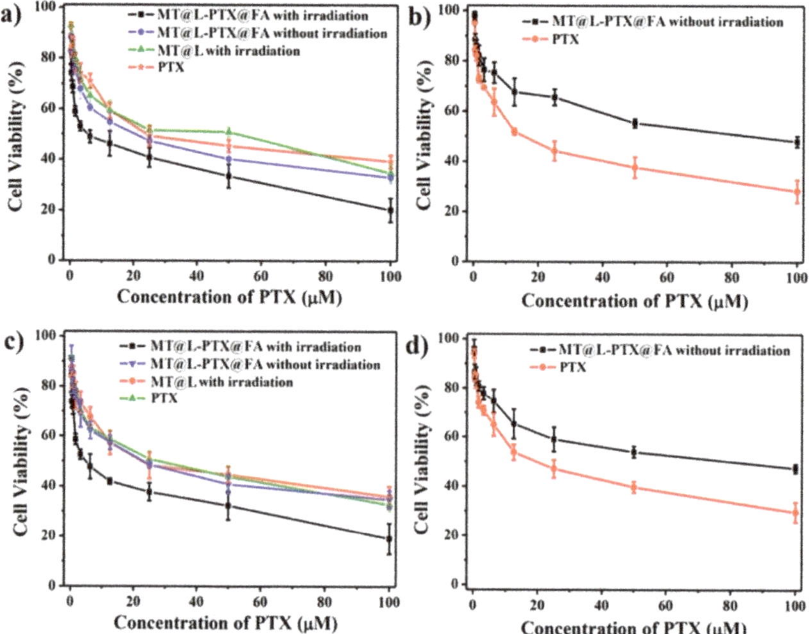

Figure 18. Experimental results showing the cytotoxicity of the hierarchical nanosystem consisting of Te NDs within Paclitaxel-Loaded MgAl LDH gated mesoporous silica NPs towards (**a**) HepG2 cells, and (c) HeLa cells for 48 h with or without irradiation. Cytotoxicity of MT@L-PTX@FA without irradiation and free Paclitaxel on (**b**) HL7702 cells, and (**d**) 293T cells for 48 h. Reprinted with permission from Ref. [233]. Copyright (2019) American Chemical Society.

CaAl LDHs have shown useful properties as a drug delivery system for methotrexate to MG-63 human osteosarcoma cell line [234]. Remarkably, CaAl LDHs alone have been recently demonstrated as an active anticancer agent, owing to the involvement of Ca^{2+} ion with the CAMKIIα protein and associated SOD activity in cancer cell [235]. The results by Bhattacharjeeet al. exhibited significant down regulation of CAMKIIα and SOD gene by CaAl LDHs at cellular level, leading to apoptosis of the cancer cell. Exfoliated ultrathin MgAl LDHs prepared by a single-step procedure in presence of formamide [236] can be used as a confinement matrix of carbon nanodots (CDs), allowing for an enhancement of fluorescence lifetime of CDs. In turn, this leads to an improvement of their efficacy in photoacoustic imaging performance and tumor inhibition under the 808 nm irradiation [237] in Hela tumor-models consisting of male nude Balb/c mice.

LDHs have also been demonstrated as an effective drug delivery system of etoposide, a chemotherapeutic agent for the treatment of a number of diseases, including glioblastoma, one of the most common and lethal intrinsic brain tumors. Wang et al. [238] incorporated etoposide into LDH nanoparticles in order to overcome the inherent drug resistant problem, such as poor tumor selectivity, and low solubility in water, permitting to significantly facilitate uptake by glioblastoma cells and enhance apoptosis efficiency, self-renewal repression, and epithelial-mesenchymal transition program reversion. The authors conducted a profound transcriptomic analysis to find the role of different protein pathways involved in the delivery of etoposide into glioblastoma cells, including PI3K–AKt, downstream mTOR, Nuclear factor kappa B (NF-κB), mitochondrial Bcl-2, activated caspase, and Wnt-β-catenin signaling (see Figure 19).

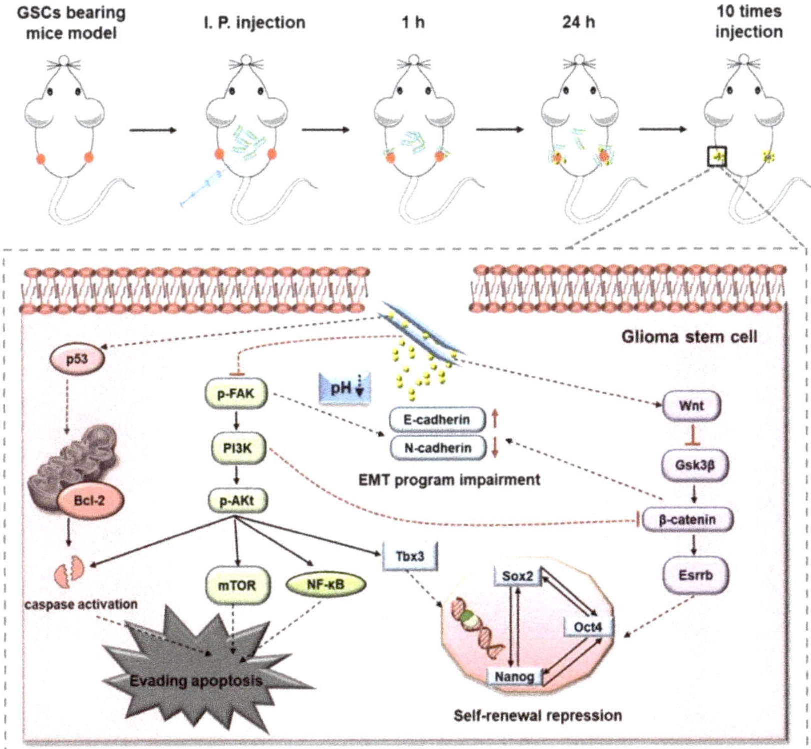

Figure 19. Molecular mechanisms involved in etoposide LDH NPs reversing chemoresistance and eradicating human glioma stem cells. Republished with permission from the Royal Society of Chemistry, from Ref. [238]; permission conveyed through Copyright Clearance Center, Inc.

A different scenario has been elicited as it concerns antibiotics loading. Model antibiotics such as tetracycline and oxytetracyline [239] were shown not to be intercalated into LDHs, but more likely to be adsorbed on their external surface and released according to a Fickian diffusion model. The LDHs adsorbed antibiotics are still active towards *E. coli* and *S. epidermis*, showing a decrease of efficacy in comparison to the free molecules. Among the recent applications in anticancer drug delivery, Asiabi et al. [240] prepared a novel biocompatible pseudo-hexagonal NaCa-layered double metal hydroxides (see Figure 20), which allowed for a sustained pH-triggered release of dacarbazine. This system allowed for enhanced anticancer activity in malignant melanoma cells (malignant A-375 melanoma and breast cancer MCF-7 cell lines) at higher values in comparison to the free drug [240]. Similarly, Shahabadi et al. [241] demonstrated the suitability of Fe_3O4@CaAl-LDHs as levodopa delivery systems for Mel-Rm Cells Melanoma (NCIt: C3224) cells, permitting improvement in the efficacy of the free drug, owing to a pH-triggered release mechanism.

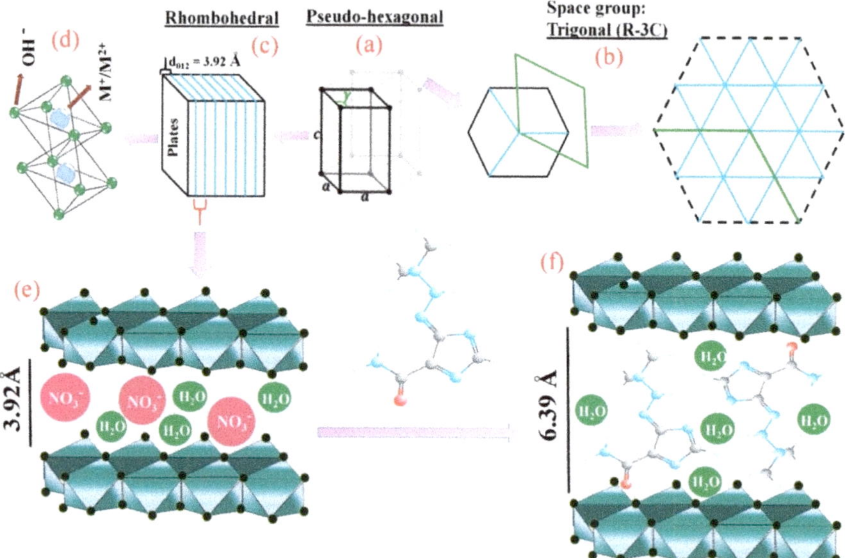

Figure 20. Schematic illustration of the structure of NaCa LDH for dacarbazine drug loading. NaCa LDH has (**a**) pseudo-hexagonal (P-3m1) structure with (**b**) a trigonal crystal structure (R-3C) and (c) rhombohedral lattice system. (**d**) The M^{2+}/M^{+} ions are surrounded approximately octahedrally by hydroxide ions, (**e**) structure of NO_3LDH, and (**f**) structure of the LDH intercalated with dacarbazine drug, showing an enlargement of the inter-layer spacing. Reprinted from Ref. [240], with permission from Elsevier.

7. Conclusions

The aim of this review is to provide the reader with a fresh view in the field of LDH applications in chemistry and biology. As for the chemistry-related applications, LDHs are a definitely promising platform for organics catalysis and photocatalysis. Their main advantages lie in the low cost, mild, eco-compatible, low temperature synthetic approaches in controlled reaction environments (i.e., the interlamellar space), along with the possibility to recycle the LDHs after their employment. However, a specific issue of these materials is the "trails and error" approach that typically characterizes the literature of these materials for chemistry-applications. In the future, a more predictive approach towards the understanding of the reaction mechanisms involving the metal centers forming the LDHs would highly benefit the development of this research field. As for the biology-related applications,

LDHs have proved to be a suitable material for drug and nucleic acid delivery, due to their partially positive charge that greatly facilitates cellular uptake by electrostatic forces. We have focused on the novel still poor explored applications in DNA and siRNA delivery, where LDHs are able to play a key role given their biocompatibility and easy synthesis and loading approaches. Still, many new investigations are needed in the field of interactions with cells (especially stem cells and also cancer cells), with the aim to precisely understand the intracellular molecular pathways that can be activated by the interaction of LDHs with cells.

This overview on LDHs considers only a portion of the numerous organic reactions protocols which could benefit of LDHs catalytic properties as well as only some remarkable examples among the different applications of these materials in the field of drug delivery and advanced medical therapies. While this study does not pretend to detail all the literature actually available, the authors hope that the most important and recent achievements have been reviewed, with the ultimate goal to offer a panoramic vision of the importance and potential of this branch of chemistry in between material science, organic synthesis and biology.

Author Contributions: Methodology, G.P., A.O., A.M., G.A., and A.B.; writing—original draft preparation, G.A., A.B.; writing—review and editing, G.P. and E.M.; supervision, E.M., B.P. and P.G.M.

Funding: This research received no external funding.

Acknowledgments: The authors acknowledge ATeN center (Unipa) for hospitality and support.

Conflicts of Interest: The authors declare no conflict of interest.

References

1. Tiwari, J.N.; Tiwari, R.N.; Kim, K.S. Zero-dimensional, one-dimensional, two-dimensional and three-dimensional nanostructured materials for advanced electrochemical energy devices. *Prog. Mater. Sci.* **2012**, *57*, 724–803. [CrossRef]
2. Sciortino, A.; Mauro, N.; Buscarino, G.; Sciortino, L.; Popescu, R.; Schneider, R.; Giammona, G.; Gerthsen, D.; Cannas, M.; Messina, F. β-C_3N_4 Nanocrystals: Carbon Dots with Extraordinary Morphological, Structural, and Optical Homogeneity. *Chem. Mater.* **2018**, *30*, 1695–1700. [CrossRef]
3. Miccichè, C.; Arrabito, G.; Amato, F.; Buscarino, G.; Agnello, S.; Pignataro, B. Inkjet printing Ag nanoparticles for SERS hot spots. *Anal. Methods* **2018**, *10*, 3215–3223. [CrossRef]
4. Cataldo, S.; Salice, P.; Menna, E.; Pignataro, B. Carbon nanotubes and organic solar cells. *Energy Environ. Sci.* **2012**, *5*, 5919–5940. [CrossRef]
5. Fang, Y.; Jiang, Y.; Acaron Ledesma, H.; Yi, J.; Gao, X.; Weiss, D.E.; Shi, F.; Tian, B. Texturing Silicon Nanowires for Highly Localized Optical Modulation of Cellular Dynamics. *Nano Lett.* **2018**, *18*, 4487–4492. [CrossRef]
6. Arrabito, G.; Errico, V.; Zhang, Z.; Han, W.; Falconi, C. Nanotransducers on printed circuit boards by rational design of high-density, long, thin and untapered ZnO nanowires. *Nano Energy* **2018**, *46*, 54–62. [CrossRef]
7. Errico, V.; Arrabito, G.; Plant, S.R.; Medaglia, P.G.; Palmer, R.E.; Falconi, C. Chromium inhibition and size-selected Au nanocluster catalysis for the solution growth of low-density ZnO nanowires. *Sci. Rep.* **2015**, *5*, 12336. [CrossRef]
8. Kim, K.S.; Zhao, Y.; Jang, H.; Lee, S.Y.; Kim, J.M.; Kim, K.S.; Ahn, J.-H.; Kim, P.; Choi, J.-Y.; Hong, B.H. Large-scale pattern growth of graphene films for stretchable transparent electrodes. *Nature* **2009**, *457*, 706–710. [CrossRef]
9. Pradhan, D.; Leung, K.T. Vertical growth of two-dimensional zinc oxide nanostructures on ITO-coated glass: Effects of deposition temperature and deposition time. *J. Phys. Chem. C* **2008**, *112*, 1357–1364. [CrossRef]
10. Jeevanandam, J.; Barhoum, A.; Chan, Y.S.; Dufresne, A.; Danquah, M.K. Review on nanoparticles and nanostructured materials: History, sources, toxicity and regulations. *Beilstein J. Nanotechnol.* **2018**, *9*, 1050–1074. [CrossRef]
11. Wang, Q.; O'Hare, D. Recent Advances in the Synthesis and Application of Layered Double Hydroxide (LDH) Nanosheets. *Chem. Rev.* **2012**, *112*, 4124–4155. [CrossRef] [PubMed]
12. Crepaldi, E.L.; Pavan, P.C.; Valim, J.B. Comparative Study of the Coprecipitation Methods for the Preparation of Layered Double Hydroxides. *J. Braz. Chem. Soc.* **2000**, *11*, 64–70. [CrossRef]

13. Bukhtiyarova, M.V. A review on effect of synthesis conditions on the formation of layered double hydroxides. *J. Solid State Chem.* **2019**, *269*, 494–506. [CrossRef]
14. Mishra, G.; Dash, B.; Pandey, S. Layered double hydroxides: A brief review from fundamentals to application as evolving biomaterials. *Appl. Clay Sci.* **2018**, *153*, 172–186. [CrossRef]
15. Guo, X.; Zhang, F.; Evans, D.G.; Duan, X. Layered double hydroxide films: Synthesis, properties and applications. *Chem. Commun.* **2010**, *46*, 5197–5210. [CrossRef] [PubMed]
16. Liu, B.J.; Li, Y.; Huang, X.; Li, G.; Li, Z. Layered Double Hydroxide Nano- and Microstructures Grown Directly on Metal Substrates and Their Calcined Products for Application as Li-Ion Battery Electrodes. *Adv. Funct. Mater.* **2008**, *18*, 1448–1458. [CrossRef]
17. Liu, J.; Huang, X.; Li, Y.; Sulieman, K.M.; He, X.; Sun, F. Facile and Large-Scale Production of ZnO/Zn−Al Layered Double Hydroxide Hierarchical Heterostructures. *J. Phys. Chem. B* **2006**, *110*, 21865–21872. [CrossRef] [PubMed]
18. Guo, X.; Xu, S.; Zhao, L.; Lu, W.; Zhang, F.; Evans, D.G.; Duan, X. One-Step Hydrothermal Crystallization of a Layered Double Hydroxide/Alumina Bilayer Film on Aluminum and Its Corrosion Resistance Properties. *Langmuir* **2009**, *25*, 9894–9897. [CrossRef]
19. Scarpellini, D.; Leonardi, C.; Mattoccia, A.; Di Giamberardino, L.; Medaglia, P.G.; Mantini, G.; Gatta, F.; Giovine, E.; Foglietti, V.; Falconi, C.; et al. Solution-Grown Zn/Al Layered Double Hydroxide Nanoplatelets onto Al Thin Films: Fine Control of Position and Lateral Thickness. *J. Nanomater.* **2015**, *2015*, 1–8. [CrossRef]
20. Polese, D.; Mattoccia, A.; Giorgi, F.; Pazzini, L.; Ferrone, A.; Di Giamberardino, L.; Maiolo, L.; Pecora, A.; Convertino, A.; Fortunato, G.; et al. Layered Double Hydroxides intercalated with chlorine used as low temperature gas sensors. *Procedia Eng.* **2015**, *120*, 1175–1178. [CrossRef]
21. Meng, F.; Li, L.; Liang, H.; Xiu, L.; Forticaux, A.; Wang, Z.; Cabán-Acevedo, M.; Jin, S. Hydrothermal Continuous Flow Synthesis and Exfoliation of NiCo Layered Double Hydroxide Nanosheets for Enhanced Oxygen Evolution Catalysis. *Nano Lett.* **2015**, *15*, 1421–1427.
22. Zhou, Y.; Li, X.; Wang, K.; Hu, F.; Zhai, C.; Wang, X. Enhanced photoluminescence emission and surface fluorescence response of morphology controllable nano porous anodize alumina Zn-Al LDH film. *J. Alloys Compd.* **2019**, *770*, 6–16. [CrossRef]
23. Teixeira, A.C.; Morais, A.F.; Silva, I.G.N.; Breynaert, E.; Mustafa, D. Luminescent Layered Double Hydroxides Intercalated with an Anionic Photosensitizer via the Memory Effect. *Crystals* **2019**, *9*, 153. [CrossRef]
24. Richetta, M.; Digiamberardino, L.; Mattoccia, A.; Medaglia, P.G.; Montanari, R.; Pizzoferrato, R.; Scarpellini, D.; Varone, A.; Kaciulis, S.; Mezzi, A.; et al. Surface spectroscopy and structural analysis of nanostructured multifunctional (Zn, Al) layered double hydroxides. *Surf. Interface Anal.* **2016**, *48*, 514–518. [CrossRef]
25. Kruissink, E.C.; Van Reijen, L.L.; Ross, J.R.H. Coprecipitated nickel-alumina catalysts for methanation at high temperature. Part 1.—Chemical composition and structure of the precipitates. *J. Chem. Soc. Faraday Trans. 1 Phys. Chem. Condens. Phases* **1981**, *77*, 649–663. [CrossRef]
26. Reichle, W.T. Catalytic Reactions by Thermally Activated, Synthetic, Anionic Clay Minerals. *J. Catal.* **1985**, *94*, 547–557. [CrossRef]
27. Martin, K.J.; Pinnavaia, T.J. Layered Double Hydroxides as Supported Anionic Reagents. Halide Ion Reactivity in [$Zn_2Cr(OH)6X.nH_2O$] (X = Cl, I). *J. Am. Chem. Soc.* **1986**, *108*, 541–542.
28. Zhao, M.Q.; Zhang, Q.; Huang, J.Q.; Wei, F. Hierarchical nanocomposites derived from nanocarbons and layered double hydroxides-properties, synthesis, and applications. *Adv. Funct. Mater.* **2012**, *22*, 675–694. [CrossRef]
29. Fan, G.; Li, F.; Evans, D.G.; Duan, X. Catalytic applications of layered double hydroxides: Recent advances and perspectives. *Chem. Soc. Rev.* **2014**, *43*, 7040–7066. [CrossRef] [PubMed]
30. Feng, J.; He, Y.; Liu, Y.; Du, Y.; Li, D. Supported catalysts based on layered double hydroxides for catalytic oxidation and hydrogenation: General functionality and promising application prospects. *Chem. Soc. Rev.* **2015**, *44*, 5291–5319. [CrossRef]
31. Newman, S.P.; Jones, W. Synthesis, characterization and applications of layered double hydroxides containing organic guests. *New J. Chem.* **1998**, *22*, 105–115. [CrossRef]
32. Leroux, F.; Taviot-Guého, C. Fine tuning between organic and inorganic host structure: New trends in layered double hydroxide hybrid assemblies. *J. Mater. Chem.* **2005**, *15*, 3628–3642. [CrossRef]
33. Rives, V.; Ulibarri, M.A. Layered double hydroxides (LDH) intercalated with metal coordination compounds and oxometalates. *Coord. Chem. Rev.* **1999**, *181*, 61–120. [CrossRef]

34. Omwoma, S.; Chen, W.; Tsunashima, R.; Song, Y.F. Recent advances on polyoxometalates intercalated layered double hydroxides: From synthetic approaches to functional material applications. *Coord. Chem. Rev.* **2014**, *258–259*, 58–71. [CrossRef]
35. Vicente, R. Characterisation of layered double hydroxides and their decomposition products. *Mater. Chem. Phys.* **2002**, *75*, 19–25.
36. Xu, Z.P.; Zhang, J.; Adebajo, M.O.; Zhang, H.; Zhou, C. Catalytic applications of layered double hydroxides and derivatives. *Appl. Clay Sci.* **2011**, *53*, 139–150. [CrossRef]
37. Auer, S.M.; Gredig, S.V.; Köppel, R.A.; Baiker, A. Synthesis of methylamines from CO_2, H_2 and NH_3 over Cu-Mg-Al mixed oxides. *J. Mol. Catal. A Chem.* **1999**, *141*, 193–203. [CrossRef]
38. Christensen, K.O.; Chen, D.; Lødeng, R.; Holmen, A. Effect of supports and Ni crystal size on carbon formation and sintering during steam methane reforming. *Appl. Catal. A Gen.* **2006**, *314*, 9–22. [CrossRef]
39. Liu, H.; Wierzbicki, D.; Debek, R.; Motak, M.; Grzybek, T.; Da Costa, P.; Gálvez, M.E. La-promoted Ni-hydrotalcite-derived catalysts for dry reforming of methane at low temperatures. *Fuel* **2016**, *182*, 8–16. [CrossRef]
40. Li, P.; Yu, F.; Altaf, N.; Zhu, M.; Li, J.; Dai, B.; Wang, Q. Two-dimensional layered double hydroxides for reactions of methanation and methane reforming in C1 chemistry. *Materials* **2018**, *11*, 221. [CrossRef]
41. Gawande, M.B.; Pandey, R.K.; Jayaram, R.V. Role of mixed metal oxides in catalysis science—Versatile applications in organic synthesis. *Catal. Sci. Technol.* **2012**, *2*, 1113–1125. [CrossRef]
42. Tkatchenko, I. Synthesis with Carbon Monoxide and a Petroleum Product. *Compr. Organomet. Chem.* **2006**, *8*, 101–223.
43. Marchi, A.J.; Apesteguía, C.R. Impregnation-induced memory effect of thermally activated layered double hydroxides. *Appl. Clay Sci.* **1998**, *13*, 35–48. [CrossRef]
44. Rocha, J.; Del Arco, M.; Rives, V.; Ulibarri, M.A. Reconstruction of layered double hydroxides from calcined precursors: A powder XRD and ^{27}Al MAS NMR study. *J. Mater. Chem.* **1999**, *9*, 2499–2503. [CrossRef]
45. Stanimirova, T.S.; Kirov, G.; Dinolova, E. Mechanism of hydrotalcite regeneration. *J. Mater. Sci. Lett.* **2001**, *20*, 453–455. [CrossRef]
46. Erickson, K.L.; Bostrom, T.E.; Frost, R.L. A study of structural memory effects in synthetic hydrotalcites using environmental SEM. *Mater. Lett.* **2005**, *59*, 226–229. [CrossRef]
47. Duan, X.; Lu, J.; Evans, D.G.; Wei, X.; Chen, J.S. Functional Host-Guest Materials. In *Modern Inorganic Synthetic Chemistry: Second Edition*; Elsevier: Amsterdam, The Netherlands, 2017; pp. 493–543. ISBN 9780444635914.
48. Meyn, M.; Beneke, K.; Lagaly, G. Anion-Exchange Reactions of Layered Double Hydroxides. *Inorg. Chem.* **1990**, *29*, 5201–5207. [CrossRef]
49. Radha, A.V.; Kamath, P.V.; Shivakumara, C. Mechanism of the anion exchange reactions of the layered double hydroxides (LDHs) of Ca and Mg with Al. *Solid State Sci.* **2005**, *7*, 1180–1187. [CrossRef]
50. Prasanna, S.V.; Kamath, P.V. Anion-exchange reactions of layered double hydroxides: Interplay between coulombic and h-bonding interactions. *Ind. Eng. Chem. Res.* **2009**, *48*, 6315–6320. [CrossRef]
51. Debecker, D.P.; Gaigneaux, E.M.; Busca, G. Exploring, tuning, and exploiting the basicity of hydrotalcites for applications in heterogeneous catalysis. *Chem. Eur. J.* **2009**, *15*, 3920–3935. [CrossRef]
52. Evans, D.G.; Slade, R.C.T. Structure & Bonding, Chapter Structural Aspects of Layered Double Hydroxides. *Struct. Bond.* **2006**, *119*, 1–87.
53. Sideris, P.J.; Nielsen, U.G.; Gan, Z.; Grey, C.P. Mg/Al Ordering in Layered Double Hydroxides Revealed by Multinuclear NMR Spectroscopy. *Science* **2008**, *321*, 113–117. [CrossRef] [PubMed]
54. Cadars, S.; Layrac, G.; Gérardin, C.; Deschamps, M.; Yates, J.R.; Tichit, D.; Massiot, D. Identification and Quantification of Defects in the Cation Ordering in Mg/Al Layered Double Hydroxides. *Chem. Mater.* **2011**, *23*, 2821–2831. [CrossRef]
55. Sideris, P.J.; Blanc, F.; Gan, Z.; Grey, C.P. Identification of Cation Clustering in Mg–Al Layered Double Hydroxides Using Multinuclear Solid State Nuclear Magnetic Resonance Spectroscopy. *Chem. Mater.* **2012**, *24*, 2449–2461. [CrossRef]
56. Manohara, G.V.; Prasanna, S.V.; Kamath, P.V. Structure and Composition of the Layered Double Hydroxides of Mg and Fe: Implications for Anion-Exchange Reactions. *Eur. J. Inorg. Chem.* **2011**, *2011*, 2624–2630. [CrossRef]
57. Xu, M.; Wei, M. Layered Double Hydroxide-Based Catalysts: Recent Advances in Preparation, Structure, and Applications. *Adv. Funct. Mater.* **2018**, *28*, 1802943. [CrossRef]

58. Hernández, W.Y.; Lauwaert, J.; Van Der Voort, P.; Verberckmoes, A. Recent advances on the utilization of layered double hydroxides (LDHs) and related heterogeneous catalysts in a lignocellulosic-feedstock biorefinery scheme. *Green Chem.* **2017**, *19*, 5269–5302. [CrossRef]
59. Kim, D.; Huang, C.; Lee, H.; Han, I.; Kang, S.; Kwon, S.; Lee, J.; Han, Y.; Kim, H. Hydrotalcite-type catalysts for narrow-range oxyethylation of 1-dodecanol using ethyleneoxide. *Appl. Catal. A Gen.* **2003**, *249*, 229–240. [CrossRef]
60. Constantino, V.R.L.; Pinnavaia, T.J. Basic Properties of $Mg^{2+}1{-}xAl^{3+}x$ Layered Double Hydroxides Intercalated by Carbonate, Hydroxide, Chloride, and Sulfate Anions. *Inorg. Chem.* **1995**, *34*, 883–892. [CrossRef]
61. Lü, Z.; Zhang, F.; Lei, X.; Yang, L.; Xu, S.; Duan, X. In situ growth of layered double hydroxide films on anodic aluminum oxide/aluminum and its catalytic feature in aldol condensation of acetone. *Chem. Eng. Sci.* **2008**, *63*, 4055–4062. [CrossRef]
62. Sels, B.F.; De Vos, D.E.; Jacobs, P.A. Hydrotalcite-like anionic clays in catalytic organic reactions. *Catal. Rev. Sci. Eng.* **2001**, *43*, 443–488. [CrossRef]
63. Cirujano, F.G. Engineered MOFs and Enzymes for the Synthesis of Active Pharmaceutical Ingredients. *ChemCatChem* **2019**. [CrossRef]
64. Krow, G.R. The Baeyer-Villiger Oxidation of Ketones and Aldehydes. *Org. React.* **2004**, *43*, 251–798.
65. Baeyer, A.; Villiger, V. Ueber die Einwirkung des Caro'schen Reagens auf Ketone. *Z. Anal. Chem.* **1902**, *41*, 765–766. [CrossRef]
66. Renz, M.; Meunier, B. MICROREVIEW 100 Years of Baeyer–Villiger Oxidations. *Eur. Hournal Org. Chem.* **1999**, *1999*, 737–750. [CrossRef]
67. Fried, J.; Thoma, R.W.; Klingsberg, A. Oxidation of steroids by micro örganisms. iii. side chain degradation, ring d-cleavage and dehydrogenation in ring a. *J. Am. Chem. Soc.* **1953**, *75*, 5764–5765. [CrossRef]
68. Mihovilovic, M.D.; Rudroff, F.; Winninger, A.; Schneider, T.; Schulz, F.; Reetz, M.T. Microbial Baeyer–Villiger Oxidation: Stereopreference and Substrate Acceptance of Cyclohexanone Monooxygenase Mutants Prepared by Directed Evolution. *Org. Lett.* **2006**, *8*, 1221–1224. [CrossRef]
69. ten Brink, G.-J.; Arends, I.W.C.E.; Sheldon, R.A. The Baeyer–Villiger Reaction: New Developments toward Greener Procedures. *Chem. Rev.* **2004**, *104*, 4105–4124. [CrossRef] [PubMed]
70. Leisch, H.; Morley, K.; Lau, P.C.K. Correction to Baeyer-Villiger Monooxygenases: More Than Just Green Chemistry. *Chem. Rev.* **2013**, *113*, 5700. [CrossRef]
71. Kaneda, K.; Ueno, S.; Imanaka, T. Heterogeneous Baeyer-Villiger Oxidation of Ketones Using an Oxidant Consisting of Molecular Oxygen and Aldehydes in the Presence of Hydrotalcite Catalysts. *J. Chem. Soc. Chem. Commun.* **1994**, *25*, 797–798. [CrossRef]
72. Kaneda, K.; Ueno, S.; Imanaka, T. Catalysis of transition metal-functionalized hydrotalcites for the Baeyer-Villiger oxidation of ketones in the presence of molecular oxygen and benzaldehyde. *J. Mol. Catal. A. Chem.* **1995**, *102*, 135–138. [CrossRef]
73. Kawabata, T.; Fujisaki, N.; Shishido, T.; Nomura, K.; Sano, T.; Takehira, K. Improved Fe/Mg-Al hydrotalcite catalyst for Baeyer-Villiger oxidation of ketones with molecular oxygen and benzaldehyde. *J. Mol. Catal. A Chem.* **2006**, *253*, 279–289. [CrossRef]
74. Kaneda, K.; Yamashita, T. Heterogeneous Baeyer-Viiliger Oxidation of Ketones Using m-Chloroperbenzoic Acid Catalyzed by Hydrotaicites. *Tetrahedron Lett.* **1996**, *37*, 4555–4558. [CrossRef]
75. Pillai, U.R.; Sahle-Demessie, E. Sn-exchanged hydrotalcites as catalysts for clean and selective Baeyer-Villiger oxidation of ketones using hydrogen peroxide. *J. Mol. Catal. A Chem.* **2003**, *191*, 93–100. [CrossRef]
76. Jiménez-Sanchidrián, C.; Hidalgo, J.M.; Llamas, R.; Ruiz, J.R. Baeyer-Villiger oxidation of cyclohexanone with hydrogen peroxide/benzonitrile over hydrotalcites as catalysts. *Appl. Catal. A Gen.* **2006**, *312*, 86–94. [CrossRef]
77. Olszówka, J.; Karcz, R.; Napruszewska, B.D.; Duraczyńska, D.; Gaweł, A.; Bahranowski, K.; Serwicka, E.M. Baeyer-Villiger oxidation of cyclohexanone with H_2O_2/acetonitrile over hydrotalcite-like catalysts: Effect of Mg/Al ratio on the ε-caprolactone yield. *Catal. Commun.* **2017**, *100*, 196–201. [CrossRef]
78. Olszówka, J.E.; Karcz, R.; Napruszewska, B.D.; Michalik-Zym, A.; Duraczyńska, D.; Kryściak-Czerwenka, J.; Niecikowska, A.; Bahranowski, K.; Serwicka, E.M. Effect of Mg–Al hydrotalcite crystallinity on catalytic Baeyer-Villiger oxidation of cyclohexanone with H_2O_2/acetonitrile. *Catal. Commun.* **2018**, *107*, 48–52. [CrossRef]

79. Olszówka, J.E.; Karcz, R.; Michalik-Zym, A.; Napruszewska, B.D.; Bielańska, E.; Kryściak-Czerwenka, J.; Socha, R.P.; Nattich-Rak, M.; Krzan, M.; Klimek, A.; et al. Effect of grinding on the physico-chemical properties of Mg-Al hydrotalcite and its performance as a catalyst for Baeyer-Villiger oxidation of cyclohexanone. *Catal. Today* **2018**, *333*, 147–153. [CrossRef]
80. Jones, G. The Knoevenagel Condensation. In *Organic Reactions*; Major Reference Works; John Wiley & Sons, Inc.: New York, NY, USA, 2011; ISBN 9780471264187.
81. Touré, B.B.; Hall, D.G. Natural Product Synthesis Using Multicomponent Reaction Strategies. *Chem. Rev.* **2009**, *109*, 4439–4486. [CrossRef]
82. Yamashita, S.; Iso, K.; Kitajima, K.; Himuro, M.; Hirama, M. Total Synthesis of Cortistatins A and J. *J. Org. Chem.* **2011**, *76*, 2408–2425. [CrossRef]
83. Rousselot, I.; Taviot-Guého, C.; Besse, J.P. Synthesis and characterization of mixed Ga/Al-containing layered double hydroxides: Study of their basic properties through the Knoevenagel condensation of benzaldehyde and ethyl cyanoacetate, and comparison to other LDHs. *Int. J. Inorg. Mater.* **1999**, *1*, 165–174. [CrossRef]
84. Choudary, B.M.; Lakshmi Kantam, M.; Neeraja, V.; Koteswara Rao, K.; Figueras, F.; Delmotte, L. Layered double hydroxide fluoride: A novel solid base catalyst for C-C bond formation. *Green Chem.* **2001**, *3*, 257–260. [CrossRef]
85. Costantino, U.; Curini, M.; Montanari, F.; Nocchetti, M.; Rosati, O. Hydrotalcite-like compounds as catalysts in liquid phase organic synthesis I. Knoevenagel condensation promoted by $[Ni_{0.73}Al_{0.27}(OH)_2]$ $(CO_3)_{0.135}$. *J. Mol. Catal. A Chem.* **2003**, *195*, 245–252. [CrossRef]
86. Li, F.; Jiang, X.; Evans, D.G.; Duan, X. Structure and basicity of mesoporous materials from Mg/Al/In layered double hydroxides prepared by separate nucleation and aging steps method. *J. Porous Mater.* **2005**, *12*, 55–63. [CrossRef]
87. Lei, X.; Zhang, F.; Yang, L.; Guo, X.; Tian, Y.; Fu, S.; Li, F.; Evans, D.G.; Duan, X. Highly crystalline activated layered double hydroxides as solid acid-base catalysts. *AIChE J.* **2007**, *53*, 932–940. [CrossRef]
88. Jia, H.; Zhao, Y.; Niu, P.; Lu, N.; Fan, B.; Li, R. Amine-functionalized MgAl LDH nanosheets as efficient solid base catalysts for Knoevenagel condensation. *Mol. Catal.* **2018**, *449*, 31–37. [CrossRef]
89. Khan, F.A.; Dash, J.; Satapathy, R.; Upadhyay, S.K. Hydrotalcite catalysis in ionic liquid medium: A recyclable reaction system for heterogeneous Knoevenagel and nitroaldol condensation. *Tetrahedron Lett.* **2004**, *45*, 3055–3058. [CrossRef]
90. Li, T.; Zhang, W.; Chen, W.; Miras, H.N.; Song, Y.F. Layered double hydroxide anchored ionic liquids as amphiphilic heterogeneous catalysts for the Knoevenagel condensation reaction. *Dalt. Trans.* **2018**, *47*, 3059–3067. [CrossRef]
91. Zhou, W.; Zhai, S.; Pan, J.; Cui, A.; Qian, J.; He, M.; Xu, Z.; Chen, Q. Bifunctional NiGa Layered Double Hydroxide for the Aerobic Oxidation/Condensation Tandem Reaction between Aromatic Alcohols and Active Methylene Compounds. *Asian J. Org. Chem.* **2017**, *6*, 1536–1541. [CrossRef]
92. Bergmann, E.D.; Ginsburg, D.; Pappo, R. The Michael Reaction. In *Organic Reactions*; Major Reference Works; John Wiley & Sons, Inc.: New York, NY, USA, 2011; pp. 179–270. ISBN 9780471264187.
93. Michael, A. Ueber die Addition von Natriumacetessig- und Natriummalonsaureathern zu den Aethern ungesattigter Sauren. *J. Prakt. Chem.* **1887**, *35*, 349–356. [CrossRef]
94. Mather, B.D.; Viswanathan, K.; Miller, K.M.; Long, T.E. Michael addition reactions in macromolecular design for emerging technologies. *Prog. Polym. Sci.* **2006**, *31*, 487–531. [CrossRef]
95. Noomen, A. Applications of Michael addition chemistry in coatings technology. *Prog. Org. Coat.* **1997**, *32*, 137–142. [CrossRef]
96. Choudary, B.M.; Lakshmi Kantam, M.; Venkat Reddy, C.R.; Koteswara Rao, K.; Figueras, F. The first example of Michael addition catalysed by modified Mg-Al hydrotalcite. *J. Mol. Catal. A Chem.* **1999**, *146*, 279–284. [CrossRef]
97. Choudary, B.M.; Lakshmi Kantam, M.; Kavita, B.; Venkat Reddy, C.; Figueras, F. Catalytic C-C bond formation promoted by Mg-Al-O-t-Bu hydrotalcite. *Tetrahedron* **2000**, *56*, 9357–9364. [CrossRef]
98. Naciuk, F.F.; Vargas, D.Z.; D'oca, C.R.M.; Moro, C.C.; Russowsky, D. One pot domino reaction accessing γ-nitroesters: Synthesis of GABA derivatives. *New J. Chem.* **2015**, *39*, 1643–1653. [CrossRef]
99. Prescott, H.A.; Li, Z.J.; Kemnitz, E.; Trunschke, A.; Deutsch, J.; Lieske, H.; Auroux, A. Application of calcined Mg-Al hydrotalcites for Michael additions: An investigation of catalytic activity and acid-base properties. *J. Catal.* **2005**, *234*, 119–130. [CrossRef]

100. Ebitani, K.; Motokura, K.; Mori, K.; Mizugaki, T.; Kaneda, K. Reconstructed hydrotalcite as a highly active heterogeneous base catalyst for carbon-carbon bond formations in the presence of water. *J. Org. Chem.* **2006**, *71*, 5440–5447. [CrossRef]
101. Kantam, M.L.; Neelima, B.; Reddy, C.V. A recyclable protocol for aza-Michael addition of amines to α,β-unsaturated compounds using Cu-Al hydrotalcite. *J. Mol. Catal. A Chem.* **2005**, *241*, 147–150. [CrossRef]
102. Vijaikumar, S.; Dhakshinamoorthy, A.; Pitchumani, K. l-Proline anchored hydrotalcite clays: An efficient catalyst for asymmetric Michael addition. *Appl. Catal. A Gen.* **2008**, *340*, 25–32. [CrossRef]
103. Otera, J. Transesterification. *Chem. Rev.* **1993**, *93*, 1449–1470. [CrossRef]
104. Porter, R.S.; Wang, L.H. Compatibility and transesterification in binary polymer blends. *Polymer* **1992**, *33*, 2019–2030. [CrossRef]
105. Petrovic, Z.S. Polyurethanes from vegetable oils. *Polym. Rev.* **2008**, *48*, 109–155. [CrossRef]
106. Ma, F.; Hanna, M.A. Biodiesel production: A review. *Bioresour. Technol.* **2003**, *70*, 1–15. [CrossRef]
107. Cantrell, D.G.; Gillie, L.J.; Lee, A.F.; Wilson, K. Structure-reactivity correlations in MgAl hydrotalcite catalysts for biodiesel synthesis. *Appl. Catal. A Gen.* **2005**, *287*, 183–190. [CrossRef]
108. Zeng, H.Y.; Feng, Z.; Deng, X.; Li, Y.Q. Activation of Mg-Al hydrotalcite catalysts for transesterification of rape oil. *Fuel* **2008**, *87*, 3071–3076. [CrossRef]
109. Xie, W.; Peng, H.; Chen, L. Calcined Mg-Al hydrotalcites as solid base catalysts for methanolysis of soybean oil. *J. Mol. Catal. A Chem.* **2006**, *246*, 24–32. [CrossRef]
110. Liu, Y.; Lotero, E.; Goodwin, J.G.; Mo, X. Transesterification of poultry fat with methanol using Mg-Al hydrotalcite derived catalysts. *Appl. Catal. A Gen.* **2007**, *331*, 138–148. [CrossRef]
111. Navajas, A.; Campo, I.; Moral, A.; Echave, J.; Sanz, O.; Montes, M.; Odriozola, J.A.; Arzamendi, G.; Gandía, L.M. Outstanding performance of rehydrated Mg-Al hydrotalcites as heterogeneous methanolysis catalysts for the synthesis of biodiesel. *Fuel* **2018**, *211*, 173–181. [CrossRef]
112. Kondawar, S.; Rode, C. Solvent-Free Glycerol Transesterification with Propylene Carbonate to Glycerol Carbonate over a Solid Base Catalyst. *Energy Fuels* **2017**, *31*, 4361–4371. [CrossRef]
113. Yagiz, F.; Kazan, D.; Akin, A.N. Biodiesel production from waste oils by using lipase immobilized on hydrotalcite and zeolites. *Chem. Eng. J.* **2007**, *134*, 262–267. [CrossRef]
114. Giannelis, E.P.; Nocera, D.G.; Pinnavaia, T.J. Anionic Photocatalysts Supported in Layered Double Hydroxides: Intercalation and Photophysical Properties of a Ruthenium Complex Anion in Synthetic Hydrotalcite. *Inorg. Chem.* **1987**, *26*, 203–205. [CrossRef]
115. Kwon, T.; Tsigdinos, G.A.; Pinnavaia, T.J. Pillaring of Layered Double Hydroxides (LDH's) by Polyoxometalate Anions. *J. Am. Chem. Soc.* **1988**, *110*, 3653–3654. [CrossRef]
116. Seftel, E.M.; Popovici, E.N.; Mertens, M.; Stefaniak, E.A.; Van Grieken, R.; Cool, P.; Vansant, E.F. SnIV-containing layered double hydroxides as precursors for nano-sized ZnO/SnO_2 photocatalysts. *Appl. Catal. B Environ.* **2008**, *84*, 699–705. [CrossRef]
117. Seftel, E.M.; Popovici, E.N.; Mertens, M.; De Witte, K.; Van Tendeloo, G.; Cool, P.; Vansant, E.F. Zn-Al layered double hydroxides: Synthesis, characterization and photocatalytic application. *Microporous Mesoporous Mater.* **2008**, *113*, 296–304. [CrossRef]
118. Jingshan, L.; Jeong-Hyeok, I.; Mayer, M.T.; Marcel, S.; Mohammad Khaja, N.; Nam-Gyu, P.; S David, T.; Jin, F.H.; Michael, G. Water photolysis at 12.3% efficiency via perovskite photovoltaics and Earth-abundant catalysts. *Science* **2014**, *345*, 1593–1596.
119. Mohapatra, L.; Parida, K. A review on the recent progress, challenges and perspective of layered double hydroxides as promising photocatalysts. *J. Mater. Chem. A* **2016**, *4*, 10744–10766. [CrossRef]
120. Greene, J.C.; Baughman, G.L. Effects of 46 Dyes on Population Growth of Freshwater Green Alga Se/eksfrum capricornutum. *Text. Chem. Color.* **1996**, *28*, 23–30.
121. Gregory, P. 3—Toxicology of textile dyes. In *Woodhead Publishing Series in Textiles*; Christie, R.M., Ed.; Woodhead Publishing: Cambridge, UK, 2007; pp. 44–73. ISBN 978-1-84569-115-8.
122. Hatch, K.L.; Maibach, H. Dyes as Contact Allergens: A Comprehensive Record. *Text. Chem. Color. Am. Dyest. Report.* **1999**, *1*, 53–59.
123. Kuo, W.G. Decolorizing dye wastewater with Fenton's reagent. *Water Res.* **1992**, *26*, 881–886. [CrossRef]
124. Rai, H.S.; Bhattacharyya, M.S.; Singh, J.; Bansal, T.K.; Vats, P.; Banerjee, U.C. Removal of dyes from the effluent of textile and dyestuff manufacturing industry: A review of emerging techniques with reference to biological treatment. *Crit. Rev. Environ. Sci. Technol.* **2005**, *35*, 219–238. [CrossRef]

125. Gupta, V.K. Suhas Application of low-cost adsorbents for dye removal—A review. *J. Environ. Manage.* **2009**, *90*, 2313–2342. [CrossRef]
126. Seftel, E.M.; Popovici, E.N.; Beyers, E.; Mertens, M.; Zhu, H.; Vansant, E.F.; Cool, P. New TiO_2/MgAl-LDH Nanocomposites for the Photocatalytic Degradation of Dyes. *J. Nanosci. Nanotechnol.* **2010**, *10*, 8227–8233. [CrossRef] [PubMed]
127. Parida, K.; Mohapatra, L. Carbonate intercalated Zn/Fe layered double hydroxide: A novel photocatalyst for the enhanced photo degradation of azo dyes. *Chem. Eng. J.* **2012**, *179*, 131–139. [CrossRef]
128. Abderrazek, K.; Najoua, F.S.; Srasra, E. Synthesis and characterization of [Zn-Al] LDH: Study of the effect of calcination on the photocatalytic activity. *Appl. Clay Sci.* **2016**, *119*, 229–235. [CrossRef]
129. Ao, Y.; Wang, D.; Wang, P.; Wang, C.; Hou, J.; Qian, J. Enhanced photocatalytic properties of the 3D flower-like Mg-Al layered double hydroxides decorated with Ag_2CO_3 under visible light illumination. *Mater. Res. Bull.* **2016**, *80*, 23–29. [CrossRef]
130. Dinari, M.; Momeni, M.M.; Ghayeb, Y. Photodegradation of organic dye by ZnCrLa-layered double hydroxide as visible-light photocatalysts. *J. Mater. Sci. Mater. Electron.* **2016**, *27*, 9861–9869. [CrossRef]
131. Suárez-Quezada, M.; Romero-Ortiz, G.; Suárez, V.; Morales-Mendoza, G.; Lartundo-Rojas, L.; Navarro-Cerón, E.; Tzompantzi, F.; Robles, S.; Gómez, R.; Mantilla, A. Photodegradation of phenol using reconstructed Ce doped Zn/Al layered double hydroxides as photocatalysts. *Catal. Today* **2016**, *271*, 213–219. [CrossRef]
132. Wu, X.; Zhang, D.; Jiao, F.; Wang, S. Visible-light-driven photodegradation of Methyl Orange using Cu_2O/ZnAl calcined layered double hydroxides as photocatalysts. *Colloids Surf. A Physicochem. Eng. Asp.* **2016**, *508*, 110–116. [CrossRef]
133. Carja, G.; Grosu, E.F.; Mureseanu, M.; Lutic, D. A family of solar light responsive photocatalysts obtained using Zn^{2+} Me^{3+} (Me = Al/Ga) LDHs doped with Ga_2O_3 and In_2O_3 and their derived mixed oxides: A case study of phenol/4-nitrophenol decomposition. *Catal. Sci. Technol.* **2017**, *7*, 5402–5412. [CrossRef]
134. Hadnadjev-Kostic, M.; Vulic, T.; Marinkovic-Neducin, R.; Lončarević, D.; Dostanić, J.; Markov, S.; Jovanović, D. Photo-induced properties of photocatalysts: A study on the modified structural, optical and textural properties of TiO_2–ZnAl layered double hydroxide based materials. *J. Clean. Prod.* **2017**, *164*, 1–18. [CrossRef]
135. Mohamed, F.; Abukhadra, M.R.; Shaban, M. Removal of safranin dye from water using polypyrrole nanofiber/Zn-Fe layered double hydroxide nanocomposite (Ppy NF/Zn-Fe LDH) of enhanced adsorption and photocatalytic properties. *Sci. Total Environ.* **2018**, *640–641*, 352–363. [CrossRef] [PubMed]
136. Timár, Z.; Varga, G.; Muráth, S.; Kónya, Z.; Kukovecz, Á. Synthesis, characterization and photocatalytic activity of crystalline Mn (II) Cr (III) -layered double hydroxide. *Catal. Today* **2017**, *284*, 195–201. [CrossRef]
137. Fu, Y.; Ning, F.; Xu, S.; An, H.; Shao, M.; Wei, M. Terbium doped ZnCr-layered double hydroxides with largely enhanced visible light photocatalytic performance. *J. Mater. Chem. A* **2016**, *4*, 3907–3913. [CrossRef]
138. Silva, C.G.; Bouizi, Y.; Fornés, V.; García, H. Layered double hydroxides as highly efficient photocatalysts for visible light oxygen generation from water. *J. Am. Chem. Soc.* **2009**, *131*, 13833–13839. [CrossRef] [PubMed]
139. Lee, J.M.; Yang, J.H.; Kwon, N.H.; Jo, Y.K.; Choy, J.H.; Hwang, S.J. Intercalative hybridization of layered double hydroxide nanocrystals with mesoporous g-C_3N_4 for enhancing visible light-induced H_2 production efficiency. *Dalt. Trans.* **2018**, *47*, 2949–2955. [CrossRef] [PubMed]
140. Wang, R.; Luo, L.; Zhu, X.; Yan, Y.; Zhang, B.; Xiang, X.; He, J. Plasmon-Enhanced Layered Double Hydroxide Composite $BiVO_4$ Photoanodes: Layering-Dependent Modulation of the Water-Oxidation Reaction. *ACS Appl. Energy Mater.* **2018**, *1*, 3577–3586. [CrossRef]
141. Vo, T.G.; Tai, Y.; Chiang, C.Y. Multifunctional ternary hydrotalcite-like nanosheet arrays as an efficient co-catalyst for vastly improved water splitting performance on bismuth vanadate photoanode. *J. Catal.* **2019**, *370*, 1–10. [CrossRef]
142. Ma, Y.; Wang, X.; Jia, Y.; Chen, X.; Han, H.; Li, C. Titanium Dioxide-Based Nanomaterials for Photocatalytic Fuel Generations. *Chem. Rev.* **2014**, *114*, 9987–10043. [CrossRef]
143. Tu, W.; Zhou, Y.; Zou, Z. Photocatalytic conversion of CO_2 into renewable hydrocarbon fuels: State-of-the-art accomplishment, challenges, and prospects. *Adv. Mater.* **2014**, *26*, 4607–4626. [CrossRef]
144. White, J.L.; Baruch, M.F.; Pander, J.E.; Hu, Y.; Fortmeyer, I.C.; Park, J.E.; Zhang, T.; Liao, K.; Gu, J.; Yan, Y.; et al. Light-Driven Heterogeneous Reduction of Carbon Dioxide: Photocatalysts and Photoelectrodes. *Chem. Rev.* **2015**, *115*, 12888–12935. [CrossRef]

145. Sharma, U.; Tyagi, B.; Jasra, R.V. Synthesis and Characterization of Mg-Al-CO_3 Layered Double Hydroxide for CO_2 Adsorption. *Ind. Eng. Chem. Res.* **2008**, *47*, 9588–9595. [CrossRef]
146. Flores-Flores, M.; Luévano-Hipólito, E.; Martínez, L.M.T.; Morales-Mendoza, G.; Gómez, R. Photocatalytic CO_2 conversion by MgAl layered double hydroxides: Effect of Mg^{2+} precursor and microwave irradiation time. *J. Photochem. Photobiol. A Chem.* **2018**, *363*, 68–73. [CrossRef]
147. Iguchi, S.; Teramura, K.; Hosokawa, S.; Tanaka, T. Photocatalytic conversion of CO_2 in water using fluorinated layered double hydroxides as photocatalysts. *Appl. Catal. A Gen.* **2016**, *521*, 160–167. [CrossRef]
148. Anton Wein, L.; Zhang, H.; Urushidate, K.; Miyano, M.; Izumi, Y. Optimized photoreduction of CO_2 exclusively into methanol utilizing liberated reaction space in layered double hydroxides comprising zinc, copper, and gallium. *Appl. Surf. Sci.* **2018**, *447*, 687–696. [CrossRef]
149. Iguchi, S.; Hasegawa, Y.; Teramura, K.; Hosokawa, S.; Tanaka, T. Preparation of transition metal-containing layered double hydroxides and application to the photocatalytic conversion of CO_2 in water. *J. CO2 Util.* **2016**, *15*, 6–14. [CrossRef]
150. Tokudome, Y.; Fukui, M.; Iguchi, S.; Hasegawa, Y.; Teramura, K.; Tanaka, T.; Takemoto, M.; Katsura, R.; Takahashi, M. A nanoLDH catalyst with high CO_2 adsorption capability for photo-catalytic reduction. *J. Mater. Chem. A* **2018**, *6*, 9684–9690. [CrossRef]
151. Zhao, H.; Xu, J.; Liu, L.; Rao, G.; Zhao, C.; Li, Y. CO_2 photoreduction with water vapor by Ti-embedded MgAl layered double hydroxides. *J. CO2 Util.* **2016**, *15*, 15–23. [CrossRef]
152. Tonda, S.; Kumar, S.; Bhardwaj, M.; Yadav, P.; Ogale, S. g-C_3N_4/NiAl-LDH 2D/2D Hybrid Heterojunction for High-Performance Photocatalytic Reduction of CO_2 into Renewable Fuels. *ACS Appl. Mater. Interfaces* **2018**, *10*, 2667–2678. [CrossRef]
153. Gao, L.G.; Gao, Y.Y.; Song, X.L.; Ma, X. rong A novel La^{3+}–Zn^{2+}–Al^{3+}–$MoO_4{}^{2-}$ layered double hydroxides photocatalyst for the decomposition of dibenzothiophene in diesel oil. *Pet. Sci. Technol.* **2018**, *36*, 850–855. [CrossRef]
154. Ristovski, Z.D.; Jayaratne, R.E.; Lim, M.; Ayoko, G.A.; Morawska, L. Influence of diesel fuel sulfur on nanoparticle emissions from city buses. *Environ. Sci. Technol.* **2006**, *40*, 1314–1320. [CrossRef]
155. Postgate, J.R. Biological nitrogen fixation: Fundamentals. *Philos. Trans. R. Soc. Lond. B. Biol. Sci.* **1982**, *296*, 375–385. [CrossRef]
156. Appl, M. Chemical Reactions and Uses of Ammonia. In *Ammonia: Principles and Industrial Practice;* Wiley Online Books: New York, NY, USA, 1999; pp. 231–234. ISBN 9783527613885. Available online: https://onlinelibrary.wiley.com/doi/book/10.1002/9783527613885 (accessed on 3 June 2019).
157. Erisman, J.W.; Sutton, M.A.; Galloway, J.; Klimont, Z.; Winiwarter, W. How a century of ammonia synthesis changed the world. *Nat. Geosci.* **2008**, *1*, 636–639. [CrossRef]
158. Zhao, Y.; Zhao, Y.; Waterhouse, G.I.N.; Zheng, L.; Cao, X.; Teng, F.; Wu, L.; Tung, C.; Hare, D.O.; Zhang, T. Layered-Double-Hydroxide Nanosheets as Efficient Visible-Light-Driven Photocatalysts for Dinitrogen Fixation. *Adv. Mater.* **2017**, *29*, 1703828. [CrossRef] [PubMed]
159. Cnrs, U.M.R.; Lyon, E.C. De Urea Biosensors Based on Immobilization of Urease into Two Oppositely Charged Clays (Laponite and Zn-Al Layered Double Hydroxides) the effect of the buffer capacity of the outer medium. *Anal. Chem.* **2002**, *74*, 4037–4043.
160. Taviot-Guého, C.; Prévot, V.; Forano, C.; Renaudin, G.; Mousty, C.; Leroux, F. Tailoring Hybrid Layered Double Hydroxides for the Development of Innovative Applications. *Adv. Funct. Mater.* **2018**, *28*, 1703868. [CrossRef]
161. Wang, J. Electrochemical Glucose Biosensors. *Chem. Rev.* **2008**, *108*, 814–825. [CrossRef] [PubMed]
162. Aleeva, Y.; Maira, G.; Scopelliti, M.; Vinciguerra, V.; Scandurra, G.; Cannatà, G.; Giusi, G.; Ciofi, C.; Figà, V.; Occhipinti, L.G.; et al. Amperometric Biosensor and Front-End Electronics for Remote Glucose Monitoring by Crosslinked PEDOT-Glucose Oxidase. *IEEE Sens. J.* **2018**, *18*, 4869–4878. [CrossRef]
163. Toghill, K.E.; Compton, R.G. Electrochemical Non-enzymatic Glucose Sensors: A Perspective and an Evaluation. *Int. J. Electrochem. Sci* **2010**, *5*, 1246–1301.
164. Cui, J.; Li, Z.; Liu, K.; Li, J.; Shao, M. A bifunctional nonenzymatic flexible glucose microsensor based on CoFe-Layered double hydroxide. *Nanoscale Adv.* **2019**, *1*, 948–952. [CrossRef]
165. Ai, H.; Huang, X.; Zhu, Z.; Liu, J.; Chi, Q.; Li, Y.; Li, Z.; Ji, X. A novel glucose sensor based on monodispersed Ni/Al layered double hydroxide and chitosan. *Biosens. Bioelectron.* **2008**, *24*, 1048–1052. [CrossRef]

166. Li, X.; Liu, J.; Ji, X.; Jiang, J.; Ding, R.; Hu, Y.; Hu, A.; Huang, X. Ni/Al layered double hydroxide nanosheet film grown directly on Ti substrate and its application for a nonenzymatic glucose sensor. *Sens. Actuators B Chem.* **2010**, *147*, 241–247. [CrossRef]
167. Fu, S.; Fan, G.; Yang, L.; Li, F. Non-enzymatic glucose sensor based on Au nanoparticles decorated ternary Ni-Al layered double hydroxide/single-walled carbon nanotubes/graphene nanocomposite. *Electrochim. Acta* **2015**, *152*, 146–154. [CrossRef]
168. Hassanein, A.; Salahuddin, N.; Matsuda, A.; Hattori, T.; Elfiky, M. Fabrication of Electrochemical Sensor Based on Layered Double Hydroxide/Polypyrrole/Carbon Paste for Determination of an Alpha-adrenergic Blocking Agent Terazosin. *Electroanalysis* **2018**, *30*, 459–465. [CrossRef]
169. Ma, Y.; Wang, Y.; Xie, D.; Gu, Y.; Zhang, H.; Wang, G.; Zhang, Y.; Zhao, H.; Wong, P.K. NiFe-Layered Double Hydroxide Nanosheet Arrays Supported on Carbon Cloth for Highly Sensitive Detection of Nitrite. *ACS Appl. Mater. Interfaces* **2018**, *10*, 6541–6551. [CrossRef] [PubMed]
170. Asif, M.; Aziz, A.; Wang, H.; Wang, Z.; Wang, W.; Ajmal, M.; Xiao, F.; Chen, X.; Liu, H. Superlattice stacking by hybridizing layered double hydroxide nanosheets with layers of reduced graphene oxide for electrochemical simultaneous determination of dopamine, uric acid and ascorbic acid. *Microchim. Acta* **2019**, *186*, 61. [CrossRef] [PubMed]
171. Liu, J.; Xu, X.; Chen, Z.; Li, R.; Kang, L.; Yao, J. A fluorometric displacement assay for adenosine triphosphate using layered cobalt(II) double hydroxide nanosheets. *Microchim. Acta* **2019**, *186*, 263. [CrossRef] [PubMed]
172. Abdolmohammad-Zadeh, H.; Zamani-Kalajahi, M. A turn-on/off fluorescent sensor based on nano-structured Mg-Al layered double hydroxide intercalated with salicylic acid for monitoring of ferric ion in human serum samples. *Anal. Chim. Acta* **2019**, *1061*, 152–160. [CrossRef] [PubMed]
173. Arrabito, G.; Galati, C.; Castellano, S.; Pignataro, B. Luminometric sub-nanoliter droplet-to-droplet array (LUMDA) and its application to drug screening by phase I metabolism enzymes. *Lab Chip* **2013**, *13*, 68–72. [CrossRef] [PubMed]
174. Wang, Z.; Liu, F.; Lu, C. Chemiluminescence flow biosensor for glucose using Mg-Al carbonate layered double hydroxides as catalysts and buffer solutions. *Biosens. Bioelectron.* **2012**, *38*, 284–288. [CrossRef]
175. Pourfaraj, R.; Kazemi, S.Y.; Fatemi, S.J.; Biparva, P. Synthesis of α- and β-CoNi binary hydroxides nanostructures and luminol chemiluminescence study for H_2O_2 detection. *J. Photochem. Photobiol. A Chem.* **2018**, *364*, 534–541. [CrossRef]
176. Pan, F.; Zhang, Y.; Yuan, Z.; Lu, C. Sensitive and Selective Carmine Acid Detection Based on Chemiluminescence Quenching of Layer Doubled Hydroxide–Luminol–H_2O_2 System. *ACS Omega* **2018**, *3*, 18836–18842. [CrossRef]
177. Haroone, M.S.; Hu, Y.; Zhang, P.; Ma, R.; Zhang, X.; Kaleem, Q.M.; Khan, M.B.; Lu, J. Fluorescence enhancement strategy for evaluation of the minor groove binder DAPI to complementary ssDNA sequence including telomere mimics in $(ssDNA@DAPI/LDH)_n$ ultrathin films. *Dyes Pigments* **2019**, *166*, 422–432. [CrossRef]
178. Fu, L.; Yan, L.; Wang, G.; Ren, H.; Jin, L. Photoluminescence enhancement of silver nanoclusters assembled on the layered double hydroxides and their application to guanine detection. *Talanta* **2019**, *193*, 161–167. [CrossRef] [PubMed]
179. Aziz, A.; Asif, M.; Azeem, M.; Ashraf, G.; Wang, Z.; Xiao, F.; Liu, H. Self-stacking of exfoliated charged nanosheets of LDHs and graphene as biosensor with real-time tracking of dopamine from live cells. *Anal. Chim. Acta* **2019**, *1047*, 197–207. [CrossRef] [PubMed]
180. Asif, M.; Aziz, A.; Wang, Z.; Ashraf, G.; Wang, J.; Luo, H.; Chen, X.; Xiao, F.; Liu, H. Hierarchical CNTs@CuMn layered double hydroxide nanohybrid with enhanced electrochemical performance in H_2S detection from live cells. *Anal. Chem.* **2019**, *91*, 3912–3920. [CrossRef]
181. Chatterjee, A.; Bharadiya, P.; Hansora, D. Layered double hydroxide based bionanocomposites. *Appl. Clay Sci.* **2019**, *177*, 19–36. [CrossRef]
182. Andrea, K.A.; Wang, L.; Carrier, A.J.; Campbell, M.; Buhariwalla, M.; Mutch, M.; MacQuarrie, S.L.; Bennett, C.; Mkandawire, M.; Oakes, K.; et al. Adsorption of Oligo-DNA on Magnesium Aluminum-Layered Double-Hydroxide Nanoparticle Surfaces: Mechanistic Implication in Gene Delivery. *Langmuir* **2017**, *33*, 3926–3933. [CrossRef] [PubMed]

183. Lu, M.; Shan, Z.; Andrea, K.; Macdonald, B.; Beale, S.; Curry, D.E.; Wang, L.; Wang, S.; Oakes, K.D.; Bennett, C.; et al. Chemisorption Mechanism of DNA on Mg/Fe Layered Double Hydroxide Nanoparticles: Insights into Engineering Effective SiRNA Delivery Systems. *Langmuir* **2016**, *32*, 2659–2667. [CrossRef] [PubMed]
184. Choy, J.-H.; Kwak, S.-Y.; Park, J.-S.; Jeong, Y.-J.; Portier, J. Intercalative Nanohybrids of Nucleoside Monophosphates and DNA in Layered Metal Hydroxide. *J. Am. Chem. Soc.* **1999**, *121*, 1399–1400. [CrossRef]
185. Desigaux, L.; Belkacem, M.B.; Richard, P.; Cellier, J.; Léone, P.; Cario, L.; Leroux, F.; Taviot-Guého, C.; Pitard, B. Self-assembly and characterization of layered double hydroxide/DNA hybrids. *Nano Lett.* **2006**, *6*, 199–204. [CrossRef] [PubMed]
186. Choy, J.-H.; Kwak, S.-Y.; Jeong, Y.-J.; Park, J.-S. Inorganic Layered Double Hydroxides as Nonviral Vectors. *Angew. Chem. Int. Ed.* **2000**, *39*, 4041–4045. [CrossRef]
187. Yazdani, P.; Mansouri, E.; Eyvazi, S.; Yousefi, V.; Kahroba, H.; Hejazi, M.S.; Mesbahi, A.; Tarhriz, V.; Abolghasemi, M.M. Layered double hydroxide nanoparticles as an appealing nanoparticle in gene/plasmid and drug delivery system in C2C12 myoblast cells. *Artif. Cells Nanomed. Biotechnol.* **2019**, *47*, 436–442. [CrossRef] [PubMed]
188. Li, Y.; Bao, W.; Wu, H.; Wang, J.; Zhang, Y.; Wan, Y.; Cao, D.; O'Hare, D.; Wang, Q. Delaminated layered double hydroxide delivers DNA molecules as sandwich nanostructure into cells via a non-endocytic pathway. *Sci. Bull.* **2017**, *62*, 686–692. [CrossRef]
189. Ádok-Sipiczki, M.; Szilagyi, I.; Pálinkó, I.; Pavlovic, M.; Sipos, P.; Nardin, C. Design of nucleic acid-layered double hydroxide nanohybrids. *Colloid Polym. Sci.* **2017**, *295*, 1463–1473. [CrossRef]
190. Wong, Y.; Markham, K.; Xu, Z.P.; Chen, M.; Max Lu, G.Q.; Bartlett, P.F.; Cooper, H.M. Efficient delivery of siRNA to cortical neurons using layered double hydroxide nanoparticles. *Biomaterials* **2010**, *31*, 8770–8779. [CrossRef]
191. Li, L.; Gu, W.; Chen, J.; Chen, W.; Xu, Z.P. Co-delivery of siRNAs and anti-cancer drugs using layered double hydroxide nanoparticles. *Biomaterials* **2014**, *35*, 3331–3339. [CrossRef]
192. Park, D.H.; Cho, J.; Kwon, O.J.; Yun, C.O.; Choy, J.H. Biodegradable Inorganic Nanovector: Passive versus Active Tumor Targeting in siRNA Transportation. *Angew. Chem. Int. Ed.* **2016**, *55*, 4582–4586. [CrossRef]
193. Acharya, R.; Alsharabasy, A.M.; Saha, S.; Rahaman, S.H.; Bhattacharjee, A.; Halder, S.; Chakraborty, M.; Chakraborty, J. Intercalation of shRNA-plasmid in Mg–Al layered double hydroxide nanoparticles and its cellular internalization for possible treatment of neurodegenerative diseases. *J. Drug Deliv. Sci. Technol.* **2019**, *52*, 500–508. [CrossRef]
194. Rahaman, S.H.; Bhattacharjee, A.; Saha, S.; Chakraborty, M.; Chakraborty, J. shRNA intercalation in CaAl-LDH nanoparticle synthesized at two different pH conditions and its comparative evaluation. *Appl. Clay Sci.* **2019**, *171*, 57–64. [CrossRef]
195. Li, L.; Zhang, R.; Gu, W.; Xu, Z.P. Mannose-conjugated layered double hydroxide nanocomposite for targeted siRNA delivery to enhance cancer therapy. *Nanomed. Nanotechnol. Biol. Med.* **2018**, *14*, 2355–2364. [CrossRef]
196. Wu, Y.; Gu, W.; Li, L.; Chen, C.; Xu, Z. Enhancing PD-1 Gene Silence in T Lymphocytes by Comparing the Delivery Performance of Two Inorganic Nanoparticle Platforms. *Nanomaterials* **2019**, *9*, 159. [CrossRef]
197. Wu, Y.; Gu, W.; Chen, C.; Do, S.T.; Xu, Z.P. Optimization of Formulations Consisting of Layered Double Hydroxide Nanoparticles and Small Interfering RNA for Efficient Knockdown of the Target Gene. *ACS Omega* **2018**, *3*, 4871–4877. [CrossRef] [PubMed]
198. Santos, A.C.; Ferreira, C.; Veiga, F.; Ribeiro, A.J.; Panchal, A.; Lvov, Y.; Agarwal, A. Halloysite clay nanotubes for life sciences applications: From drug encapsulation to bioscaffold. *Adv. Colloid Interface Sci.* **2018**, *257*, 58–70. [CrossRef] [PubMed]
199. Massaro, M.; Cavallaro, G.; Colletti, C.G.; Lazzara, G.; Milioto, S.; Noto, R.; Riela, S. Chemical modification of halloysite nanotubes for controlled loading and release. *J. Mater. Chem. B* **2018**, *6*, 3415–3433. [CrossRef]
200. Arrabito, G.; Cavaleri, F.; Montalbano, V.; Vetri, V.; Leone, M.; Pignataro, B. Monitoring few molecular binding events in scalable confined aqueous compartments by raster image correlation spectroscopy (CADRICS). *Lab Chip* **2016**, *16*, 4666–4676. [CrossRef] [PubMed]
201. Kumar, B.V.V.S.P.; Patil, A.J.; Mann, S. Enzyme-powered motility in buoyant organoclay/DNA protocells. *Nat. Chem.* **2018**, *10*, 1154–1163. [CrossRef] [PubMed]
202. Choi, G.; Eom, S.; Vinu, A.; Choy, J.H. 2D Nanostructured Metal Hydroxides with Gene Delivery and Theranostic Functions; A Comprehensive Review. *Chem. Rec.* **2018**, *18*, 1033–1053. [CrossRef] [PubMed]

203. Forano, C.; Bruna, F.; Mousty, C.; Prevot, V. Interactions between Biological Cells and Layered Double Hydroxides: Towards Functional Materials. *Chem. Rec.* **2018**, *18*, 1150–1166. [CrossRef]
204. Kura, A.U.; Hussein, M.Z.; Fakurazi, S.; Arulselvan, P. Layered double hydroxide nanocomposite for drug delivery systems; bio-distribution, toxicity and drug activity enhancement. *Chem. Cent. J.* **2014**, *8*, 47. [CrossRef]
205. Choi, S.-J.; Choy, J.-H. Layered double hydroxide nanoparticles as target-specific delivery carriers: Uptake mechanism and toxicity. *Nanomedicine* **2011**, *6*, 803–814. [CrossRef]
206. Saifullah, B.; Hussein, M.Z.; Hussein-Al-Ali, S.H.; Arulselvan, P.; Fakurazi, S. Antituberculosis nanodelivery system with controlled-release properties based on para-amino salicylate—Zinc aluminum-layered double-hydroxide nanocomposites. *Drug Des Devel Ther.* **2013**, *7*, 1365–1375.
207. da Costa Fernandes, C., Jr.; Pinto, T.S.; Kang, H.R.; Magalhães Padilha, P.; Koh, I.H.J.; Constantino, V.R.L.; Zambuzzi, W.F. Layered Double Hydroxides Are Promising Nanomaterials for Tissue Bioengineering Application. *Adv. Biosyst.* **2019**, 1800238. [CrossRef]
208. Cunha, V.R.R.; De Souza, R.B.; Da Fonseca Martins, A.M.C.R.P.; Koh, I.H.J.; Constantino, V.R.L. Accessing the biocompatibility of layered double hydroxide by intramuscular implantation: Histological and microcirculation evaluation. *Sci. Rep.* **2016**, *6*, 30547. [CrossRef] [PubMed]
209. Li, H.; Peng, F.; Wang, D.; Qiao, Y.; Xu, D.; Liu, X. Layered double hydroxide/poly-dopamine composite coating with surface heparinization on Mg alloys: Improved anticorrosion, endothelialization and hemocompatibility. *Biomater. Sci.* **2018**, *6*, 1846–1858. [CrossRef] [PubMed]
210. Peng, F.; Li, H.; Wang, D.; Tian, P.; Tian, Y.; Yuan, G.; Xu, D.; Liu, X. Enhanced Corrosion Resistance and Biocompatibility of Magnesium Alloy by Mg–Al-Layered Double Hydroxide. *ACS Appl. Mater. Interfaces* **2016**, *8*, 35033–35044. [CrossRef]
211. Errico, V.; Arrabito, G.; Fornetti, E.; Fuoco, C.; Testa, S.; Saggio, G.; Rufini, S.; Cannata, S.; Desideri, A.; Falconi, C.; et al. High-Density ZnO Nanowires as a Reversible Myogenic–Differentiation Switch. *ACS Appl. Mater. Interfaces* **2018**, *10*, 14097–14107. [CrossRef]
212. Hansel, C.S.; Crowder, S.W.; Cooper, S.; Gopal, S.; João Pardelha da Cruz, M.; de Oliveira Martins, L.; Keller, D.; Rothery, S.; Becce, M.; Cass, A.E.G.; et al. Nanoneedle-Mediated Stimulation of Cell Mechanotransduction Machinery. *ACS Nano* **2019**, *13*, 2913–2926. [CrossRef]
213. Ramanathan, G.; Liji, L.S.; Fardim, P.; Sivagnanam, U.T. Fabrication of 3D dual-layered nanofibrous graft loaded with layered double hydroxides and their effects in osteoblastic behavior for bone tissue engineering. *Process Biochem.* **2018**, *64*, 255–259. [CrossRef]
214. Lee, S.S.; Choi, G.E.; Lee, H.J.; Kim, Y.; Choy, J.H.; Jeong, B. Layered Double Hydroxide and Polypeptide Thermogel Nanocomposite System for Chondrogenic Differentiation of Stem Cells. *ACS Appl. Mater. Interfaces* **2017**, *9*, 42668–42675. [CrossRef]
215. Kang, H.R.; da Costa Fernandes, C.J.; da Silva, R.A.; Constantino, V.R.L.; Koh, I.H.J.; Zambuzzi, W.F. Mg–Al and Zn–Al Layered Double Hydroxides Promote Dynamic Expression of Marker Genes in Osteogenic Differentiation by Modulating Mitogen-Activated Protein Kinases. *Adv. Healthc. Mater.* **2018**, *7*, 1700693. [CrossRef]
216. da Silva Feltran, G.; da Costa Fernandes, C.J.; Rodrigues Ferreira, M.; Kang, H.R.; de Carvalho Bovolato, A.L.; de Assis Golim, M.; Deffune, E.; Koh, I.H.J.; Constantino, V.R.L.; Zambuzzi, W.F. Sonic hedgehog drives layered double hydroxides-induced acute inflammatory landscape. *Colloids Surf. B Biointerfaces* **2019**, *174*, 467–475. [CrossRef]
217. Li, B.; Gu, Z.; Kurniawan, N.; Chen, W.; Xu, Z.P. Manganese-Based Layered Double Hydroxide Nanoparticles as a T1-MRI Contrast Agent with Ultrasensitive pH Response and High Relaxivity. *Adv. Mater.* **2017**, *29*, 1700373. [CrossRef]
218. Saha, S.; Ray, S.; Acharya, R.; Chatterjee, T.K.; Chakraborty, J. Magnesium, zinc and calcium aluminium layered double hydroxide-drug nanohybrids: A comprehensive study. *Appl. Clay Sci.* **2017**, *135*, 493–509. [CrossRef]
219. Naz, S.; Shamoon, M.; Wang, R.; Zhang, L.; Zhou, J.; Chen, J. Advances in therapeutic implications of inorganic drug delivery nano-platforms for cancer. *Int. J. Mol. Sci.* **2019**, *20*, 965. [CrossRef]
220. Lv, F.; Xu, L.; Zhang, Y.; Meng, Z. Layered Double Hydroxide Assemblies with Controllable Drug Loading Capacity and Release Behavior as well as Stabilized Layer-by-Layer Polymer Multilayers. *ACS Appl. Mater. Interfaces* **2015**, *7*, 19104–19111. [CrossRef]

221. Choi, G.; Lee, J.H.; Oh, Y.J.; Choy, Y.B.; Park, M.C.; Chang, H.C.; Choy, J.H. Inorganic-polymer nanohybrid carrier for delivery of a poorly-soluble drug, ursodeoxycholic acid. *Int. J. Pharm.* **2010**, *402*, 117–122. [CrossRef]
222. Gu, Z.; Zuo, H.; Li, L.; Wu, A.; Xu, Z.P. Pre-coating layered double hydroxide nanoparticles with albumin to improve colloidal stability and cellular uptake. *J. Mater. Chem. B* **2015**, *3*, 3331–3339. [CrossRef]
223. Oh, J.-M.; Choi, S.-J.; Kim, S.-T.; Choy, J.-H. Cellular Uptake Mechanism of an Inorganic Nanovehicle and Its Drug Conjugates: Enhanced Efficacy Due To Clathrin-Mediated Endocytosis. *Bioconjug. Chem.* **2006**, *17*, 1411–1417. [CrossRef]
224. Bao, W.; Wang, J.; Wang, Q.; O'Hare, D.; Wan, Y. Layered double hydroxide nanotransporter for molecule delivery to intact plant cells. *Sci. Rep.* **2016**, *6*, 26738. [CrossRef]
225. Kong, X.; Jin, L.; Wei, M.; Duan, X. Antioxidant drugs intercalated into layered double hydroxide: Structure and in vitro release. *Appl. Clay Sci.* **2010**, *49*, 324–329. [CrossRef]
226. Rojas, R.; Palena, M.C.; Jimenez-Kairuz, A.F.; Manzo, R.H.; Giacomelli, C.E. Modeling drug release from a layered double hydroxide-ibuprofen complex. *Appl. Clay Sci.* **2012**, *62–63*, 15–20. [CrossRef]
227. Yasaei, M.; Khakbiz, M.; Ghasemi, E.; Zamanian, A. Synthesis and characterization of ZnAl-$NO_3(CO_3)$ layered double hydroxide: A novel structure for intercalation and release of simvastatin. *Appl. Surf. Sci.* **2019**, *467–468*, 782–791. [CrossRef]
228. Bouaziz, Z.; Djebbi, M.A.; Soussan, L.; Janot, J.M.; Amara, A.B.H.; Balme, S. Adsorption of nisin into layered double hydroxide nanohybrids and in-vitro controlled release. *Mater. Sci. Eng. C* **2017**, *76*, 673–683. [CrossRef]
229. Mallakpour, S.; Hatami, M. Fabrication and characterization of pH-sensitive bio-nanocomposite beads havening folic acid intercalated LDH and chitosan: Drug release and mechanism evaluation. *Int. J. Biol. Macromol.* **2019**, *122*, 157–167. [CrossRef]
230. Piao, H.; Kim, M.H.; Cui, M.; Choi, G.; Choy, J.H. Alendronate-anionic clay nanohybrid for enhanced osteogenic proliferation and differentiation. *J. Korean Med. Sci.* **2019**, *34*, e37. [CrossRef]
231. Khorsandi, K.; Hosseinzadeh, R.; Shahidi, F.K. Photodynamic treatment with anionic nanoclays containing curcumin on human triple-negative breast cancer cells: Cellular and biochemical studies. *J. Cell. Biochem.* **2019**, *120*, 4998–5009. [CrossRef]
232. Liu, C.G.; Kankala, R.K.; Liao, H.Y.; Chen, A.Z.; Wang, S.B. Engineered pH-responsive hydrazone-carboxylate complexes-encapsulated 2D matrices for cathepsin-mediated apoptosis in cancer. *J. Biomed. Mater. Res. Part A* **2019**, 1184–1194. [CrossRef]
233. Wen, J.; Yang, K.; Ding, X.; Li, H.; Xu, Y.; Liu, F.; Sun, S. In Situ Formation of Homogeneous Tellurium Nanodots in Paclitaxel-Loaded MgAl Layered Double Hydroxide Gated Mesoporous Silica Nanoparticles for Synergistic Chemo/PDT/PTT Trimode Combinatorial Therapy. *Inorg. Chem.* **2019**, *58*, 2987–2996. [CrossRef]
234. Chakraborty, M.; Mitra, M.K.; Chakraborty, J. One-pot synthesis of CaAl-layered double hydroxide–methotrexate nanohybrid for anticancer application. *Bull. Mater. Sci.* **2017**, *40*, 1203–1211. [CrossRef]
235. Bhattacharjee, A.; Rahaman, S.H.; Saha, S.; Chakraborty, M.; Chakraborty, J. Determination of half maximal inhibitory concentration of CaAl layered double hydroxide on cancer cells and its role in the apoptotic pathway. *Appl. Clay Sci.* **2019**, *168*, 31–35. [CrossRef]
236. Yu, J.; Martin, B.R.; Clearfield, A.; Luo, Z.; Sun, L. One-step direct synthesis of layered double hydroxide single-layer nanosheets. *Nanoscale* **2015**, *7*, 9448–9451. [CrossRef]
237. Weng, Y.; Guan, S.; Lu, H.; Meng, X.; Kaassis, A.Y.; Ren, X.; Qu, X.; Sun, C.; Xie, Z.; Zhou, S. Confinement of carbon dots localizing to the ultrathin layered double hydroxides toward simultaneous triple-mode bioimaging and photothermal therapy. *Talanta* **2018**, *184*, 50–57. [CrossRef]
238. Wang, Z.; Liang, P.; He, X.; Wu, B.; Liu, Q.; Xu, Z.; Wu, H.; Liu, Z.; Qian, Y.; Wang, S.; et al. Etoposide loaded layered double hydroxide nanoparticles reversing chemoresistance and eradicating human glioma stem cells: In vitro and in vivo. *Nanoscale* **2018**, *10*, 13106–13121. [CrossRef]
239. Bouaziz, Z.; Soussan, L.; Janot, J.M.; Jaber, M.; Ben Haj Amara, A.; Balme, S. Dual role of layered double hydroxide nanocomposites on antibacterial activity and degradation of tetracycline and oxytetracyline. *Chemosphere* **2018**, *206*, 175–183. [CrossRef]

240. Asiabi, H.; Yamini, Y.; Alipour, M.; Shamsayei, M.; Hosseinkhani, S. Synthesis and characterization of a novel biocompatible pseudo-hexagonal NaCa-layered double metal hydroxides for smart pH-responsive drug release of dacarbazine and enhanced anticancer activity in malignant melanoma. *Mater. Sci. Eng. C* **2019**, *97*, 96–102. [CrossRef]
241. Shahabadi, N.; Razlansari, M.; Zhaleh, H.; Mansouri, K. Antiproliferative effects of new magnetic pH-responsive drug delivery system composed of Fe_3O_4, CaAl layered double hydroxide and levodopa on melanoma cancer cells. *Mater. Sci. Eng. C* **2019**, *101*, 472–486. [CrossRef]

MDPI
St. Alban-Anlage 66
4052 Basel
Switzerland
Tel. +41 61 683 77 34
Fax +41 61 302 89 18
www.mdpi.com

Crystals Editorial Office
E-mail: crystals@mdpi.com
www.mdpi.com/journal/crystals

www.ingramcontent.com/pod-product-compliance
Lightning Source LLC
LaVergne TN
LVHW072009150826
845671LV00011B/307

* 9 7 8 3 0 3 6 5 0 3 0 6 6 *